PERSPECTIVES ON SCIENCE EDUCATION

A LEADERSHIP SEMINAR

Claire Reinburg, Director
Wendy Rubin, Managing Editor
Rachel Ledbetter, Associate Editor
Amanda Van Beuren, Associate Editor
Donna Yudkin, Book Acquisitions Coordinator

Art and Design
Will Thomas Jr., Director
Rashad Muhammad, Cover and Interior Design

Printing and Production
Catherine Lorrain, Director

National Science Teachers Association
David L. Evans, Executive Director
David Beacom, Publisher

1840 Wilson Blvd., Arlington, VA 22201
www.nsta.org/store
For customer service inquiries, please call 800-277-5300.

20 19 18 17 4 3 2 1

NSTA is committed to publishing material that promotes the best in inquiry-based science education. However, conditions of actual use may vary, and the safety procedures and practices described in this book are intended to serve only as a guide. Additional precautionary measures may be required. NSTA and the authors do not warrant or represent that the procedures and practices in this book meet any safety code or standard of federal, state, or local regulations. NSTA and the authors disclaim any liability for personal injury or damage to property arising out of or relating to the use of this book, including any of the recommendations, instructions, or materials contained therein.

Cataloging-in-Publication Data for this book are available from the Library of Congress.
ISBN: 978-1-941316-30-6
e-ISBN: 978-1-941316-41-2

CONTENTS

SECTION VII
CLASSROOM PRACTICES FOR SCIENCE TEACHING

SECTION VIII
PROFESSIONAL DEVELOPMENT FOR TEACHERS OF SCIENCE

SECTION IX
ASSESSMENT AND ACCOUNTABILITY IN SCIENCE EDUCATION

SECTION X
REFORMS, POLICIES, AND POLITICS IN SCIENCE EDUCATION

SECTION XI
LEADERSHIP AND EDUCATION

SECTION XII
LEADERSHIP IN CONTEMPORARY SCIENCE EDUCATION

ACKNOWLEDGMENTS

Our gratitude extends to the many individuals with whom we have interacted and from whom we have learned about the various realms of science education. To our teachers, college faculty in both science and education, and professional colleagues, we extend our appreciation for your many and varied contributions to our perspectives on science education. We also have held positions of leadership, and in all of these situations, numerous individuals have made decisions and supported our opportunities to lead. Finally, we have had thousands of conversations in classes and meetings and at conferences that have informed our professional knowledge and abilities. We thank all of the individuals who shared their views with us.

Rodger: I would like to thank Janet Carlson, Stanford University, for her insights and recommendations, especially about curriculum and instruction. Ann Rivet, Teachers College, Columbia University, offered her syllabus and personal insights about an introductory course on science education. Kathy Stiles, WestEd, provided ideas and support on professional development, including the book she wrote with Susan Mundry, Nancy Love, Peter Hewson, and the late Susan Loucks-Horsley. Edys Quellmalz and her colleagues at WestEd included me as an advisor on several assessment projects. Jim Short, now at Carnegie Corporation of New York and formerly at the American Museum of Natural History, provided me with several opportunities to advise on his projects and numerous discussions, all of which enriched my understanding of professional development, curriculum, and teaching.Dr. Robert Pletka, Superintendent, Fullerton California School District, and Corey Bess, Assistant Principal, both provided insights about administrative leadership at the district and school levels.

Stephen: My first acknowledgement has to be to my co-author and very good friend, Rodger Bybee. He has probably forgotten more than I will ever learn, and he has been a great friend and mentor throughout the development of the *Next Generation Science Standards* (*NGSS*) and this book. I have had an opportunity to meet so many great people through my work with *NGSS* and am thankful for them all. However, it was Rodger who kept me sane, reminded me of history, allowed me to vent, and pushed me intellectually at every turn to do something special for the students of our country. I am honored to be his friend.

My second acknowledgement is to my two children, Samuel and Abby. It is for you I work hard to made education better for future generations. You both inspire me.

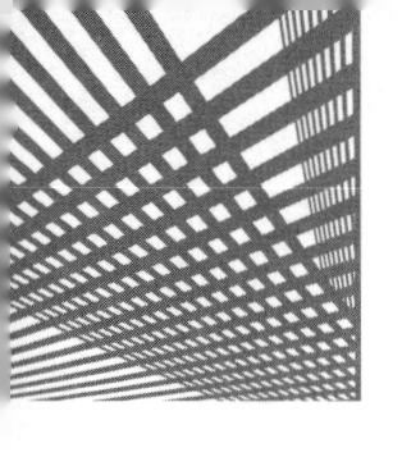

ACKNOWLEDGMENTS

Finally, and most important, my greatest acknowledgment is to my wife, Cecelia. She has made my life a dream. She has believed in me far more than I believed in myself. She pushed me to do great things for our kids, the kids of the country, and now the kids of Kentucky. I am not being sentimental when I say I owe all my success to her. She saw abilities in me I never did. She pushed me to use them, and as a result I have had a special career. Thank you for your love, support, and encouragement.

Both of us acknowledge colleagues at ACHIEVE. Your understanding and permission to use *NGSS*-related material is deeply appreciated. We also thank Claire Reinburg at NSTA for her continual encouragement of this project, and Wendy Rubin, our editor at NSTA, for the improvements she has made in our book.

The manuscript for this book was reviewed by Harold Pratt, Peter McLaren, and Cary Sneider. We attended to your comments and suggestions. The book is improved because of your reviews. Thank you.

Rodger's wife Kathryn carefully reviewed, read, and edited the entire book. She identified spelling and grammatical errors and provided insightful recommendations and resources from a district science coordinator's view. This book is much improved due to her time, effort, and understanding of American science education. We deeply appreciate and fully acknowledge her contributions to this book.

Once again, Byllee Simon assisted with her thorough support. She found errors, did research, and asked insightful questions about various features of the book. Our debt to Byllee is deep, and we thank her for all she did.

Finally, we recognize our families for their understanding and support. They know only too well what it takes to write a book.

Rodger W. Bybee
Stephen L. Pruitt

PREFACE

Depending on the policy, report, or event, one can claim that science education reform has been on the national agenda for days, months, years, decades, or even centuries. Today's media regularly report on the poor achievement of American students on national and international assessments. From January 2001 until December 2015, educators heard about issues associated with the No Child Left Behind Act (NCLB); now there are new challenges posed by the Every Student Succeeds Act (ESSA). Business leaders look to science education to prepare a 21st-century workforce. Finally, there are continuous references throughout this book to health, environment, climate, and other issues that require significant levels of scientific literacy for all citizens.

Whether the means for improvement resides with curriculum materials, teachers' professional education, assessments, or assorted other initiatives, responsibility for improvement ultimately rests with the science education community. This community includes classroom teachers, science coordinators, district administrators, state science supervisors, college and university researchers, curriculum developers, science assessment specialists, administrators of national organizations, and federal agencies.

Looking at the science education community, there is a clear and—we think—compelling need to develop a new generation of leaders who understand science education and are willing to confront the challenges of reform. This book is our response to those ready to face the challenges and provide leadership for education reform.

The general idea for this book originated about a decade ago. F. James Rutherford and Rodger Bybee had a series of discussions about the need for a new generation of leaders in science education. The discussions included many of the themes in this book, such as the goals of science education, standards, and assessments. These conversations were rich in content, drew on professional experiences, and capitalized on different perspectives. However, like many such discussions, they were not fully developed and the ideas never evolved into a book.

The notion of a book on contemporary perspectives and leadership re-emerged with an invitation to both of us to make a presentation at the 2014 National Science Teachers Association (NSTA) national conference. We decided to engage in a dialogue about science education standards through the years.

PREFACE

While the *Next Generation Science Standards* (*NGSS*) were the primary focus of that presentation, we thought it important to identify earlier reforms of science education. Such a discussion naturally centered on aims and goals, standards and benchmarks, curriculum and instruction, assessment and accountability, and teacher education and professional development. These, after all, are topics at the core of science education and central concerns of the science education community in general and science teachers in particular.

This book is not about the need to reform science education. Others have made that argument and undoubtedly will continue making it in the future. This book is about science education and what one needs to know, value, and be able to do as a leader initiating and sustaining reform. The book serves as an introduction to purposes, policies, programs, and practices that science education leaders should understand and be able to apply. Beyond an introduction to science education, we have presented some of the contemporary challenges and controversies that leaders will face: Is the purpose of science education to prepare scientists and engineers, a 21st-century workforce, or scientifically literate citizens? What is the role of federal, state, and local governments in setting standards for science education? To what degree should the curriculum include science-related social issues? What are the roles of politics and policies in science education?

We conceived the book as a seminar, one that begins with an introduction to themes that unify the presentations—perspectives, challenges, standards, and leadership. This introduction is followed by personal introductions. These brief autobiographies present our backgrounds and experiences in science education.

The book (and seminar) continues with a brief history of science education and a close look at the Sputnik era. These two sections of the book set the stage for the central topics of the purposes and goals of science education, national standards, state standards and district leadership, curriculum programs, classroom practices, professional development, and assessment and accountability. These are followed by reform, policies, politics, and two concluding sections on leadership.

Throughout the book, we use an informal, conversational style, as we would in an actual seminar. Most sections of the book include suggested readings that have historical or contemporary significance, personal perspectives, our common perspectives and leadership opportunities, and issues and questions for discussion.

Who is the audience for this book? This book is for those individuals already in leadership positions at national, state, district, and school levels; those enrolled in courses on curriculum and instruction; those participating in professional development; and those teachers of science who want to broaden and deepen their understanding of the foundations and dynamics of science education. Some individuals know much of what we present in the following chapters. They also are probably the ones who are teaching courses or arranging continued professional development. We hope our insights and discussions serve as the basis for continuing discussions. There are others who just want to understand more about their profession. This book is for all of you.

As mentioned, we conceptualized and developed this book as though we were presenting a seminar. Both of us have a broad set of experiences that range from teaching in science classrooms

to formulating and implementing national and international policies and programs. We have careers that include the successes and failures of leadership. In a seminar, we would have the place to present our own scholarly work and that of others, the chance to express our perspectives on issues, and opportunities to challenge the participants with questions, situations, and issues.

Ultimately, science education leaders have to make decisions and set directions based on their positions and opportunities. In our careers, we have done just that. This book is not so much about answers; it is more about questions. It is not about persuading you of the need to reform; it is more about developing your understanding of science education and recognizing the challenges and opportunities of leadership. We present some of the perennial issues to structure the conversations so ideas are exchanged and individuals develop the understanding and abilities to lead. To the best of our knowledge, values, skills, and experiences, we try to begin professional conversations that will contribute to a deeper understanding of science education in general and your leadership in particular.

SECTION I

PERSPECTIVES ON SCIENCE EDUCATION

INTRODUCTIONS

Individuals in science education have perspectives based on their past and present experiences. Their perspectives influence the decisions they make and the priorities they assign to education policies, school curriculum, teaching practices, teachers' professional development, and classroom, state, and national assessments.

You might have some questions. First, what is this seminar about? Chapter 1 introduces the seminar's themes: perspectives, challenges, standards, and leadership. Second, who are these guys, and what have they done that qualifies them to lead this seminar? Personal introductions to Rodger Bybee and Stephen Pruitt are the subject of Chapter 2.

SUGGESTED READINGS

American Association for the Advancement of Science (AAAS). 2013. Grand challenges in science education. Special issue, *Science* 340 (6130).

Atkin, J. M. 1994. Developing world-class education standards: Some conceptual and political dilemmas. In *The future of education: Perspectives on standards in America*, ed. N. Cobb, 61–84. New York: College Entrance Examination Board.

Banilower, E., S. Smith, I. Weiss, K. Malzahn, K. Campbell, and A. Weis. 2013. *Report of the 2012 Survey of Science and Mathematics Education.* Chapel Hill, NC: Horizon Research. A great resource for basic data on science education.

National Research Council (NRC). 2002. *Investigating the influence of standards.* Washington, DC: National Academies Press. A framework for research on national standards.

National Research Council (NRC). 2012. *A framework for K–12 science education: Practices, crosscutting concepts, and core ideas.* Washington, DC: National Academies Press. A fundamental report influencing all dimensions of contemporary science education.

NGSS Lead States. 2013. *Next Generation Science Standards: For states, by states.* Washington, DC: National Academies Press. *www.nextgenscience.org/next-generation-science-standards.*

Pedersen, J., K. Finson, B. Spector, and P. Jablon. 2013. *Going back for our future: Carrying forward the spirit of pioneers of science education.* Charlotte, NC: Information Age Publishing. A valuable set of autobiographical and biographical chapters. Many individual leaders (i.e., pioneers) discussed in Pedersen et al.'s book are referred to in the following chapters of this book. You likely will find insights and enjoy reading about leaders such as F. James Rutherford, Paul DeHart Hurd, J. Myron Atkin, and others.

CHAPTER 1

INTRODUCING THE SEMINAR

This chapter provides an orientation and overview for the book. We designed the book to have the feel of a seminar. In a seminar, one studies a subject—in this case, different perspectives on dimensions, dynamics, topics, challenges, and issues in science education. This seminar covers topics such as standards, challenges such as improving achievement for all students, and issues such as understanding the politics of education reform. Our perspectives are historical and contemporary, personal and shared, and we evaluate different components of the science education system—purposes and goals, policies, programs, and practices.

American science education presents diverse challenges for groups ranging from classroom teachers to national policy makers. Regardless of your place and work, we are in a time that requires a broader and deeper understanding of science education. A view of science education as one component of a dynamic system helps one understand that most initiatives for reform and improvement originate from combinations of contextual forces—some internal forces, such as research on student learning, and some external social forces, such as economics and politics. Contemporary examples include the National Research Council report *How People Learn* (Bransford, Brown, and Cocking 1999) as an internal force and the economic influence of calls from business and industry to prepare a 21st-century workforce as an external one.

The chapter continues with discussions of our main themes: perspectives, challenges, standards, and leadership.

PERSPECTIVES

In the world of art, the term *perspective* refers to the representation of three-dimensional objects and depth relationships, often on two-dimensional surfaces. This is a nice metaphor for our use of the term in the title of the seminar and as a prominent theme of the book. Perspectives give orientation, depth, and relationships to components of a subject such as science education.

The seminar includes historical and contemporary perspectives. We also present our personal perspectives, as well as positions, challenges, and opportunities on which we have common perspectives. In addition, we introduce four perspectives that represent different dimensions and orientations in the science education system: purposes, policies, programs, and practices. As you will see, these terms are used for the general titles of sections and chapters in the book. We introduce these four perspectives in the following paragraphs, and the next sections provide detailed descriptions of each.

You may find it useful to recognize different perspectives on initiatives in education. The perspectives may be expressed by individuals, groups, or organizations within the science education community. Some, for example, express views about goals, while others state concerns about school curricula and effective teaching strategies. Our position should be clear: The perspectives are not good or bad; they are different.

The different perspectives usually express the individual's or group's professional orientation and concerns. Science teachers have an understandable classroom and teaching perspective, while some educators have a philosophical or policy view as they describe major goals such as scientific literacy. Unfortunately, there are occasional statements by individuals holding one perspective while being critical of another's perspective. We caution against this dismissive view and encourage the recognition and appreciation of different perspectives in science education.

Because there are different perspectives, a "map" will help you recognize where you are and where others are coming from. Such a map will help you identify the location, means of movement, and direction and difficulties of navigation in science education and in this seminar. In the science education community, numerous reports, standards, and studies on science education present what may be a confusing array of destinations and directions, all of which may also be discussed at national, state, or district levels. A simple map is quite helpful for locating and clarifying different efforts in the geography of contemporary science education. The map we propose uses the terms *purposes, policies, programs*, and *practices*—the 4 *P*s—to identify various important perspectives in science education. The 4*P*s were described in an earlier publication (Bybee 1997) and will be used as themes in this seminar. The following discussion introduces the 4*P*s, and a "map" is shown in Figure 1.1 (p. 7).

The Purposes of Science Education

Science educators have expressed many aims, goals, and objectives in various documents, such as national standards, state frameworks, school syllabi, and teaching lessons (DeBoer 1991; Bybee 1993; Bybee and DeBoer 1994; DeBoer and Bybee 1995). For this perspective, we use the term *purposes* as it refers to universal aims and goals of what K–12 science education should achieve. Such statements are abstract and apply to all concerned within science education.

Achieving scientific literacy is a statement of purposes for science education. The strength of a purpose statement lies in its widespread acceptance and general agreement within the science education community. One weakness lies within the statement's ambiguity concerning specific situations in science education. For example, what does the goal of achieving scientific literacy mean for an elementary school teacher, a high school chemistry teacher, a teacher educator, a policy maker, or a curriculum developer? The answers, of course, vary for individuals and situations, hence the need for more concrete statements and the adaptation of the purpose statements for various factions of the science education community. These statements are based on the general purpose but address more specific situations, and they will introduce the role of policies.

Policies for Science Education

Policy statements are concrete translations of the purposes for various constituents within the science education community. Documents that give direction and guidance but are not actual programs serve as policies. Examples of policy documents include a course outline for a high school Earth science class, district syllabi for K–12 science courses, state frameworks, standards, and federal legislation such as the 2015 Every Student Succeeds Act (ESSA). Likewise, college or university requirements for undergraduate teacher education and state and national frameworks for assessing science also fall into the category of policies. At the national level, examples of policy documents include the *Next Generation Science Standards* (*NGSS*; NGSS Lead States 2013). We suggest that contemporary reform is poised at the critical point between completing the *NGSS* and implementing programs that align with those standards.

Programs for Science Education

Major categories of programs include curriculum materials, assessments, teacher education courses, and professional development. School programs, for example, include the actual curriculum materials, textbooks, and coursework based on the policies. Programs are unique to grade levels, disciplines, and aspects of science education such as teacher education or a middle school integrated science curriculum.

School science programs may be developed by organizations and marketed commercially, or they may be developed by states or local school districts. Who develops the materials is not the defining characteristic; the fact that schools, colleges, state agencies, and national organizations have programs aligned with policies such as the national or state standards is the important feature of this aspect of science education.

Practices in Science Education

Practice, as we use the term in this seminar, refers to the specific actions and processes of teaching science in schools, colleges, universities, museums, science centers, and other informal settings. The practices of science education include the personal interactions between teachers and students and among students, as well as the roles and uses of assessment, educational technologies, laboratories, and myriad strategies for teaching science. In the perspectives described here, implementing new classroom practices implies they would be consistent with policies, and programs would be designed to achieve purposes of scientific literacy, for example. (*Note:* This seminar also introduces scientific and engineering practices that, in some cases, may be specific examples of generic classroom practices.)

Improving the practices of teaching science centers on the most individual, unique, and fundamental aspect of the map. Science educators can propose new goals; design new state standards, syllabi, and scope and sequence charts; and develop new curriculum materials, but the critical aspect of any reform is improving teaching and enhancing learning in science classrooms.

Recognizing these perspectives in science education is relatively new. As you will see, for decades science educators have recognized the importance of goals, curriculum programs, and classroom practices. Little attention has been directed to questions of policy. Typical discussions have been framed by challenges of topics such as "research to practice" or "theory into action" (see, e.g., NSTA 1964). The purposes, programs, and practices of science education do not exist independently. There is a need to recognize the place and influence of policies such as national and state standards.

Here is a note of caution. Although one may be tempted to think of the 4*P*s as a sequence of stages, such that the ideal would be to begin with purposes and develop policies followed by programs and practices, history indicates that reforms in science education do not reflect this ideal. Different reforms of science education have emphasized different elements of the 4*P*s.

To conclude this discussion, one can view discussions, articles, and reports on science education in at least four connected and interconnected perspectives: purposes, policies, programs, and practices (see Figure 1.1). Each perspective has advocates and audiences, problems and solutions, and roles and responsibilities in science education. The challenge that leaders face is to recognize and accept these perspectives, as they all have a place in this seminar.

CHALLENGES

The first decades of the 21st century present a world with a variety of challenges. As the global society advances, America's education system lags behind many countries in providing our students with the knowledge and skills that will enable them to sustain and build the technology-rich culture of the 21st century. The K–12 system of science and technology education is one vehicle for ensuring a society that possesses essential skills for higher-order thinking; evidence-based decision making; rational argument; and scientific, technological, and quantitative literacy.

Grand Challenges?

The April 2013 issue of *Science*, the journal of the American Association for the Advancement of Science (AAAS), devoted a special section to the theme "Grand Challenges in Science Education." Educators addressed 20 challenges, which are summarized in Figure 1.2 (p. 8). Take a few minutes to review these challenges. Do you agree they are "grand challenges"? Would you eliminate any of the challenges? How about adding challenges? What is your list of "grand challenges" in science education?

In the spirit of provoking discussion in this seminar, we ask these questions:

- Do you agree there are grand challenges in science education?

- How would you differentiate between grand challenges and those problems that are important but not large and dominating enough to be considered grand?
- What are the size, scope, and persistence of challenges that would define them as grand?

Figure 1.1. Perspectives on Initiatives in Science Education

PURPOSE

Purpose statements include aims, goals, and rationales. These statements tend to be universal and abstract and apply to all components of the science education system (e.g., teacher education, curriculum, instruction, and assessment). Although it presents elements of both purpose and policy, *A Framework for K–12 Science Education: Practices, Crosscutting Concepts, and Core Ideas* (NRC 2012) has served as the purpose in this era of standards-based reform.

POLICY

Policies are more specific statements of standards, benchmarks, syllabi, frameworks, and strategic plans based on the defined purposes. Policy statements are concrete translations of the purpose and apply to specific components such as teacher education, K–12 curriculum, and assessments. The *NGSS* (NGSS Lead States 2013) provide the policy statement most applicable to this contemporary discussion.

PROGRAM

Programs are the actual instructional materials used in states, schools, and classrooms. Programs are unique to disciplines, K–12 grades, and different components of the education system. Curriculum materials for K–12 science, state assessments, and undergraduate teacher education are different examples of programs. Programs are a translation of policies into the unique requirements of the contexts in which they must be implemented. As mentioned above, the requirements of teacher education and assessments differ when one is developing a program based on new policies.

PRACTICE

Practice refers to the specific actions of educators as they implement the program. Classroom teaching of science is an example of practice. Practice is the most unique and fundamental level at which the purposes of science education are translated to classrooms.

Other Challenges?

As the seminar progresses, you will encounter other challenges. We present some in our personal perspectives. Others are presented in different sections of this book. We do propose that there are challenges for each individual in the science education community—for science teachers, department chairs, district coordinators, state supervisors, policy makers at all levels, and people in national organizations. Although the challenges may not be grand, do they feel that way for the individual?

As you will see, we propose that the importance and influence of national and state standards and their implementation by district leaders and classroom teachers present some of the most immediate

Figure 1.2. Grand Challenges in Science Education

- Use technology to improve pedagogy, management, and accountability.
- Improve access to and quality of pre- and post-primary education.
- Develop appropriate policies for regulating and supporting the private sector in education.
- Develop an understanding of how individual differences in brain development interact with formal education.
- Adapt learning pathways to individual needs.
- Create online environments that use stored data from individual students to guide them to virtual experiments that are appropriate for their stage of understanding.
- Determine the ideal balance between virtual and physical investigations.
- Identify the skills and strategies that teachers need to implement a science curriculum featuring virtual and physical laboratories.
- Identify the underlying mechanisms that make some teacher professional development (PD) programs more effective than others.
- Identify the kind of PD that will best prepare teachers to implement the *NGSS*.
- Harness new technologies and social media to make high-quality science PD available to all teachers.
- Help students explore the personal relevance of science and integrate scientific knowledge into complex practical solutions.
- Develop students' understanding of the social and institutional basis of scientific credibility.
- Enable students to build on their own enduring science-related interests.
- Shift incentives to encourage education research on the real problems of practice.
- Create a set of school districts where long-standing, multidisciplinary teams work together to identify effective improvements.
- Create a culture in school systems that allows for meaningful experimentation.
- Design valid and reliable assessments that reflect the integration of practices, crosscutting concepts, and core ideas in science.
- Use assessment results to establish an empirical evidence base regarding progressions in science proficiency across K–12.
- Build and test tools and information systems that help teachers effectively use assessments to promote learning in the classroom.

Source: Adapted from AAAS 2013.

and significant challenges in American science education. Among the challenges related to standards, we focus on those related to topics in the seminar—namely, curriculum programs, classroom practices, assessments and accountability, and professional development. Many of the topics listed in Figure 1.1 are likely to be included in this challenge. Standards will be an important theme in this seminar, as they influence science educators' perspectives and opportunities for leadership.

STANDARDS

The American system of public education involves a network of 50 state agencies (or more, if one counts the District of Columbia, Department of Defense schools, and education systems in U. S. territories), thousands of school districts, and several million teachers. The system also includes community and four-year colleges, as well as media, museums, science and technology centers, botanical gardens, and a variety of outdoor and environmental activities that also educate the public. In all of these institutions, educators face a difficult but achievable challenge: to improve the way they teach and what students learn about science and technology. How can the science education community effect change, sustain reform, and improve education across such a diverse system? The answer lies in a common vision founded on a cohesive set of ideas, communicated broadly, and implemented with consistency and commitment. We thus introduce the role of standards for education.

If one considers the statements of individuals, committees, and reports from national organizations, the story of standards in American education is long and varied. However, most of these statements did not include the word *standards* in the title. The contemporary story of standards began with the 1983 report *A Nation at Risk* (National Commission on Excellence in Education 1983). To reduce the perceived risk, American educators were challenged to meet two of the report's recommendations: (1) Strengthen content of the core curriculum, and (2) raise expectations using measurable standards.

Six years later, President George H. W. Bush worked with 49 of the nation's governors to agree on the need for national goals for education. This bipartisan agreement (yes, bipartisan agreements can happen!) clarified the need for national standards and assessments.

This seminar addresses several common uses of the term *standard* in education contexts. The following definitions introduce the different forms and functions of standards.

Content Standards

These standards describe what teachers are expected to teach and students are expected to learn. These are sometimes referred to as *curriculum standards*. Content standards should be clear and detailed descriptions of the knowledge, skills, and abilities that should be taught and learned. The knowledge includes the enduring ideas, major concepts, and fundamental information of a discipline such as science. The skills and abilities may include ways of thinking, communicating, reasoning, and conducting investigations using processes unique to a discipline such as science. The *National Science Education Standards* (NRC 1996) are content standards, as are most state standards. A final and important note on content standards: They are not the instructional materials that teachers actually use in their teaching, but they directly influence what teachers will teach.

Performance Standards

Performance standards describe the levels and types of competencies or mastery students are expected to attain. These standards are usually related to content standards. If content standards indicate what is to be learned, performance standards inform assessments that identify degrees of attainment using terms such as *basic, proficient,* or *advanced.* The *NGSS* use performance expectations as a central feature of the standards. These standards incorporate science and engineering practices, disciplinary core ideas, and crosscutting concepts. The National Assessment of Educational Progress (NAEP) periodically sends America a report card with grades identifying students' science achievement at grades 4, 8, and 12.

Clearly, much more can and will be said about standards and American education. Next, we describe standards for science education.

Standards for Science Education

The publication of *Curriculum and Evaluation Standards for School Mathematics* (NCTM 1989) introduced an era of standards-based reform in American science and mathematics education. We note that *Science for All Americans* (Rutherford and Ahlgren 1989) was published in the same year. The work of F. James Rutherford and Andrew Ahlgren and the AAAS Project 2061 continued with the development and publication of *Benchmarks for Science Literacy* (AAAS 1993). In science education, *Benchmarks for Science Literacy* was a forerunner to the National Research Council's *National Science Education Standards,* published in 1996, and the *Standards for Technological Literacy,* published by the International Technology Education Association (ITEA) in 2000.

Early in this era of standards-based reform for science education, politicians, leaders in business and industry, and the American public viewed the United States as a nation at risk, requiring clarity of learning outcomes and measures of student achievement. Standards and assessments have served as descriptions of learning outcomes and, thus, the basis for education reform. In time, standards and assessments have been subject to political attacks, and assessments in particular have been criticized for their frequency, duration, and lack of consequences.

A new generation of national standards began when the National Research Council released *A Framework for K–12 Science Education: Practices, Crosscutting Concepts, and Core Ideas* (*Framework*; NRC 2012). This document set forth a new vision for science education by identifying fundamental science and engineering practices; essential core ideas for physical, life, and Earth and space science; and key crosscutting concepts that all students should learn during their K–12 education. Additionally, and perhaps most important, the *Framework* provided a vision for science education that centered on students performing at the nexus of these three dimensions.

The *Framework* served as the basis for new standards that would replace those developed in the 1990s—that is, *Benchmarks for Science Literacy* and *National Science Education Standards.* In April 2013, the *NGSS* were released at the national conference of the National Science Teachers Association (NSTA). These standards were developed with leadership provided by 26 states and a team that included scientists, engineers, science teachers, and other science educators. (We should also note that no federal money supported the development of *NGSS.*)

Whether your immediate perspectives and concerns center on the national, state, district, or classroom level, standards for science education will influence your work. It is important to understand how standards influence the greater science education systems, so in this seminar, we will discuss standards and their implications for science education.

Understanding the Influence of Standards

Because we are emphasizing contemporary perspectives, from the early 1980s until the present, we have appealed to the National Research Council report *Investigating the Influence of Standards: A Framework for Research in Mathematics, Science, and Technology Education* (NRC 2002) and a subsequent workshop review of evidence for the influence of standards (NRC 2003). In particular, we draw your attention to the general framework for understanding the influence of standards on the education system (see Figure 1.3).

Figure 1.3. A Framework for Understanding the Influence of Standards on the Education System

CONTEXTUAL FORCES

- Politicians and policy makers
- Public
- Business and industry
- Professional organizations

CHANNELS OF INFLUENCE WITHIN THE EDUCATION SYSTEM

Curriculum

- State and district policy decisions
- Instructional materials development
- Text and materials selection

Teacher Development

- Initial preparation
- Certification
- Professional development

Assessment and Accountability

- Accountability systems
- Classroom assessment
- State and district assessment
- College entrance and placement practices

→ **Teachers and Teaching Practice**

- In classroom and school contexts

→ **Student Learning**

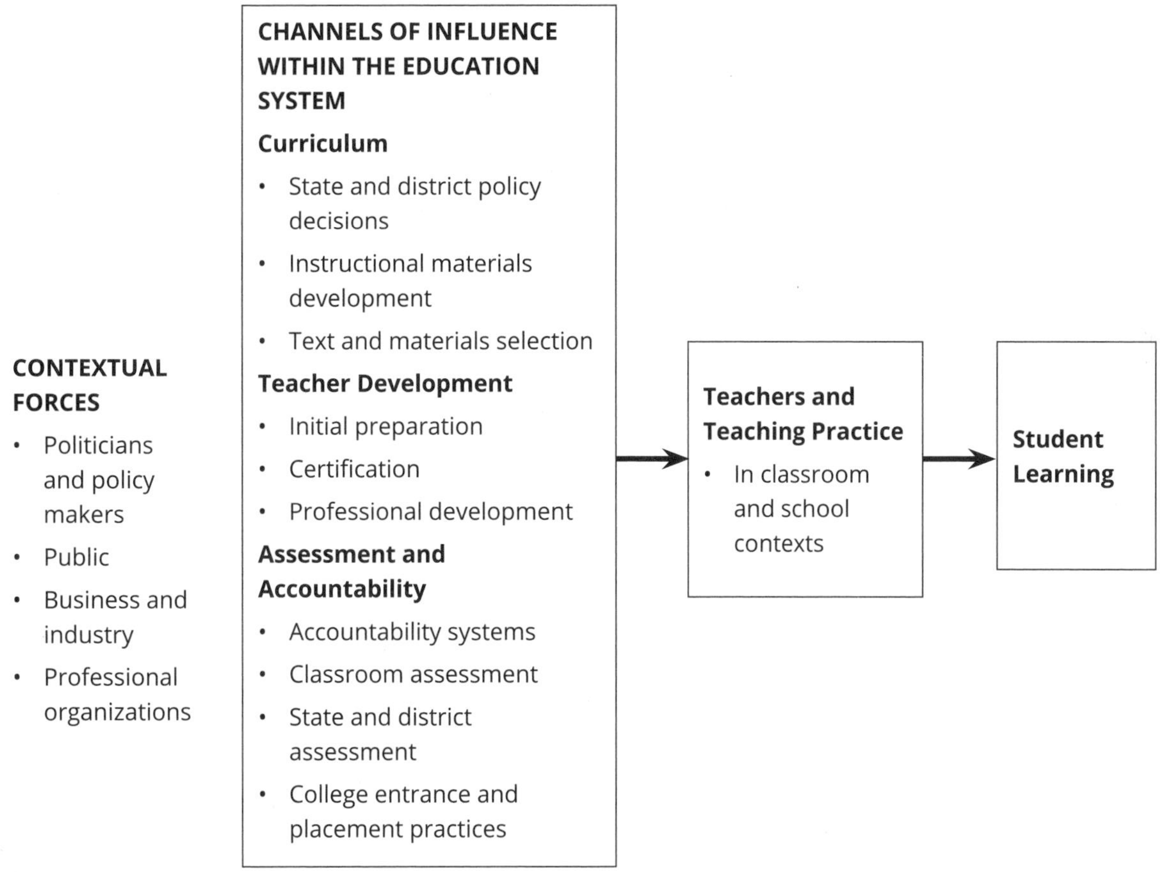

Source: Adapted from NRC 2002.

CHAPTER 1

LEADERSHIP

In recent decades, the literature on leadership in education has developed with, for example, views of person-centered school leaders (Combs, Miser, and Whitaker 1999), building leadership capacity in schools (Lambert 1998; Patterson 1993), using systems thinking to develop leadership and sustainability (Fullan 2005), and understanding the instructional core (City et al. 2009). As one considers leadership in this era, we believe it is important to begin with a definition from the classic book *Leadership* by James MacGregor Burns (1978, p. 19):

> *I define leadership as leaders inducing followers to act for certain goals that represent the values and motivations—the wants and needs, the aspirations and expectations—of both leaders and followers.*

One should note the term *inducing* is not as strong as other terms such as *require*; rather, the term implies persuasion or influence. To be clear, Burns was, for the most part, writing about political, military, and other leaders in powerful (often national) positions. Most leaders in science education are not in such positions. For them, the needs, aspirations, and expectations center on improving student achievement by reforming standards, implementing new school programs, providing professional development, and improving classroom teaching. Still, the leader is asked to consider the purposes, concerns, and values of those being led, whether from national organizations, state agencies, or the classroom. Burns went on to say, "And the genius of leadership lies in the manner in which leaders see and act on their own and their followers' values and motivations" (1978, p. 19). In the context of this discussion, a leader in a school district may have the expectation of reforming the science curriculum, and it is important, for example, to acknowledge the perspective of those who will be influenced by the new curriculum. With this general introduction, we turn to leadership in the context of science education.

Leaders Must Have a Vision

One of the consistent requirements of leadership is that leaders have a vision. Leaders may, for example, have a longer-term perspective, see larger systemic issues, present future scenarios, or discern fundamental problems and present possible solutions rather than spending time and energy assigning blame for problems. Depending on their situation, leaders in science education have diverse ways of clarifying their visions. Some may do so in speeches, others in articles, still others in policies, and others—most importantly—may lead every day in their classrooms. One vision may unify a group, organization, or community, while another may set priorities or resolve conflicts among constituencies. The emergence of the leader's vision likely will have many sources and result from extensive review and careful thought. This is especially true in today's complex education environment.

Leadership in science education extends from classroom teachers to the highest levels of government. Science education consists of numerous systems and subsystems, all with individuals who have power, constituents, and goals that contribute to better science education for students. Not all in the science education community can or should be involved in constructing international assessments, developing curriculum materials, presenting the arguments for scientific practices, defending the integrity of science, or providing professional development. But all of us do have our roles and responsibilities that relate to these and many other leadership opportunities, and that is what will ultimately make a difference for students.

For most, the fundamental purpose of science education is expressed by the following aim—achieving high levels of scientific literacy for all students. Such a broad—and, we would argue, deep—perspective touches critical components of the science education system; national, state, and district school science programs; and classroom practices of curriculum, instruction, and assessments. Indeed, the fundamental purpose is comprehensive and inclusive. This is the vision required of science education leadership in the 21st century.

Contemporary justification for a new vision of science education is expressed in themes such as economic stability, basic skills for the 21st-century workforce, and college and career readiness. Such themes differ from earlier justifications such as the space race and responding to a nation at risk. The economic rationale emerged from a significant recession, the realization that the U.S. economy is part of a global economy, and the acknowledgment that the educational levels of the broad public influence the rate and direction of a country's economic progress.

Concretely, a contemporary vision for science education resides in national and state standards and the implied reform of curriculum, instruction, and assessments. Providing leadership requires communicating a vision for those one wishes to lead. To state the obvious, the vision need not be complex and complicated, but it must be different from the status quo. The vision must be new and substantial and look to the future. The title *Science for All Americans* (Rutherford and Ahlgren 1989) is an excellent example of a vision from history. In this era, the title *Next Generation Science Standards: For States, By States* expresses a vision and identifies the importance of states in the reform.

Communicating the vision often requires a translation from the abstract—for example, "achieving scientific literacy"—to contexts that have meaning for a specific community. An example of a vision with context would be "translating state standards to curriculum programs and classroom practices at the local level."

All of this said, leaders must recognize the power of the past as a force against a new vision. Those in the science education community will evaluate the leaders' visions with their own visions grounded in the past or current situation. We cannot say what the evaluation will entail, but we can guarantee that it will be grounded in the past: "Where is my topic in the new standards?" "Will the practices of scientific inquiry be on the state test?" Being able to illustrate the vision with examples that constituents recognize will be most helpful for leaders.

Leaders Must Have a Plan

A plan must complement the vision. It is one thing to express a new vision for science education; many do this regularly. It is quite another to provide a plan for achieving that vision. The plan must have clear examples for constituents. Having a vision for science education is fine, but what does the vision mean for the school curriculum in general and the unique situations in teachers' classrooms in particular?

So, leadership requires both a vision and a plan. If you only have a vision, you are like a utopian thinker. If you only have a plan, you likely are a manager, not a leader.

Politics and Criticisms

Among the essentials of leadership is a leader's ability to recognize and address the political realities of the work. The insight here is that the leader has to recognize that initiating changes means addressing the political issues within the system. Not all issues are solely educational. Indeed, it may be the case that all education issues ultimately have political components. The paradox embedded here can be understood as the balance between achieving education goals while addressing political realities. Leaders recognize that *either/or* thinking often expresses the paradox, while *both/and* thinking provides insights into the plans for resolutions.

Experience teaches one more lesson for those in leadership positions. If you are leading, you cannot avoid conflict, criticism, and controversy. The larger the system and greater the change, the more controversy you will experience. Indeed, this is a paradox in that achieving your goals requires enduring criticism, and the criticism often is unfair, constant, and personal.

What lessons have we learned that might help you in your leadership? We divide this discussion into two categories. First, we mention several general ideas of leadership. Second, we discuss specific ideas related to standards-based reform in science education.

Be a Leader

With regard to leadership in science education, our first suggestion is to be a leader; do not wait for others to show the way. As opportunities for leadership emerge, such as new standards for science education and the need for changing the school science program, get involved. Learn about the standards and develop an understanding of what the new standards may mean for your state, district, or school.

Your leadership should include both a vision and a plan or strategies for change. We underscore the need for *both* a vision and a plan. If you only have a vision, there may be excitement and support, but the changes will flounder and likely fail based on questions such as, "What do we do?" "What should change?" and "What does this mean for my students and me?" On the other hand, having only a plan produces problems with questions such as, "Why are we doing this?" and "How does this change relate to others at the national, state, or district level?" The effective leader must have answers to questions such as these.

REFERENCES

American Association for the Advancement of Science (AAAS). 1993. *Benchmarks for science literacy*. New York: Oxford University Press.

American Association for the Advancement of Science (AAAS). 2013. Grand challenges in science education. Special issue, *Science* 340 (6130).

Bransford, J. D., A. L. Brown, and R. R. Cocking, eds. 1999. *How people learn: Brain, mind, experience, and school*. Expanded ed. Washington, DC: National Academies Press.

Burns, J. M. 1978. *Leadership*. New York: Harper & Row Publishers.

Bybee, R. W. 1993. *Reforming science education: Social perspectives and personal reflections*. New York: Teachers College Press, Columbia University.

Bybee, R. W. 1997. *Achieving scientific literacy: From purposes to practices*. Portsmouth, NH: Heinmann.

Bybee, R., and G. DeBoer. 1994. Research on goals for the science curriculum. In *Handbook of research on science teaching and learning*, ed. D. L. Gabel, 357–387. New York: Macmillan.

City, E., R. Elmore, S. Fiarman, and L. Teitel. 2009. *Instructional rounds in education*. Cambridge, MA: Harvard Education Press.

Combs, A., A. Miser, and K. Whitaker. 1999. *On becoming a school leader*. Alexandria, VA: Association for Supervision and Curriculum Development (ASCD).

DeBoer, G. 1991. *A history of ideas in science education*. New York: Teachers College Press, Columbia University.

DeBoer, G., and R. Bybee. 1995. The goals of science curriculum. In *Redesigning the science curriculum: A report on the implications of standards and benchmarks for science education*, ed. R.W. Bybee and J. D. McInerney, 71–74. Colorado Springs, CO: BSCS.

Fullan, M. 2005. *Leadership and sustainability*. Thousand Oaks, CA: Corwin Press.

International Technology Education Association (ITEA). 2000. *Standards for technological literacy: Content for the study of technology*. Reston, VA: ITEA

Lambert, L. 1998. *Building leadership capacity in schools*. Alexandria, VA: Association for Supervision and Curriculum Development (ASCD).

National Commission on Excellence in Education (NCEE). 1983. *A nation at risk: The imperative for educational reform*. Washington, DC: U.S. Government Printing Office.

National Council of Teachers of Mathematics (NCTM). 1989. *Curriculum and evaluation standards for school mathematics*. Reston, VA: NCTM. *www.standards.nctm.orglindex.htm*

National Research Council (NRC). 1996. *National science education standards*. Washington, DC: National Academies Press.

National Research Council (NRC). 2002. *Investigating the influence of standards: A framework for research in mathematics, science, and technology education*. Washington, DC: National Academies Press.

National Research Council (NRC). 2003. *What is the influence of the* National Science Education Standards? *Reviewing the evidence, a workshop summary*. Washington, DC: National Academies Press.

National Research Council (NRC). 2012. *A framework for K–12 science education: Practices, crosscutting concepts, and core ideas*. Washington, DC: National Academies Press.

National Science Teachers Association (NSTA). 1964. *Theory into action … in science curriculum development*. Washington, DC: NSTA.

NGSS Lead States. 2013. *Next Generation Science Standards: For states, by states*. Washington, DC: National Academies Press. *www.nextgenscience.org/next-generation-science-standards*.

Patterson, J. 1993. *Leadership in tomorrow's schools.* Alexandria, VA: Association for Supervision and Curriculum Development (ASCD).

Rutherford, F. J., and A. Ahlgren. 1989. *Science for all Americans: A Project 2061 report on literacy goals in science, mathematics, and technology.* Washington, DC: American Association for the Advancement of Science.

CHAPTER 2

PERSONAL INTRODUCTIONS

Teachers, whether they are in K–12 classrooms or graduate seminars, bring more to a course than their knowledge of subject matter. Yes, they have a background of knowledge in the subject, and they also have perspectives based on their experiences. Because the perspectives have developed from life experiences and surely influence the topics, contexts, and orientation of a course, it is helpful for seminar participants to have some knowledge of their teachers' backgrounds and experiences.

This chapter will help you answer questions such as, "Have the leaders of this seminar been in classrooms? Helped set policy?" and "What leadership experiences have shaped their knowledge, values, abilities, and perspectives that will be directly and indirectly expressed in this seminar?" To answer questions such as these, at least in part, we present brief summaries of our experiences in science education. To be clear, these are informal and meant as introductions to the views and opinions we will express in this seminar.

A CAREER OF CHALLENGES AND OPPORTUNITIES

Rodger Bybee

I have had a rich and varied career in science education, one that has provided numerous opportunities and wonderful mentors. In the late 1960s, I began teaching 9th grade Earth science and high school biology. During this period, I also had the opportunity to teach elementary school (K–6) science. These opportunities were in the Greeley, Colorado, school district and at the Laboratory School at the University of Northern Colorado (UNC), also in Greeley. These experiences provided my first view of excellent curriculum materials when I used the Earth Science Curriculum Project (ESCP), Biological Sciences Curriculum Study (BSCS), and several elementary school programs—most important, the Science Curriculum Improvement Study (SCIS). During this period, Robert B. Sund, a professor at UNC, was a mentor advising me through a master's degree and guiding me into a career as a professional science educator.

CHAPTER 2

My next milestone was graduate study and work in a Trainer of Teacher Trainers (TTT Project), a federally funded program supporting graduate study for individuals who planned to work at the college level in teacher education. Beyond the usual course work, during my second year, I worked in an elementary school in Manhattan (P.S. 33), collaborating with teachers on improving their curriculum and instruction. My dissertation was a synthesis of science education history, philosophy of education, and education psychology, particularly the psychology of motivation and development. The title of my dissertation—*Implications of Abraham H. Maslow's Philosophy and Psychology for Science Education in the United States*—expresses the comprehensive nature of the work. I grew immeasurably from the advice of faculty such as Darrell Barnard, Morris Shamos, and Janice Gorn, but the mentorship of F. J. Rutherford was surpassed by none. He brought the history of science education, an understanding of science, and experience with curriculum reform to my education. Mr. Rutherford's experience included teaching high school science in California, earning graduate degrees from Stanford and Harvard, and serving as a division director at the National Science Foundation (NSF) and deputy secretary at the U.S. Department of Education. He later directed Project 2061 at the American Association for the Advancement of Science (AAAS).

My next career opportunity included 15 years of teaching in the education department at Carleton College, a small, liberal arts college in Northfield, Minnesota. Being a faculty member opened a range of opportunities—for example, teaching a course on sustainable society with theologian Ian Barbour, as well as broadening and deepening my understanding of educational psychology, philosophy, and science and technology policy.

In 1985, I left Carleton and joined BSCS as associate director. The next decade entailed writing proposals and assuming responsibility for designing and developing innovative school science programs for elementary, middle, and high school students and undergraduate nonmajors. With the leadership of Joseph McInerney, the director of BSCS and a leader in biology education, I also learned about organizational management.

In the early 1990s, during my tenure at BSCS, I worked on national standards and chaired the science content group for this initiative. This work was centered at the National Research Council (NRC) in Washington, DC. The *National Science Education Standards* (NSES) were released in late 1995 (with a 1996 copyright). Shortly before the release of the standards, I joined the NRC as executive director of the Center for Science, Mathematics, and Engineering Education (CSMEE). My position at the NRC introduced me to leadership at the national level, as there were interactions with federal agencies such as the NSF and the U.S. Department of Education, congressional committees, and the executive branch. During this time, I worked closely with Gerry Wheeler, then executive director at the National Science Teachers Association (NSTA). Working with Gerry and NSTA meant that I also maintained a view of science programs and science teachers, something that can be lost when one is working in Washington, DC. Donald Kennedy, past president of Stanford University, and Bruce Alberts, then president of the National Academy of Sciences, were invaluable mentors during this work at the national level.

After five years at NRC, I returned to BSCS as executive director. I remained at BSCS until I retired in 2007, but my work has continued with writing books such as *The Teaching of Science: 21st-Century Perspectives* (Bybee 2010), *The Case for STEM Education: Challenges and Opportunities* (Bybee 2013a), *Translating the* NGSS *for Classroom Instruction* (Bybee 2013b), and *The BSCS 5E Instructional Model: Creating Teachable Moments* (Bybee 2015), as well as work on *A Framework for K–12 Science Education* (NRC 2012) and the *Next Generation Science Standards* (NGSS Lead States 2013).

My career in science education has included publishing in refereed journals such as the *Journal of Research in Science Teaching, Science Education, International Journal of Science Education, American Biology Teacher,* and all of the NSTA journals. In addition, my publications include monographs, methods textbooks, and other reports.

Because this is a personal introduction, I will mention several books I am proud of, even if they did not sell well or change science education. My first book was *Becoming a Better Elementary Science Teacher* (Sund and Bybee 1973). This was a book of readings that developed my understanding of the profession and introduced me to publishing. While attending a summer seminar at the Aspen Institute, I met E. Gordon Gee, who went on to become president of West Virginia University, University of Colorado, Brown University, Ohio State University, and West Virginia University (a second time). We wrote *Violence, Values, and Justice in the Schools* (Bybee and Gee 1982). Written for school administrators, the book applied concepts of justice to classrooms and school discipline problems.

In the field of science education, I published *Reforming Science Education: Social Perspectives and Personal Reflections* (Bybee 1993) and *Achieving Scientific Literacy: From Purposes to Practices* (Bybee 1997). Both of these works established several themes that are clearly evident in this seminar: scientific literacy, reform, and different education perspectives.

Finally, I will briefly summarize some of the influences and insights from my career, which may be of interest to future leaders.

- I have continually experienced and learned about facing challenges, developing standards, and providing leadership through my work with many mentors.
- My work has broadened from its original exclusive perspective on science in education to include technology, engineering, and mathematics.
- I clearly benefitted from perspectives that included classroom teaching at the elementary, middle, and high school levels, as well as at the undergraduate and graduate levels. My understanding of the history of science education, curriculum development, national policy, and academic psychology and philosophy has influenced my view of science education.
- A career that ranged from teaching science in classrooms to developing international assessments has, to say the least, contributed to my broad and deep perspectives of education leadership.
- I have learned that the work of leadership requires resolving what appear to be contradictions as the aims of science education shift with changing social and political priorities.

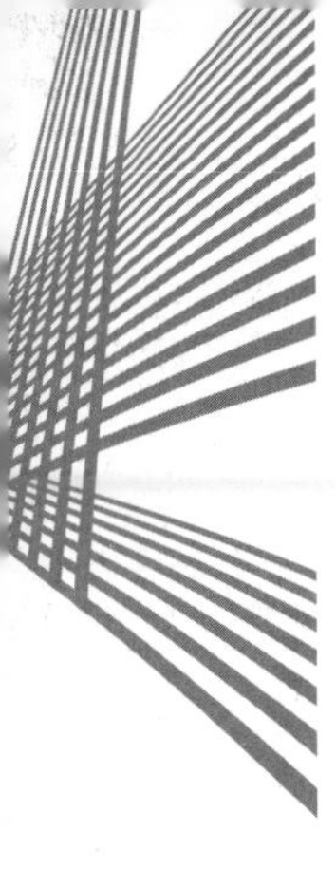

CHAPTER 2

A CAREER OF SCIENCE TEACHING, STANDARDS, AND EDUCATION POLICY—SO FAR

Stephen Pruitt

First and foremost, I spent 12 years as a high school science teacher. I am very proud of my time in the classroom. I left the classroom not for money or titles, but because I felt I could support more students by working with teachers. Serving as a seminar leader in this book gives me an opportunity to support teachers and science education as a whole.

In my previous capacity in state government, I had the opportunity to manage educational change and lead people. When I started at the Georgia Department of Education (GaDOE), the condition of science education was insufficient for serving all students. In 2003, only 36% of African American students passed the Georgia High School Graduation Test in science; those students who did not pass were unable to receive a high school diploma. We instituted programs meant to increase the success of students in science, including the largest change to the state standards in Georgia's history and a program to support science teachers in struggling schools. As a result, the state passage rate for all students went from 68% to 90% during my tenure at the agency. Passage rates for African American students went up to 84% (Georgia Department of Education 2002, 2010). In addition, during my time at the GaDOE as curriculum director, I oversaw the implementation of new math and language arts standards. Finally, as chief of staff, I oversaw the implementation of some of the largest budget cuts in education and to the GaDOE in Georgia's history. These changes required a solid knowledge of state and federal regulations and the creativity to help districts and the agency literally do more with less funding. I was also responsible for moving the strategic plan of the agency forward. I focused on ensuring that all efforts supported the organization's goal of leading the nation in improving student achievement.

My role as chief of staff also involved coordinating and working with our state board of education and staff. I also spent a great deal of time interacting with board members to answer their questions and prepare them for meetings. I always felt that some of the most important tasks involved working with the board. The more I worked with them, the more smoothly our board meetings proceeded. I was often able to deal with issues long before they became public or or needed to be addressed in the actual board meetings. I believe they trusted me, and that led to a great working relationship. This facet of my experience shaped my understanding of the need to work with diverse groups to develop a common vision and goal around the whole enterprise of education.

While serving as science supervisor at the GaDOE, I had a staff of 18 individuals who were charged with working with struggling schools and teachers in science. That group expanded during my time at the GaDOE until I assumed a position as chief of staff at the GaDOE and began overseeing the day-to-day operations of an organization with more than 600 employees. I had a hand in hiring all personnel and addressing issues that arose among agency employees. At Achieve, I oversaw a division of individuals focused on the implementation of the *Next Generation Science Standards* (*NGSS*). It is my fervent belief that people within an organization can contribute if they are put in the correct positions and given appropriate direction and support.

In my role at Achieve, I led one of the biggest reform efforts in science education in the early 21st century. The development of the *NGSS* required me to understand not only the science and science education research needed for new standards but also the politics around the standards. I worked with many groups and individuals to build the best set of standards possible. The ability to operate in a political environment and rally support from associated organizations and partners is critical in an era of standards-based reform. We are in a time of tremendous change, and leaders need to be able to support all stakeholders through the changes required to improve science education. In the development of the *NGSS*, I had the honor of working with the best and brightest from around the country in science, science education, and policy. How did this shape my perspective? Perhaps the most significant effect on me was the irrefutable reality that leadership matters. The leadership provided by the teachers, administrators, science researchers, policy makers, and engineers on this project was an incredible example of what can happen when there is a common goal—in this case, to open access to a quality science education for all students.

In October 2015, I had the opportunity to become the commissioner of education for the Commonwealth of Kentucky. While leaving full-time science work was tough, the opportunity to work with and in the state of Kentucky was too great to pass up. My role allows me to work with all aspects of education, and my experiences with *NGSS* and Georgia have given me a unique perspective.

As Rodger did in the previous section, I would like to conclude with a few insights from my experiences that guide my thinking about science education:

- We should all embrace the idea that we never stop learning. Thus, mentors and leaders around us should be encouraged and sought out.
- Teachers are the key to quality education. They certainly need support, but they are among the most creative and courageous people I know.
- Teaching is a profession and should be treated like a profession by the general populace, as well as by teachers themselves. We should boldly say, "I *am* a teacher" and never say, "I am *only* a teacher."
- For any type of change in schools, administrators must be provided with as much support as teachers.
- Science is open to all students, and it should be their choice if they select a career in science. They will only make such a choice if they receive a quality science education.
- A lot can be accomplished if diverse groups of people focus on a common goal, but this requires leadership and vision.
- A varied background has shaped what I believe; the heart of a classroom teacher and the faces of children have made me who I am.

CONCLUSION

We hope these introductions provided enough information about our background and experiences to answer a reasonable question—what qualifies these two to lead this seminar? You will hear more about our ideas, experiences, perspectives, and leadership as the seminar continues.

OUR COMMON PESPECTIVE ON THE SEMINAR'S TOPICS

This section introduces the seminar's general topics and the sections of the book. This introduction to the book's topics serves as the basis for our common perspective and highlights the opportunities for leadership in science education.

Science Education in America: Historical Perspectives

The history of science education reflects shifts in the priorities among goals and changes in the fundamental purpose of science education. Understanding and expressing contemporary purposes, aims, and goals are fundamental challenges for the science education community.

The Purposes and Goals of Science Education

Addressing the contemporary challenges of science education begins with clarifying the purposes—how the national and state standards express those goals.

National Standards: Policies for Science Education

An era of standards-based reform began in the late 20th century and continues in the early decades of the 21st century. National and state standards offer the potential for greater coherence within science education. Implementing these policies, however, is often met with the countervailing challenges of politics.

State Standards and District Leadership: Policies and Politics in Science Education

Ultimately, most of the challenges in science education reside at state and district levels, especially when the challenges of reform are perceived in terms of changing classroom teaching and enhancing student learning. State policies, such as new standards, have to be addressed by district leaders. They are an essential link in a chain of education reform that includes national and state leaders, college and university faculty, district administrators and science coordinators, and ultimately classroom teachers.

Curriculum Programs for Science Education

In American schools, fragmented and generally inconsistent curriculum programs result in less-than-optimal learning and inadequate levels of student achievement. A challenge resides in the commitment to, and implementation of, standards and the potential to increase the focus, rigor, and coherence in K–12 curriculum programs. At least two challenges present themselves: (1) the viability of new commercial programs, and (2) the need for understanding and leadership to build effective K–12 curriculum programs.

Classroom Practices for Science Teaching

Changing the quality of interactions between teachers and students—the unique and varied classroom practices—is the most basic reform in science education. The ultimate challenge is improving science teaching to enhance student learning.

Professional Development for Teachers of Science

Teacher preparation and professional development processes are inadequate to support contemporary reforms. Beyond the issues of teacher shortage, recruitment, and retention, our challenge is for continuous, rigorous professional development that begins with solid undergraduate preparation in science content, educational programs, and teaching practices. That initial professional education must be carried into the induction years and continue for the duration of the science teacher's professional life.

Assessment and Accountability in Science Education

The education system is overburdened with assessments that are not aligned with our valued goals for all students. The challenge is at once both simple and complex—align classroom, district, state, and national assessments with those outcomes that the science education community accepts and values.

Reforms, Policies, and Politics in Science Education

A mixture of initiatives, policies, and politics influences the science education system. There is a clear and compelling need for leadership with appropriate visions and adequate abilities to address challenges and opportunities for reform of curriculum programs, instructional practices, and assessments for science education.

Leadership and Education

Leadership and improvement of science education cannot rely on a few prominent leaders. Science education needs leaders at all levels, from classroom teachers to the U.S. Department of Education. Helping individuals develop the knowledge, abilities, and, most important, courage to lead is a challenge for the science education community.

Leadership in Contemporary Science Education

This final section serves as a conclusion and summarizes a range of leadership opportunities.

ISSUES AND QUESTIONS FOR DISCUSSION

1. If you had to identify two or three challenges in science education, what would they be?
2. What would you propose as solutions to the challenges you identified?
3. How would you describe your current leadership as a science educator?
4. Leaders must have a vision and a plan; the two are inextricable. Explain why you do or do not agree with this idea.
5. What is your experience with standards for science education? Are you familiar with national standards? Your state standards?
6. How would you describe the response to nationally developed standards? To state standards?
7. What recent changes have you seen in science education?
8. What are your major concerns relative to contemporary science education?

REFERENCES

Bybee, R. W. 1993. *Reforming science education: Social perspectives and personal reflections.* New York: Teachers College Press, Columbia University.

Bybee, R. W. 1997. *Achieving scientific literacy: From purposes to practices.* Portsmouth, NH: Heinmann.

Bybee, R. W. 2010. *The teaching of science: 21st-century perspectives.* Arlington, VA: NSTA Press.

Bybee, R. 2013a. *The case for STEM education: Challenges and opportunities.* Arlington, VA: NSTA Press.

Bybee, R. 2013b. *Translating the* NGSS *for classroom instruction.* Arlington, VA: NSTA Press.

Bybee, R. 2015. *The BSCS 5E Instructional Model: Creating teachable moments.* Arlington, VA: NSTA Press.

Bybee, R. W., and E. G. Gee. 1982. *Violence, values, and justice in the schools.* Boston, MA: Allyn and Bacon.

National Research Council (NRC). 2012. *A framework for K–12 science education: Practices, crosscutting concepts, and core ideas.* Washington, DC: National Academies Press.

NGSS Lead States. 2013. *Next Generation Science Standards: For states, by states.* Washington, DC: National Academies Press. *www.nextgenscience.org/next-generation-science-standards.*

Sund, R. B., and R. W. Bybee, eds. 1973. *Becoming a better elementary science teacher.* Columbus, OH: Charles E. Merrill Publishing Company.

SECTION II

SCIENCE EDUCATION IN AMERICA

HISTORICAL PERSPECTIVES

SECTION II

Each generation has the challenge of learning a diverse range of ideas; developing skills and abilities; and cultivating beliefs, values, and sensibilities in the interest of productively maintaining and changing society.

We believe that those in the science education community should understand the historical foundations that underlie contemporary challenges, reforms, and opportunities for leadership. As you will see in this section, science education in the United States has a rich and varied history. A clear understanding of present goals, policies, programs, and practices has roots in the past.

In general, reforms of science education are initiated by forces outside the community of science educators. Social and economic changes, political and public priorities, and concerns of business and industry can result in calls to improve education in general and science education in particular. One contemporary result of these forces has been the development of national and state standards.

One way to introduce the reality of different forces that influence science education is by reviewing the history. Chapter 3 presents several historical examples of science education reform, with an emphasis on school programs and classroom practices. Chapter 4 discusses the launch of Sputnik as part of a significant era of reform.

For those interested in details of science education in particular periods, we refer you to Appendix A, which presents the following summaries of historical perspectives for 1635 through 1965:

- Early Education (1635–1751)
- Emergence of Science Education (1751–1821)
- Science Education Is Established (1821–1870)
- College and Committee Policies Influence Science Education (1870–1910)
- School Reorganization and Development of General Science and the Junior High School (1910–1930)
- Progressive Policies and Science Education (1930–1945)
- Post–World War II: New Policies for Preparation of Science Teachers (1945–1955)
- A Revolution in School Science Programs (1955–1965)

SUGGESTED READINGS

Atkin, J. M., and P. Black. 2003. *Inside science education reform: A history of curricular and policy change.* New York: Teachers College Press, Columbia University. Personal histories from two leaders of reform in the United States and the United Kingdom.

DeBoer, G. 1991. *A history of ideas in science education: Implications for practice.* New York: Teachers College Press, Columbia University. An excellent introduction to the history of science education.

DeBoer, G. 2014. The history of science curriculum reform in the United States. In *Handbook of research on science education, volume II,* ed. N. Lederman and S. Abell, 559–578. New York: Routledge.

DeHart Hurd, P. 1961. *Biological education in American secondary schools 1890–1960.* Washington, DC: American Institute of Biological Sciences (AIBS). This book is perhaps the definitive history of biology education.

Rudolph, J. L. 2002. *Scientists in the classroom.* New York: Palgrave. The most insightful history of the Sputnik era in science education.

Tyack, D., and L. Cuban. 1995. *Tinkering toward utopia: A century of public school reform.* Cambridge, MA: Harvard University Press. The title and subtitle tell the story.

CHAPTER 3

SCIENCE EDUCATION IN AMERICA

DIFFERENT MODELS OF PROGRAMS AND PRACTICES

In this chapter, we examine several historical models of curriculum programs and associated instructional practices. We do not attempt to provide a thorough and detailed history of science education. For that, we refer you to George DeBoer's 1991 book *A History of Ideas in Science Education: Implications for Practice.* Rather, we introduce a number of significant reports and policy statements and examine the influences of science education reform. Our emphasis is primarily on curriculum and instruction. We begin with a discussion of aims as they influence programs and practices.

QUESTIONS CENTRAL TO DIFFERENT MODELS OF SCIENCE PROGRAMS

When constructing curricula and planning instruction, science educators consider several questions either explicitly or implicitly, depending on their perspectives. Those questions include the following:

- What does the scientific community know in a specific discipline or disciplines?
- How do scientists develop new knowledge and advance their understanding of natural phenomena and the designed world?
- What does the education community know about how students learn and develop at a particular age or grade level?
- How will the science education program contribute to students' obligations as citizens?

- How will the science education program contribute to students' awareness of and preparation for science- and technology-related careers?

These questions center on historical aims of science education: scientific knowledge, scientific methods, personal development, social issues, and careers. Details of these aims are described in Chapter 6. Here, the emphasis is on the role of these aims in different models of science programs.

The organization of programs and practices in science reflects the way in which the aims are structured, and the structure and emphasis they receive are functions of science educators' perspectives and societal forces. Different aims and different arrangements of these aims account for the range of approaches to classroom science teaching. To clarify what is meant by the organization of science programs and practices, an example from the performing arts may help. Three underlying elements in the performing arts are story, gesture, and music, but the priority given to these basic elements results in three entirely different art forms. Drama emphasizes the story, followed by gesture and music; in ballet, gesture is the primary element, then music and story; opera grants primacy to music over story and gesture. Although these arts include all three elements in some form, they are different because of the ways they emphasize, modify, and vary these elements. The same is true for science programs and practices. For example, if a teacher has decided that knowledge of science concepts is *the* important aim of instruction, then the teacher may include some emphasis on method, social issues, personal development, and careers in designing the science program. Science concepts would dominate the selections of topics and emphasis of the curriculum. However, the curriculum and instruction with social issues as the primary aim would be very different. In this case, topics and experiences that focus on students' understanding of science-related social issues (e.g., addressing climate change, controlling natural hazards, and preventing and treating infectious diseases) would have priority. Science teachers would introduce concepts that expand on and complement students' experiences.

Examination and organization of these aims is a useful way to describe historical changes, present differences, and explore contemporary trends in science programs. Several examples from the history of science education in the United States should help clarify the structure of science education and the role of societal influence in forming and reforming the structure of science programs and practices.

HISTORICAL MODELS OF SCIENCE EDUCATION

In this section, we use historical examples to illustrate different models of science teaching and demonstrate how societal pressures affect the aims and structure of science education, though this discussion is not intended as a comprehensive history. One example comes from the elementary level at the turn of the century, another from the secondary level in the 1930s, and the last from the curriculum reform movements of the 1960s and 1970s.

Before exploring the models of science education, we briefly review the early educational history of America.

Education in the Colonial Period

The early education history of this country reveals that little or no science instruction occurred. Colonization began in 1630 with the Great Migration to the New World by those dissatisfied with conditions in Europe. In the American wilderness lay the promise of solving those problems, at least economically, by exploiting the natural resources of the land. The culture and institutions of the Old World were transplanted to help compensate for the unfamiliarity of the new environment. In addition, the migrants brought certain ideals that altered those old institutions for a new and different purpose. Because 90% of the migrants to the colonies were English (Cremin 1970; see also Cohen 1974, pp. 7–30], it was natural that English language, customs, and traditions predominated. Even though the English colonists' environment, religious motivation, and sense of destiny differed from that of the countries from which other colonists came, every aspect of colonial culture had a distinctly English flair. In the Puritan and Quaker communities, the education institutions replicated those in England. In the southern colonies, the plantation system was distinctly reminiscent of the old English manorial system.

It was in New England that the English educational antecedents were transplanted most successfully. Because of the poor topography and soil, New Englanders were forced to use subsistence farming and gather together in towns. Fishing and, later, manufacturing would compensate for this lack of agrarian opportunity, but the formation of towns meanwhile provided the necessary setting for the establishment of schools. Cremin (1970) points to the fact that the Puritans emigrated as families as another important factor in the development of education opportunity in New England. Puritanism itself was an important factor because knowledge, in its scheme, was a fundamental part of obtaining salvation. To the Puritans, ignorance was considered the chief obstacle in the way of salvation. Of course, this ignorance was perceived to be spiritual rather than intellectual, and education consequently had a strong religious flavor.

Educational development in the southern colonies was hindered by geographical separation from colonies in New England and the fact that the agrarian economy, based on plantations, discouraged the development of towns and, subsequently, of schools. The most dominant obstacle to educational opportunities in the south was slavery. There was a fear that education would make slaves aware of their situation, incite rebellion, and undermine the economy and class system (Cremin 1970; Cohen 1974).

The New England Puritan community established a nearly universal system of popular education. This system would later be the model for the rest of the country in the development of public education. The Massachusetts Bay School Laws of 1642 and 1647 laid the foundation for this system by requiring the compulsory maintenance of town schools and the compulsory education of all children. Almost as soon as the colonists of New England set foot on American soil, or at least as soon as the first roofs were raised over their heads, dame schools began. A dame school involved the gathering of neighborhood children to the home of one woman in the neighborhood who taught the alphabet and other rudimentary skills. Sometimes the teacher was hired as a servant to teach the children of the household and those of others. In that case, the school was called a petty school. In some towns, "free" schools were founded so that all children of the community might attend.

The Massachusetts Law of 1642 ordered the town selectmen to determine periodically whether parents and masters were complying with the obligation to educate their children, or the children under their authority, "for the calling and employment" that they must someday assume (Cohen 1974, pp. 44–45). Those found not in compliance with this command were to be fined. Enforcement was difficult, though, so in 1648 an amendment to the statute required that children be taught an Orthodox catechism, something more measureable than "calling and employment."

In 1647, the requirement for compulsory schools was established with the Old Deluder Satan Act (Cohen 1974, pp. 45–46), a statute meant to preserve religious orthodoxy and otherwise encourage the education of the young. The statute's language illustrates the religious sentiment that governed the education environment. Here is the beginning statement of the law (Cohen 1974, pp. 47–48):

> *It being one chief project of that old deluder, Satan, to keep men from the knowledge of the Scripture. …*
>
> *It is there for ordered that every township in this jurisdiction after the Lord hath increased them (in) number to fifty householders, shall then forthwith appoint one within their town to teach all such children as shall resort to him to write and read, whose wages shall be paid either by the parents or masters of such children, or by the inhabitants in general. . ."*

As you can see in this selection, primary education was the greatest concern, but the Old Deluder Satan Act also required the establishment of grammar schools. These schools were the initial form of secondary education that developed alongside the elementary schools. They corresponded closely with the English Latin grammar schools and prepared young men of the colonies for university (Cremin 1970; Cohen 1974).

The Puritan educational system had the foundation and vitality to enjoy continued development in America. Although established with the primary purposes of maintaining Protestant beliefs and values, we note that the education system also reflected concerns about social stability and community growth.

What about science in the colonial educational system? As we noted, there was the use of natural resources, the need to farm in a new environment, and, undoubtedly, concerns about health. While Puritans were confronting these science-related issues, Sir Francis Bacon was calling for a spirit of scientific inquiry, the study of nature, and experimentation with natural phenomena. During this era, William Harvey discovered the circulation of blood; Robert Boyle developed an understanding of changes in the volume and pressure of gases; and Sir Isaac Newton proposed theories of matter, motion, and space.

Although the Puritans were not opposed to science (they saw the natural world as God's creation, and this deserved to be studied), there was only a slow infusion of science into education. During the late 18th and early 19th centuries, religious indoctrination decreased in schools and utilitarian objectives increased. With this change, science slowly gained a place in American education.

The earliest forms of science instruction for children have been traced to stories and didactic literature designed for home tutoring. These materials were based on the theories of John Locke and Jean-Jacques Rousseau and accentuated the firsthand study of "things and phenomena," as well as Christian doctrine. Though originating around 1750, these materials reached their peak between 1800 and 1825. With the rise in group instruction, books for home use evolved into textbooks designed for school use. Science was included in many of the lessons, which stressed the memorization of factual knowledge, generally supporting theological concepts (Underhill 1941).

From approximately 1860 to 1880, Object Teaching was another movement in elementary schools. The primary aim of Object Teaching was mental development, including cultivating the capacities of reasoning and thinking; the learning of science subject matter was of secondary importance. A method of teaching based primarily on the ideas of Johann Pestalozzi, this movement had some influence on American education, but from its inception, it was strongly criticized and seldom implemented. The criticisms in science education came largely from educators advocating for scientific knowledge as an aim. Interestingly, contemporary standards emphasize understanding "things and phenomena" as the basis for applying scientific practices and concepts. In time, Object Teaching was replaced by science as a school subject, with specific curricula (Underhill 1941).

The mid-18th to early 19th centuries marked the period of the academy in secondary-level education. As religion ceased to dominate the instruction program, it was replaced with a more practical curriculum, which included some science in contexts such as agriculture and navigation.

Following a depression in 1873, there were severe criticisms of American schools. This is an example of contextual focus influencing changes in American education. The attack was led by citizens asking a question common to such periods: "What are we getting for our money?" As the social and economic patterns changed, the educators followed with clear demands for more education in science. The aim of science education was to give the public greater understanding of science and technology, the warp and woof of the Industrial Revolution. From this brief overview of education and the introduction of science, we turn to a model for elementary schools. (For additional details, see Appendix A, Tables A.1, A.2, and A.3).

Two Models of Science for Elementary Schools

In the late 1800s, the combination of industrial expansion and migrations from rural areas to urban centers led to two models of science teaching at the elementary level. One was a model referred to as "elementary science" that primarily answered the question, "What does the scientific community know in specific disciplines?"; the other was nature study and had as its primary question, "How can science education programs contribute to students' appreciation of nature and their future role as citizens?"

Changes in society resulting from a developing technology created an interest in science and gave impetus to the rise of elementary science: "The chief emphasis during this period (about 1880) was in terms of giving a wider knowledge and understanding of the rapidly increasing science and technology" (Underhill 1941, p. 111). There were several education leaders who contributed to this model of elementary science, namely Francis Parker, Wilbur Jackman, William T. Harris, and E. G. Howe.

Francis Parker's model used a unifying theme for the elementary school. He wanted students to understand the universe and use scientific techniques as a means of solving problems (Underhill 1941). Wilbur Jackman worked with Parker at the Cook County Normal School to develop an elementary science program that had scientific concepts as the major organizational aims and the use of observation and experimentation as methodological goals (Jackman 1891). Jackman's model had scientific knowledge as the primary aim and scientific methods as a secondary aim. It is interesting to note that Jackman gave the title "nature study" to his elementary science program, and his book (Jackman 1891) gives considerable attention to scientific methodology. In fact, his methodological ideas were later of considerable influence on the nature study movement.

William T. Harris often receives credit for formulating the first substantial elementary science program (Underhill 1941). The Harris model emphasized the relationship of ideas and used science, as opposed to other disciplines, as an organizational framework. E. G. Howe (1894) later changed the Harris science curriculum into a program titled *Systematic Science Teaching.* Howe's program was conceptually similar to the Harris model, but as a curriculum it was more elaborate. The common factor among these models during the period from 1870 to 1900 was the dominance of scientific knowledge as their aim. These models were substantially influenced by the transition of this country from agrarian to industrial-technological society.

Nature study was designed as a countervailing force to slow or stop the emigration of people from rural agricultural communities to urban industrial centers. Many people who had migrated from rural to urban environments found themselves unemployed and on the swelling relief rolls in the cities. Science education programs were seen by some as the remedy to this social problem. In 1895, the Committee for the Promotion of Agriculture in New York State and the New York Association for Improving the Condition of the Poor (Underhill 1941) called a joint conference. The conference examined the causes and possible remedies for the agricultural depression; many individuals turned to nature study as a possible solution. The nature study idea had originated in 1893 as the vision of Liberty Hyde Bailey, and the idea gained prominence in education in the context of the aforementioned needs of society. Nature study was supported as a way of helping children become interested in farming, thereby slowing emigration to cities: "Nature study was the great remedy for the alienation of man from the land and from his neighbor" (Cremin 1964, p. 77).

Liberty Hyde Bailey directed the nature study program at Cornell University. Over the years, Bailey and his associates were instrumental in developing and disseminating materials representing this second model of science teaching at the elementary level. Bailey (1903) contrasted nature study with elementary science in his book *The Nature Study Idea:*

> *Nature study is a revolt from the teaching of mere science in the elementary grades. In teaching practice, the work and the methods of the two [elementary science and nature study] integrate ... and as the high school and college are approached, nature study passes into science teaching, or*

gives way to it; but the ideals are distinct—they should be contrasted rather than compared. Nature study is not science. [It is not facts.] It is concerned with the child's outlook on the world. (p. 4)

Bailey continued, making the distinction between nature study and elementary science clear:

Nature may be studied with either of two objects: to discover new truths for the purpose of increasing the sum of human knowledge; or to put the pupil in a sympathetic attitude toward nature for the purpose of increasing his joy of living. The first object, whether pursued in a technical or elementary way, is a science-teaching movement, and its professed purpose is to make investigators and specialists. The second object is a nature study movement, and its purpose is to enable everyone to live a richer life, whatever his business or profession may be. (pp. 4–5)

These quotations clearly outline the differences between scientific knowledge and the societal issues models of science curriculum. The social conditions supported both curriculum models. The combination of an emerging industrial-technological society and the need to maintain a substantial agricultural base for society, as well as reduce unemployment in the large cities, influenced the development of different curriculum models. Thus, a nature study model, which stressed personal development and appreciation of nature, and an elementary science model, which stressed scientific knowledge and methods, were both implemented by science educators.

In summary, the slow but steady recognition of the role science and technology played in developing an industrial society inevitably resulted in a popular interest in science and, subsequently, in science education, including in elementary schools. Just before the beginning of the 20th century, there were two models of elementary science teaching: the content model and the nature study model. The first model of science curricula designed specifically for elementary schools was developed in the period between 1870 and 1900. In general, these courses of study were outlines of content from specialized science disciplines and were practical and utilitarian.

Nature study was a combination of the broader romanticism and a "new" education that was based on the theories of John Amos Comenius, Pestalozzi, Rousseau, and Friedrich Froebel. Nature study was in fact a transformation of the earlier object study movement. The nature study model emphasized the child's interest as the origin of study, increased the individual's appreciation of and attitude toward nature, and promoted the student's personal development. Nature study was mostly observational and largely biological.

We conclude this discussion of science models for elementary schools with a variation on the elementary science model. Any discussion of elementary school science would be negligent if it did not include Gerald S. Craig's contributions. In 1927, Gerald Craig completed a dissertation at Teachers College at Columbia University (Craig 1927). Craig's work provided elementary science with a teaching model that lasted until the reform movement of the 1960s, and in many respects, this model still affects elementary education today. After carefully listening to children's questions, Craig set forth the science generalizations that were fundamental to providing answers to the children's queries. Craig then placed the science generalizations in a context that also had utilitarian meaning to the students. These ideas formed the core of Craig's curriculum and a textbook series. Craig's study has had a profound effect on science education, and his later writing was a great force in directing science education at the elementary level. Textbook series dominated elementary school teaching from the 1940 to the 1960s. In the 1960s, when inquiry-oriented curriculum materials were developed, the challenge of science educators was to displace the older "science readers" with the newer programs. To date, we have only had limited success in addressing this challenge. (For more details, see Appendix A, Tables A.3 and A.4.)

Secondary Education in an Industrial Society: Reports and Recommendations

In the late 19th century, American secondary education was subject to critical review relative to its purpose and function. Students with diverse interests and needs motivated administrators to confront the challenge of providing a curriculum that was broader than one designed exclusively for college preparation. The growing industrialization and urbanization in American society increased the demand for a high school education. At the same time, the percentage of students with college aspirations decreased. This situation paved the way for a major report on secondary education and the high school science curriculum.

A report released in 1893 by the National Education Association (NEA) facilitated a new era for science education. The Committee of Ten (NEA 1893) asserted that all students should be taught the same curriculum whether they intend to attend college or not. The committee detailed such matters as the subjects to be taught and the hours per week and weeks per year to be devoted to each subject. This report helped reduce the domination of colleges over high school programs and formed a stronger connection between high school and elementary school programs. The report, however, did place stress on academic or intellectual goals as the object of teaching.

The Committee of Ten report addressed challenges centering on the goals and content of secondary education. Should high schools offer students different courses of study if they are ending their education at the secondary level and entering the workforce? The report recommended against separate courses of study for—to use contemporary terms—college and career.

By 1915, the emphasis was shifting to goals broader than college entrance. The *Central Association of Science and Mathematics Teachers Report of the Committee on a Four-Year High School Science Course* (Central Association of Science and Mathematics Teachers 1916) suggested that science should (1) give pupils such a knowledge of the world of nature as will help them get along better in everyday life; (2) stimulate students to more direct purposeful activity; (3) help students choose intelligently for future occupations; (4) give pupils methods of obtaining accurate knowledge; and (5) give pupils greater, clearer, and more intelligent enjoyment of life.

College domination of the high school science program was further eroded when the Commission on the Reorganization of Secondary Education (1918) completed its work and published *The Cardinal Principles of Secondary Education*. This report called for a shift in the goals of education from the narrower intellectual indoctrination to a broadened socialization of the student. The seven cardinal principles were health, command of fundamental processes, worthy home membership, vocation, civic education, worthy use of leisure, and ethical character. School subjects were to be reorganized so students would attain the knowledge and skills associated with these principles more effectively.

Reorganization of Science in Secondary Schools (Caldwell 1920) was also published by a subcommittee of the original commission. The report discussed the contributions science teaching could make to the cardinal principles of secondary education. In general, the report stressed the importance of organization and sequencing of secondary science, and it also pointed out social goals beyond the traditional knowledge goals usually stressed in secondary science.

The use of scientific processes as instructional strategies to teach scientific knowledge was in evidence by 1924. The American Association for the Advancement of Science Committee on the Place of the Sciences in Education (Caldwell 1924) reported on a study of the problems in science teaching. The report underscored the importance of scientific thinking as an objective of teaching. The purposes of instruction, according to the committee, should be founded on educational and scientific observation and experimentation for "a factual basis worthy of the spirit of science" (Caldwell 1924, p. 536).

A 1932 National Survey of Secondary Education included a survey of science teaching guides, courses of study, and syllabi. The study was conducted by the Office of Education and reported by Wilbur Beauchamp (1933) in the monograph *Instruction in Science*. In general, the conclusions suggested that the current programs lacked a coherent theoretical structure; there was confusion of grade-level placement of courses and chaos concerning the methods of teaching used. In contrast to the turmoil, the report suggested a variety of innovative practices that should be considered when constructing new programs, such as problem methods of teaching, interpretation of the environment, use of illustrative materials and demonstrations, coordination of laboratory and textbook work, and greater use of visual aids.

The reports cited above represent changes in values, as seen in education, from the high ideals and social unity following World War I to the disillusionment of the economic depression of the late 1920s and 1930s. The 1930s witnessed a number of committee reports that all attempted to help reform education in some way. It is as though these reports all attempted to answer George Counts's (1932) question, *Dare the Schools Build a New Social Order?* (For more details, see Appendix A, Tables A.4 and A.5.)

Models of Science for Secondary Schools

Historically, at the secondary level, the model of science teaching that has been most dominant answered the question, "What does the scientific community know in specific disciplines?" High school teachers made their position quite clear concerning the nature study model:

> *High school teachers of science must protest against a mass of so-called nature study, more or less sentimental and worthless … and must not be content simply with treating lightly this farcical science teaching, or passing it by with silent contempt. (Brownell 1903, p. 253)*

Scientific knowledge continued to be the dominant goal of secondary science teaching. This observation is supported by the aforementioned 1932 report *Instruction in Science* (Beauchamp

1933). A review of the objectives indicated that acquiring knowledge was the primary aim of science instruction, followed by an emphasis on using scientific methods. The methodological aim, as described earlier, answers the question, "How do scientists develop new knowledge and advance their understanding of natural phenomena?" This aim is most directly concerned with the methods and practices of scientific inquiry.

Another 1932 study compared the objectives of secondary science teachers and members of the National Association for Research in Science Teaching (NARST; Hunter and Knapp 1932). In 1938, NARST replicated the 1932 survey of secondary science teachers ("Report of Committee on Secondary School Science of the National Association for Research in Science Teaching"; NARST 1938) and reported the following:

> *By far, the greatest emphasis is placed by both groups (junior and senior high schools science teachers) on the acquisition of useful general information and practical understandings which come from contacts with science. (p. 224)*

These reports indicate the dominant influence of scientific knowledge as the aim of secondary science teaching during the 1930s.

The aim of introducing students to the methods and experiences of scientific investigation had been discussed in the literature of science education for some time, and during the early part of the 20th century, the aim received new support. In *Experience and Education,* John Dewey (1938) strongly criticized the knowledge model of education. His pragmatic philosophy reaffirmed the thesis that education should be functional, center on the student, and reflect the realities of the time—which included solving many social problems. One of Dewey's significant contributions to science education was his contention that the methods of science were as important as the knowledge of science.

As early as 1910, Dewey published an article called "Science as Subject Matter and as Method" (Dewey 1910), in which his general theme was that science teaching gave too much emphasis to the accumulation of knowledge and not enough to science as an effective method of inquiry and a habit of mind. Near the conclusion of his 1910 publication, Dewey makes this powerful statement:

> *One of the only two articles that remain in my creed of life is that the future of our civilization depends upon the widening spread and deepening hold of the scientific habit of mind; and that the problem of problems in our education is therefore to discover how to nurture and make effective this scientific habit. (p. 127)*

Here is a clear call for a model of science curriculum that gives emphasis to scientific inquiry and practices and unfortunately may have influenced decades of science textbooks with an introductory chapter on "the scientific method."

In the context of models for science curriculum, Dewey's aim of reflective thinking became the presentation of a *problem,* formation of a *hypothesis, experimentation, collection of data,* and, finally, reaching a *conclusion.*

The conviction that teaching the scientific method was an important aim of science education received support from other individuals during this period. Hurd (1961) reports:

> *Many scientists in this time (1920–1930) expressed the opinion that the central purpose of education in science should be the development of an understanding of the nature of science, its methods, attitudes, and cultural impact. As a result, much educational research in the succeeding years was directed toward identifying the elements of the scientific method and scientific attitudes. (p. 51)*

The aim of teaching the scientific method had further support in journals (e.g., Special Committee on the Place of Science in Education 1928) and books (e.g., National Society for the Study of Education 1932; Progressive Education Association 1938).

During the period from 1920 to 1940, the knowledge model of secondary science continued to dominate science curriculum, although an increasing emphasis was also placed on scientific methods as an important focus of that curriculum. The knowledge model reflected a back-to-basics posture often associated with times of economic austerity. The emphasis on scientific methods was an educational manifestation of the need to solve the many social problems of that time. School systems changed their fundamental aims in response to the Great Depression. Similar changes had also occurred during the earlier depressions of the 1830s, 1850s, and 1870s. In the early decades of the 21st century, America is again experiencing pressures because of economic austerity and other problems such as emerging diseases, severe weather, and natural hazards.

The final model is covered in Chapter 4, which provides a thorough examination of the Sputnik era of reform. In the late 1950s, a science manpower shortage in America initiated a curriculum reform movement that accelerated and captured the public's attention with the launch of Sputnik by the Soviet Union.

Whether Soviet superiority was real or imagined, in the late 1950s and early 1960s, the United States responded as though it was real and moved the new frontier to space. Although there are some who doubt the importance of *Sputnik 1* as an incentive for change because reform was already in progress prior to the launch (NSF 1965), without the symbol created by *Sputnik*, curriculum reform probably would have been much less robust.

Curriculum reform in the early 1960s adopted a model of science education described by Jerome Bruner (1960) in *The Process of Education*. Scientific knowledge was the dominant aim, and the scientific method was the means to achieve this aim. In Bruner's model, knowledge consisted of science concepts forming the structure of a discipline. Bruner states his aim: "The curriculum of a subject should be determined by the most fundamental understanding that can be achieved of the underlying principles that give structure to that subject" (Bruner 1960, p. 31).

By 1964, scientific knowledge and methods were firmly established as the important aims in science education. The National Science Teachers Association (NSTA) publication *Theory Into Action … in Science Curriculum Development* (NSTA 1964) made this point clear. The aims were now called "the conceptual schemes of science" and "the process of science."

There were three significant structural changes during the early 1960s: (1) The importance of the aims that emphasized personal and social development declined, particularly at the secondary level (Hurd 1969); (2) the knowledge aim was revised and updated and became the primary aim, particularly at the secondary level; and (3) scientific methods were implemented as the means to achieve the goal of knowledge. Although the aims of personal and social development were virtually eliminated at the secondary level, they remained intact at the elementary level, primarily due to the influence of Piaget's theory of cognitive development (Piaget 1973).

A 1962 essay by Ralph Tyler, "Forces Redirecting Science Teaching," makes clear the influence of society on the changes in science education during this major period of transformation. Tyler cited the major forces redirecting science education as the technological revolution and the decreasing number of students entering careers associated with science and technology. He also pointed out the vast difference between the science programs that existed and the needs of those who were entering careers aligned with science. Finally, he listed the relationship between scientists and science teachers, the knowledge explosion, and the variety of student needs as other forces redirecting science education.

A stated objective of the curriculum reform in the 1960s was to increase the number of individuals pursuing careers in science and engineering, yet this aim was to be achieved through a curriculum model dominated by the goal of scientific knowledge and lacking in materials that explicitly focused on career awareness. Early editions of programs such as Physical Sciences Study Committee (PSSC) for physics, Biological Sciences Curriculum Study (BSCS) for biology, and Chemical Bond Approach (CBA) and Chemical Education Materials Study (CHEM study) for chemistry make little or no mention of careers related to these science disciplines, not to mention engineering and technology. (For more detail, see Appendix A, Tables A.5, A.6, and A.7.)

A PERSONAL PERSPECTIVE

NCLB—NO CHILDREN LACKING BIOLOGY?

Stephen Pruitt

So far, this chapter has reviewed the history of science education and pointed out how social events spurred different models of science programs. I want to take a slightly different tact. In 2001, the No Child Left Behind Act (NCLB) was signed into law. I will not spend time here discussing the pros or cons of NCLB; however, this law is an important part of science education history in that it drove many decisions that are currently in place in our nation's schools. In 2009, all states were required to have state-level science standards and assessments. What were the implications? Beyond the focus on mathematics and English language arts, how did NCLB change perceptions and programs for science education?

First, requiring each state to have state-level standards was a big step. There were many states with standards, but this requirement was a game-changer in that science had been out of the conversation because of the focus on mathematics and English language arts. Of course, states turned to the *National Science Education Standards* (NRC 1996) and the *Benchmarks for Science Literacy* (AAAS 1993) to help them develop their state standards. These two documents were incredibly influential in and critical to shaping the nation's thoughts about science classrooms. States developed unique standards based on their own understandings of these two documents. Of course, there was a lot of overlap between the two documents, but there were differences as well. As a result, many states used one as a base document and filled in gaps with the other. As a result, many state standards were too broad to be taught.

I remember seeing Georgia writers getting out of their cars during the revision of the Georgia state standards in the late 1990s, carrying all of their notebooks, three-ring binders, and textbooks, ready to write the new standards. As a result of using so many materials, the standards ended up the proverbial mile wide and inch deep. In that same time, interesting interpretations and implementation of key issues contained in those documents became rooted and continue today. I hope we can all agree that science concepts and processes have to be experienced to be engaging and allow students to understand science more deeply. Unfortunately, when states wrote their standards, most separated inquiry from content and typically ignored unifying themes altogether. I think it is fair to say the intention of the two major documents was for teachers to use inquiry in classrooms to let students experience science in a way that allows them to place it into a cognitive structure for later use. Unfortunately, in many classrooms, science became more about the fun and experiences and much less about a deep understanding of conceptual ideas through inquiry. I know I was guilty of this, especially in my early classroom years. I loved doing labs and demonstrations—I could blow up a methane bubble with the best of them. In the name of inquiry and the scientific method, I would have my students ask questions and make claims about the phenomena. The kids loved it, but many did not see connections of the lab or demo to the world around them or anything that affected their lives. I am afraid I was to blame for that. Later in my classroom career, I began to understand the need for students to see these connections and would offer them an opportunity to apply it to something they cared about. In the early decades of the 21st century, when *A Framework for K–12 Science Education* (NRC 2012) and the *Next Generation Science Standards* (*NGSS*; NGSS Lead States 2013) were being developed, these types of issues, as well as international perspectives, were considered. As a result, the *NGSS* require the integration of practice and content to push the system toward what the research shows to be a quality science education experience.

Standards are only part of the issue that came about as a result of NCLB. Assessment became a part of our culture and has significantly affected science education. As large-scale (state-level) assessments became more prevalent, the need to keep them cheap and easy to grade trumped quality. Because of the structure of state standards, state tests gave slight nods to inquiry by developing simple, contrived connections to inquiry through skills such as using parts of a microscope or testing the steps of the scientific method. Additionally, the content portion of the assessments stayed at the fact level. It is far easier and cheaper to write assessment items that require students to identify anaphase or a physical property. Another interesting item that is overlooked regarding the assessment requirements is how much the science students were required to learn has been narrowed and, as a result, scientific literacy has been limited. Here, we could get into the equity issues again, but I will save that for another chapter. NCLB and now the Every Student Succeeds Act (ESSA) require science to be assessed once in elementary school, once in middle school, and once in high school. While I have not reviewed all 50 states' assessment blueprints, the tested grades tend to also contain all of the tested material; that is to say, an eighth-grade test assesses eighth-grade material only. I have seen in some elementary schools that science is not taught until the testing year, and the volume of material is just not possible to teach; it can be covered but not properly taught or learned. But rather than discuss anecdotes, let's take a look at how the assessment policy has limited high school.

Biology has been elevated more than other sciences. For my biology friends, I know this is a good thing, but it has limited the measure of scientific literacy to one discipline. According to the National Center for Education Statistics (Kena et al. 2016), 96% of students take biology, or some version of it. So no student is lacking biology, but the other disciplines are at an incredible deficit. NCES also reports that 70% of students take a high school chemistry course and only 36% take a high school physics course. Only 28% of graduating students take a version of an Earth and space science class, which will contain information critical to 2 of the top 10 most sought-after jobs over the next 20 years. Only 30% of students take a rigorous three-course sequence in high school. Why is biology taken by so many students? The simple reality is that biology is the most typically assessed high school course. This has allowed students to avoid the other disciplines and has perpetuated inequities in science education. It also has damaged students' abilities to acquire core knowledge that would better prepare them for college and careers in the STEM fields, not to mention make them more scientifically literate.

Now, let me discuss a few final thoughts on this historical perspective. I cannot write about this without saying there were some good things about NCLB, the greatest being that for the first time we saw attention being given to *all* students. Underserved populations could not be "hidden" in the data. Having said that, we now have an opportunity for our Sputnik moment. In a recent survey, 84% of scientists from the American Association for the Advancement of Science (AAAS) and 68% of the general public rank science and STEM education in the United States as average or below average compared to the rest of the world (Pew Research Center 2015). Historically, it has been rare that the scientific community and the general public have the same opinion. If we can agree that science education needs some change, the *NGSS* can be the beginning of that change. There is much work to do around implementation, such as instructional materials and policies, which will be discussed in a later chapter. Our Sputnik moment—what I see as one of our biggest and most exciting events—is that the *NGSS* were written to ensure all students have opportunities with all standards.

CONCLUSION

Our theme of perspectives played an important role in the discussion of different models of science programs and practices. Many have missed an important point in discussing different models of science education programs. For example, nature study is often discussed using terms such as *interlude, intrusion,* or *interruption* in the development of science curricula for elementary schools. This interpretation derives from the dominance of a content or knowledge orientation within science education as opposed to a larger social perception of change in science curricula and instruction. The contextual forces at the turn of the century supported two types of elementary science programs, each with a different character. The developing urban-industrial society required science programs that were content oriented, vocational, and utilitarian. Because of the massive migration to the cities—as well as immigration from other countries, unemployment, and swelling welfare lines—a second science program was implemented for the rural agricultural communities. This nature study program was intended to slow the emigration from rural environments by developing an enjoyment and aesthetic appreciation of nature. It is true that many failed to see the larger direction of social development toward an industrial economy, and this inevitably resulted in the diminished effectiveness of the nature study movement. The important point is that science education did respond to the larger social demands with different models of programs and practices.

Models for the secondary level also highlight different perspectives. Those views include the prevailing dominance of scientific knowledge and other models such as the one proposed by John Dewey that emphasized scientific inquiry and practices. Still other models emphasize personal and social issues.

As Stephen pointed out, the No Child Left Behind legislation placed assessment at the forefront of education, which affected science education. In fact, that effect continues with the Every Student Succeeds Act of 2015.

REFERENCES

American Association for the Advancement of Science (AAAS). 1993. *Benchmarks for science literacy.* New York: Oxford University Press.

Bailey, L. H. 1903. *The nature study idea.* New York: Doubleday.

Beauchamp, W. 1933. *Instruction in science.* Washington, DC: U.S. Government Printing Office.

Brownell, H. 1903. Science teaching preparatory for the high school. *School Science and Mathematics* 2 (2): 253.

Bruner, J. 1960. *The process of education.* New York: Vintage.

Caldwell, O. W. 1920. *Reorganization of science in secondary schools: A report of the Commission on the Reorganization of Secondary Education.* Washington, DC: U.S. Government Printing Office

Caldwell, O. W. 1924. American Association for the Advancement of Science Committee on the place of the sciences in education. *Science* 60 (1563): 536–540.

Central Association of Science and Mathematics Teachers. 1916. Central Association of Science and Mathematics Teachers report of the committee on the four-year high school science course. *School Science and Mathematics* 16 (5): 393–399.

Cohen, S. 1974. *A history of colonial education: 1607–1776. Studies in the History of American Education* series. New York: John Wiley & Sons.

Commission on the Reorganization of Secondary Education. 1918. *The cardinal principles of secondary education.* Washington, DC: U.S. Bureau of Education.

Counts, G. 1932. *Dare the schools build a new social order?* New York: John Day.

Craig, G. 1927. *Certain techniques used in developing a course of study in science for Horace Mann Elementary School.* New York: Teachers College Press, Columbia University.

Cremin, L. 1964. *The transformation of the school: Progressivism in American education: 1876–1957.* New York: Knopf.

Cremin, L. 1970. *American education: The colonial experience: 1607–1783.* New York: Harper and Row Publishers.

DeBoer, G. 1991. *A history of ideas in science education: Implications for practice.* New York: Teachers College Press, Columbia University.

Dewey, J. 1938. *Experience and education.* New York: MacMillan Company

Dewey, J. 1910. Science as subject matter and as method. *Science* 31: 122–125.

Howe, E. G. 1894. *Systematic science teaching.* New York: D. Appleton.

Hunter, G., and R. Knapp. 1932. Science objectives at the junior- and senior- high-school level. *Science Education* 16 (5): 407–416.

Hurd, P. D. 1961. *Biological education in American secondary schools 1890–1960.* Washington, DC: American Institute of Biological Sciences (AIBS).

Hurd, P. D. 1969. *New directions in teaching secondary school science.* Chicago: Rand McNally.

Jackman, J. 1891. *Nature study for the common schools.* New York: Henry Holt.

Kena, G., W. Hussar, J. McFarland, C. de Brey, L. Musu-Gillette, X. Wang, J. Zhang, A. Rathbun, S. Wilkinson-Flicker, M. Diliberti, A. Barmer, F. Bullock Mann, and E. Dunlop Velez. 2016. *The condition of education 2016* (NCES 2016-144). Washington, DC: U.S. Department of Education, National Center for Education Statistics. *http://nces.ed.gov/pubsearch.*

National Association for Research in Science Teaching (NARST). 1938. Report of Committee on Secondary School Science of the National Association for Research in Science Teaching. *Science Education* 22 (5): 223–233.

National Education Association (NEA). 1893. *Report of the Committee of Ten on secondary school studies.* Washington, DC: U.S. Government Printing Office.

National Research Council (NRC). 1996. *National science education standards.* Washington, DC: National Academies Press.

National Research Council (NRC). 2012. *A framework for K–12 science education: Practices, crosscutting concepts, and core ideas.* Washington, DC: National Academies Press.

National Science Foundation (NSF). 1965. *Science education in the schools of the United States.* Report to the House Committee on Science Research and Development. Washington, DC: U.S. Government Printing Office.

NGSS Lead States. 2013. *Next Generation Science Standards: For states, by states.* Washington, DC: National Academies Press. *www.nextgenscience.org/next-generation-science-standards.*

National Science Teachers Association (NSTA). 1964. *Theory into action … in science curriculum development.* Washington, DC: NSTA.

Pew Research Center. 2015. Public and scientists' views on science and society. Washington, DC: Pew Research Center. *www.pewinternet.org/2015/01/29/public-and-scientists-views-on-science-and-society/#*

Piaget, J. 1973. *To understand is to invent.* New York: Grossman.

Progressive Education Association. 1938. *Science in general education.* Report of the Committee on the Function of Science in General Education. New York: Appleton-Century-Crofts.

Special Committee on the Place of Science in Education. 1928. On the place of science in education. *School Science and Mathematics* 28 (6): 640–664.

Tyler, R. 1962. Forces redirecting science teaching. In *Revolution in teaching new theory, technology, and curricula,* ed. A. DeGrazin and D. Sohn, 187–193. New York: Bantam.

Underhill, O. E. 1941. *The origins and development of elementary school science.* New York: Scott Foresman.

CHAPTER 4

SPUTNIK AND SCIENCE EDUCATION

AN ERA OF MAJOR REFORM

The education reform of the 1950s and 1960s was already in progress in October 1957, when the Soviet Union placed Sputnik in orbit. In 1951, with the leadership of Max Beberman, the University of Illinois Committee on School Mathematics (UICSM) initiated a reform of the secondary school math curriculum. The stage had been set for reform in science education by Jerrold Zacharias, who in 1956 began the Physical Sciences Study Committee (PSSC; Zacharias 1956), a year before the launch of Sputnik. However, Sputnik still played a significant role in reform; indeed, Sputnik has become the symbol for a major turning point in American science education. To say the least, Sputnik was a contextual force for reform. For the public, it represented a threat to American security, our superiority in science and technology, and our progress and political freedom. Although the threat to security was not real, the United States perceived itself as scientifically, technologically, militarily, and economically weak. As a result, educators, scientists, and mathematicians broadened and accelerated education reform, the public understood and supported the effort, and policy makers increased federal funding for reform of school science programs in particular.

What is sometimes referred to as the "golden age" of science and mathematics education began in the 1950s with the development of new programs that eventually became known by their acronyms. Science programs included the Physical Sciences Study Committee, known as PSSC Physics; the Chemical Education Materials Study, known as CHEM Study; the Biological Sciences Curriculum Study, known as BSCS Biology; and the Earth Science Curriculum Project, known as ESCP Earth Science. The junior high level had introductory physical science, or IPS. At the elementary level, there was the Elementary Science Study, known as ESS; the Science Curriculum Improvement Study, known as SCIS; and Science—A Process Approach, known as SAPA.

In mathematics, the new programs included UICSM, the School Mathematics Study Group (SMSG), the Greater Cleveland Mathematics Program (GCMP), the University of Illinois Arithmetic Project, the University of Maryland Mathematics Project (UMMP), the Suppes Experimental Project in the Teaching of Elementary-School Mathematics, and the Madison Project.

CHAPTER 4

WHAT WAS EDUCATION LIKE AT THE TIME SPUTNIK WAS LAUNCHED?

After World War II, debate about the quality of American education escalated. Individuals such as Admiral Hyman Rickover (1959) and, most notably, Arthur Bestor (1953) became critics of John Dewey's ideas and the rhetoric of progressive education, especially the theme of life adjustment. The dominant theme of the critics was *back*—back to fundamentals, back to basics, back to drill and memorization, and back to facts. Bestor called for a return to past practices and argued for a restoration of learning as the theme for reform (Cremin 1961; Ravitch 1983). Several observations are worth noting about the criticism of progressive ideas and the emergence of Sputnik-era programs. First, such criticism of education was not new; for example, in the late 1800s, critics said that students were being "spoon-fed," the curriculum was too easy, and music and art took too much time from fundamentals. Second, some of what critics such as Bestor wrote included a distortion of facts. Furthermore, the critics seldom appealed to evidence in support of their arguments; they relied on personal authority and powerful rhetoric. Third, educators did not respond to the critics. There is no clear explanation for the educators' silence. However, this was the Cold War and the period of McCarthyism, so they may have been fearful to say anything. Also, progressive education was on the decline. In 1955, the Progressive Education Association closed its doors, and two years later the journal *Progressive Education* folded, so those educators disposed to counter the critics may have thought it would make no difference. Regardless, educators remained silent. Fourth, life adjustment education did not convey a message that students would learn basic concepts of mathematics, science, and other disciplines.

Progressive educators had introduced the term *life adjustment* to describe programs for secondary schools that built on the "important needs of youth" expressed in the Educational Policies Commission's report *Education for All American Youth: A Further Look* (1952). Life adjustment education focused on the needs of students in "general tracks" and proposed a curriculum of functional experiences in areas such as the practical arts, family living, and civic participation. Such rhetoric about the curriculum seemed to neglect aspects of the disciplines that critics thought vital. Finally, progressive educators lacked—or probably never developed—public support for their ideas. The critics' opinions of progressive education had a natural appeal to the public's perception of what constitutes a good education. This explains the critics' appeal to themes such as "restoration of learning," which implied students were not learning anything. The critics' ideas and recommendations for reform were aligned with the educational experiences adults had when they were in school and represented activities parents knew and could do with their children.

In October 1957, the debate about American education reached a turning point. Sputnik resolved the debate in favor of those recommending greater emphasis on higher academic standards, especially in science and mathematics, because the American public understood it would be in the country's best interest. Although the public had previously opposed federal aid to schools on the grounds that

federal aid would lead to federal control, public support for a federal response was unusually high after Sputnik, and Congress passed the National Defense Education Act (NDEA) in 1958.

Another important point: Science education leaders of the Sputnik era shared a common vision. Across disciplines and within the education community, leaders generated enthusiasm for their initiatives. They would replace the current content of topics and information with a curriculum based on conceptually fundamental ideas of science and mathematics and the modes of scientific inquiry and mathematical problem solving. The reform would replace textbooks with instructional materials that included films, activities, and readings. No longer would schools' science and mathematics programs emphasize information, terms, and applied aspects of content; rather, students would learn the structures and procedures of science and mathematics disciplines (Bruner 1960).

The leaders' vision of replacing the curriculum, combined with united political and economic support for education improvement, stimulated the reform. The Eisenhower administration (1953–1961) provided initial economic support, and the enthusiasm of the Kennedy administration (1961–1963) moved the nation forward with reform initiatives. While the Soviet Union had provided Sputnik as a symbol for the problem, President John F. Kennedy provided a manned flight to the Moon as America's scientific and technological solution to the problem.

Reformers enjoyed financial support from both public and private sources for their curriculum projects. Federal agencies, particularly the National Science Foundation (NSF), and major philanthropic foundations, such as Carnegie Corporation of New York and the Rockefeller Brothers Fund, provided ample support for the development of new programs.

The leaders included senior scholars from prestigious institutions such as the National Academy of Sciences (NAS), the National Academy of Engineering (NAE), and the American Mathematical Society (AMS). They had affiliations with Harvard University, Massachusetts Institute of Technology, Stanford University, University of Illinois, University of Maryland, and University of California. In the public's and funders' views, the scientists, mathematicians, and engineers who led projects during this era gave credibility and confidence that we could really achieve a revolution in American education. In 1963, Frances Keppel, then U.S. Commissioner of Education, commented that "more time, talent, and money than ever before in history have been invested in pushing outward the frontiers of educational knowledge, and in the next decade or two we may expect even more significant developments" (Gross and Murphy 1964, p. 1). The commissioner may have been correct about the investment and the frontiers of educational knowledge, but in the next decade, there were significant developments in education that changed Keppel's optimistic projection of the Sputnik-based revolution in American education. Again, public and political forces influenced the reform of science and mathematics education.

Americans developed a new awareness as a result of the events in the late 1950s. A social awareness of civil rights developed, the origins of which included the Supreme Court decision *Brown v. Board of Education* (1954) and Arkansas governor Orval Faubus's refusal to allow African American students to enter Little Rock Central High School. In the early 1960s, society increased its focus on civil rights, poverty, and an escalating war in Vietnam. Socially, we entered an era of protest that

education did not escape. The titles of books from this period clearly express the educational protest—*Compulsory Miseducation* (Goodman 1964), *Death at an Early Age* (Kozol 1967), *Our Children Are Dying* (Hentoff 1966), and *How Children Fail* (Holt 1964). The criticisms of this period were many and about different issues. At the same time, constructive solutions were few and did little to address the issues. Interestingly, there was a call for relevance of school programs—a call that echoed progressive ideas—although most critics did not identify them as such. Programs from the Sputnik era were included in the critics' view of what was going on in American schools. Indeed, as the new PSSC, CHEM Study, BSCS, ESCP, IPS, SCIS, ESS, and other programs were reaching students, criticisms of their elitism and lack of accommodation for disadvantaged students mounted.

Just as social and political factors had initiated and supported the Sputnik era of education reform, in the 1960s, social and political factors also arose and acted as countervailing forces to the pursuit of excellence, high academic standards, and knowledge of the conceptual and methodological basis of science and mathematics disciplines. We point out the observation that in the Sputnik era, political, social, and economic support, combined with the enthusiasm of leaders and a single focus on replacing curriculum programs, omitted a necessary aspect of education reform—establishing policies at the state and local levels that would sustain the nationally developed programs in states, districts, and local schools.

WAS CURRICULUM REFORM IN THE SPUTNIK ERA A FAILURE?

Education reform should not be subject to a pass-or-fail evaluation. Every reform effort contributes to the overall development and continuous improvement of the education system. The education community and the public learn from the experience. Often, the critics of education reform express the misconception that a particular reform will fix our education problems once and for all. This perspective may have political value, but it does not stand in the light of history. Leaders of the Sputnik era did not fail. Although the reformers made mistakes and the programs had weaknesses, the approaches they used, the groups they formed, and the programs they developed have all had positive and lasting influences on American education. Reports in the late 1970s indicated that the curriculum programs had a broad impact. The new programs were being used extensively and commercial textbooks had incorporated these approaches (Weiss 1978; Helgeson, Blosser, and Howe 1977). For example, in the academic year 1976–1977, nearly 60% of school districts were using one or more of the federally funded programs in grades 7 through 12, and 30% of school districts reported using at least one program in elementary schools. Reviews of the effect of science curricula on student performance indicated that the programs were successful (i.e., student achievement was higher in Sputnik-era programs than in programs with traditional curriculum), especially the BSCS programs (Shymansky, Kyle, and Alport 1983).

Mathematics presented a different situation. Mathematicians criticized the new programs because the content was too abstract and neglected significant applications; teachers criticized the programs because they were too difficult to teach; and parents criticized the new math because they worried

that their children would not develop fundamental computational skills. Although 30% of districts reported using NSF-supported mathematics programs in the early 1970s, only 9% reported using NSF programs during the 1976–1977 school year. Most important, mathematics teachers supported this change from Sputnik-era programs back to curriculum perceived as "basic" (Dow 1991).

An important but unrecognized outcome of the Sputnik era was the birth of education groups that specialized in the development of instructional materials. Some of the groups continue today, such as BSCS, Lawrence Hall of Science, and Education Development Center (EDC). Furthermore, new groups that serve a similar educational function have emerged since the Sputnik era, such as the National Science Resources Center (NSRC), the Technical Education Research Center (TERC), and the Concord Consortium.

The Sputnik era had other indirect but important effects on the education system. The reform in general and specific programs in particular brought a unified purpose to the science education community. Individual teachers of the Sputnik era talk about the influence that a particular science program—for example, *BSCS Biology: An Ecological Approach (Green Version)*, ESCP's *Investigating the Earth*, or CHEM Study—had on their lives. Some of these individuals are now scientists, science educators, and science teachers; many are not, but they are citizens who have an interest in science, which is an important goal of science and mathematics education.

A not-insignificant influence from the Sputnik era can be found in the many classroom activities and lessons that infuse science and mathematics education. For example, the ESS program produced activities called "Batteries and Bulbs" and "Mystery Powders." These activities and many others are still used in contemporary classrooms, undergraduate teacher education, and professional development workshops. Though these programs are not as nationally prominent as achievement scores, the science education community did effect positive changes in the teaching and learning of science.

A significant innovation of the Sputnik era was the involvement of senior scientists, mathematicians, and engineers working along with teachers and other educators in this reform. Their leadership set a precedent for future reforms of education. Also, it is significant that many educators—for example, those responsible for teacher education—were not immediately involved in the reform and were slow to support it through revision of state programs for certification and licensure, professional development for teachers, and undergraduate courses for future teachers.

The Sputnik era continued into the early 1970s. If we had to indicate an end of the era, it would be 1976, when *Man: A Course of Study* (MACOS), an anthropology program developed with NSF funds, came under scrutiny and attack from conservative critics who objected to the subject matter (Dow 1991). The combined forces of House subcommittee hearings, an NSF internal review, and the Government Accounting Office's investigation of the financial relationships between NSF and the developers signaled the end of the MACOS program and symbolized the end of this era of curriculum reform. (For more details, see Appendix A, Tables A.7 and A.8.)

CHAPTER 4

A PERSONAL PERSPECTIVE

SPUTNIK, SYMBOLISM, AND STANDARDS

Rodger Bybee

I will begin with a question based on the perspective that the United States needs a Sputnik to stimulate reform: Does the United States need another Sputnik? Clearly, there will not be another Sputnik, but we may need what Sputnik has come to symbolize—an era of significant reform of science, technology, engineering, and mathematics (STEM) education. In events of the first decades of the 21st century, one can identify similar concerns about national security, economic efficiency, use of resources, emerging and re-emerging infectious diseases, and, yes, the scientific evidence supporting climate change.

Here are several insights about Sputnik that leaders might consider when using Sputnik as a symbol for an era of reform. The competitor and venue were clear: the Soviet Union and a race to space. President Kennedy challenged the nation to respond by setting a clear goal: Send a man to the Moon and return him safely. The president also set a timeline—by the end of the decade. This goal employed a clear and visible symbol that every American could see on a regular basis—the Moon. Accomplishing the ultimate goal included approximations of success that the public could see and understand—suborbital flights, orbital flights, a flight around the Moon and back, and ultimately landing a man on the Moon. One component of the U.S. response involved curriculum reform led by the scientific and education communities. One final insight from the Sputnik era included the use of curriculum materials and science teacher institutes as the primary methods of reform. Both of these methods center on the core of teachers' effective interaction with students. This was a positive and productive way the federal government facilitated education reform.

America now confronts new challenges associated with, for example, the economy, the environment, and natural resources. The National Research Council (NRC) report *Rising Above the Gathering Storm* (NRC 2007) has become one of several major reports signaling the need for a national response. This report holds the attention of scientists, engineers, and the business community. A decade ago, Thomas Friedman engaged public attention for reform with his provocative book *The World Is Flat* (Friedman 2005). Friedman has an interesting, if not compelling, premise: The international economic playing field is level, hence the metaphor "the world is flat." The "flattening" that Friedman refers to is a result of information technologies and associated innovations that have made it technically possible and economically feasible for U.S. companies to locate work "offshore,"such as by having call centers in India. The revolution of information technologies has developed a generation of digital natives that cast many others of us as digital immigrants. The implications for STEM education are significant, to say the least.

On balance, Friedman argues that a flatter world will benefit all of us, in both developed and developing countries. Friedman does address education questions in a chapter titled "The Quiet Crisis." According to Friedman, "The American education system from kindergarten through twelfth grade just is not stimulating enough for young people to want to go into science, math, and engineering" (2005, p. 270). Friedman continues,

> *Because it takes fifteen years to create a scientist or advanced engineer, starting from when that young man or woman first gets hooked on science and math in elementary school, we should be embarking on all-hands-on-deck, no-hold-barred, crash program for science and engineering education immediately. The fact that we are not doing so is our quiet crisis. (p. 275)*

Leaders in the science education community are left with a fundamental question: What should be done to address this crisis?

The United States must continue reforming science education, in this case because we are losing our competitive edge in the global economy and clearly must attend to environmental and resource issues that often underlie economic realities. However, this era is very different from the Sputnik era. The competitors are greater in number, including countries with developed economies, such as Canada, France, Germany, and Japan. Also, we must consider the fastest-growing economies, such as China, India, Singapore, and South Korea. The primary goal is less clear and more complex—to prosper in a global economy. And the timeline for achievement is less clear—a decade? A half century? What is the symbol? The Dow Jones Industrial Average? NASDAQ? Do we have any indicators? Increased trade? A lower trade deficit?

All of the above involve the economy. What about societal issues related to energy? The environment? Natural hazards? Infectious disease? Climate change? These, too, are forces with implications for education systems, especially science education.

Finally, we turn to national standards. What role should national and state standards for science education play? They have an influence on school programs and the emphasis given to what students should know, value, and be able to do. Few would disagree with the assertion that the K–12 education system can and should play a central role in the responses.

CHAPTER 4

OUR COMMON PERSPECTIVE AND LEADERSHIP OPPORTUNITIES

An examination of history, especially the Sputnik era and, more recently, the results of the No Child Left Behind Act, reveals some significant insights about reforms of science education. Some observations are worth noting for this seminar.

School programs can vary based on the emphasis and priority given to the aims of science education. The perspectives of those preparing reports and standards influence the priorities of aims. Leaders should recognize the different perspectives and make judgments about the design, development, and emphasis of science programs.

Replacement of school science programs is a complex, difficult process at best. Although leaders in the Sputnik era used terms such as *revision* and *reform*, the intention was to replace school science programs. The leaders' zeal and confidence were great. They thought that if they built strong curriculum materials, then science teachers would adopt them, thus replacing traditional programs. This "field of dreams" perspective, however, confronted pervasive institutional resistance and raised the personal concerns of teachers.

We think the lesson for the 21st century resides in the importance of using our knowledge about education change and addressing the concerns of teachers. Not only are new programs important, but other components of the education system must themselves change and provide support for the implementation of education innovations such as those in the *Next Generation Science Standards* (*NGSS*; NGSS Lead States 2013). Those components include professional development for teachers, curriculum materials, assessments, administrators, school boards, the community, and a variety of local, state, and national policies.

Teachers' reluctance to change increases as innovations require a shift away from current school programs and classroom practices. This is especially true if the innovations lack political, social, and educational support. Teachers likely will have difficulty with the content and pedagogy of new programs. Lacking educational support within their system and experiencing political criticism from outside of education, teachers may seek security by staying with or returning to the programs they know and are comfortable using. This insight can be seen in contemporary initiatives such as the *Common Core State Standards* and *NGSS.*

The educational lesson here centers on the importance of both initial and ongoing professional development and support for the new state standards' school programs and classroom practices.

Recognize the different components and individuals within the science education community. Exclusion of those in the larger science education community (e.g., state coordinators, teacher educators, science education researchers, and the public) contributed to the slow acceptance and implementation of the programs, reduced understanding by those entering the profession, and afforded inadequate professional development for teachers in the classroom.

We have learned to involve more than teachers. Education is a system consisting of many different components. Important components include those who have some responsibility for state policies, teacher preparation, professional development, assessments, and the implementation of school science programs. We suggest it is best to work from a perspective that attempts to unify and coordinate

efforts among teachers, educators, and scientists, all of whom have strengths and weaknesses in their respective contributions to reform efforts.

State policies have a critical and essential role in reforms of science education. History reveals that many commissions, reports, and reforms paid very little attention to state-level policies relative to science education. Such a perspective may not have stopped reforms, but it certainly did not facilitate the improvement of science education by overlooking the need to update high school graduation requirements, teacher certification, financial support for instruction materials, and state assessments.

With this observation, as various reforms are proposed in the 21st century, those proposing fundamental changes must attend to state-level policies and their consequences. Those who developed and are working on the implementation of the *NGSS* (NGSS Lead States 2013) recognize the basic issue and have continually acknowledged the important role of the states.

Remember that there are realities of state and local school districts. In the Sputnik era, for example, support from federal agencies and national foundations freed developers from the political and educational constraints of state and local agencies and the power and influence of commercial publishers.

In this era of standards, educators acknowledge this lesson and direct their attention to a broader, more systemic view of education, one that includes a variety of policies. One view of education suggests that it involves policies, programs, and practices. Usually, individuals, science educators, professional organizations, and state and local agencies contribute in various ways to the formulation of policy, development of school programs, or implementation of classroom practices; however, there must be coordination and consistency among the various efforts. Designing and developing new programs without attending to a larger educational context to support those programs and change classroom practices to align with the innovative program surely marginalize the success of the initiative. Now, a majority of states have new *standards* for science education but they do not have science *programs*. At best, this lag in curriculum materials will slow the initiative, and at worst it will stop the reform.

Restricting initiatives to curriculum for specific groups of students (e.g., science-prone and college-bound students) resulted in criticism of Sputnik-era reforms as inappropriate for other groups, such as average and disadvantaged students. To the degree that school systems implemented the new programs, teachers found that the materials were inappropriate for some populations of students and too difficult for others. Restricting policies or targeting programs opens the door to criticism on the grounds of equity. Proposing initiatives for *all* students also often results in criticism from those who maintain there is a need for a specific program for students inclined toward science and mathematics careers, as well as from those who argue that programs for all do not adequately meet the needs of either college or career.

Examining the nature and lessons of the Sputnik era, as well as reform initiatives that came before and after, clearly demonstrates that education reforms differ. Although this may seem obvious, leaders have not always paid attention to some of the common themes and general lessons that may benefit the steady work of improving science education. Stated succinctly, those lessons are that we should use what we know about educational change; include all the key stakeholders in the education community; align policies, programs, and practices with the stated purposes of education; work on improving education for all students; and attend to the support and continuous professional development of classroom teachers because they are the most essential resource in the system of science education.

ISSUES AND QUESTIONS FOR DISCUSSION

1. Do you think it is the role of the science education community to respond to the variety of social force seen in our history? Why? Why not?
2. With the exception of the Sputnik era, few of the changes recommended by individuals and committees have ever been implemented in science classrooms. Is this observation true? How would you explain your response?
3. How does the era of NCLB, ESSA, and *NGSS* differ from earlier periods of reforms? What are the implications of those differences?
4. What lessons do you take from the history of science education?
5. The Sputnik era stimulated education research and curriculum implementation, the concerns of teachers, and educational change. What is your understanding of this research? How does it apply to contemporary science education?

REFERENCES

Bestor, A. 1953. *Educational wastelands: A retreat from learning in our public schools.* Urbana, IL: University of Illinois Press.

Bruner, J. 1960. *The process of education.* New York: Vintage.

Cremin, L. 1961. *The genius of American education.* New York: Vintage.

Dow, P. 1991. *Schoolhouse politics.* Cambridge, MA: Harvard University Press.

Educational Policies Commission. 1952. *Education for all American youth: A further look.* Washington, DC: National Education Association and the American Association of School Administrators.

Friedman, T. 2005. *The world is flat.* New York: Farrar, Straus and Giroux.

Goodman, P. 1964. *Compulsory miseducation.* New York: Horizon Press.

Gross, R., and J. Murphy, eds. *The revolution in the schools.* New York: Harcourt, Brace & World.

Helgeson, S. L., P. E. Blosser, and R. W. Howe. 1977. *The status of pre-college science, mathematics, and social science education: 1955–1975.* Volume 1. Columbus, OH: Center on Science and Mathematics Education, Ohio State University.

Hentoff, N. 1966. *Our children are dying.* New York: Viking Press.

Holt, J. 1964. *How children fail.* New York: Delacorte.

Kozol, J. 1967. *Death at an early age.* Boston, MA: Houghton-Mifflin.

National Research Council (NRC). 2007. *Rising above the gathering storm.* Washington, DC: National Academies Press.

NGSS Lead States. 2013. *Next Generation Science Standards: For states, by states.* Washington, DC: National Academies Press. *www.nextgenscience.org/next-generation-science-standards.*

Ravitch, D. 1983. *The troubled crusade: American education 1945–1980.* New York: Basic.

Rickover, H. 1959. *Education and freedom.* New York: E. P. Dutton.

Shymansky, J. A., W. C. Kyle, and J. M. Alport. 1983. The effects of new science curricula on student performance. *Journal of Research in Science Teaching* 20 (5): 387–404.

Weiss, I. 1978. *Report of the 1977 National Survey of Science, Mathematics, and Social Studies Education.* Washington, DC: U.S. Government Printing Office.

Zacharias, J. 1956. *Physical Sciences Study Committee.* Boston, MA: MIT.

SECTION III

THE PURPOSES AND GOALS OF SCIENCE EDUCATION

Science education has a limited number of goals. Students in grades K–12 should acquire scientific knowledge, develop methods of scientific inquiry, apply the knowledge and methods to personal and social situations, and gain an understanding of careers related to science and technology. Historically, the terms used to describe these purposes have changed, as has the emphasis among the goals.

The term *purpose* suggests the aims or goals for which something—in this case, science education—exists. Having a purpose implies intended outcomes for science education. Asking and answering questions about the purposes of science education are among the most important actions one can take because clarifying the purposes and goals for educating in science gives direction for the creation of curriculum, instruction, assessment, and the professional development of teachers. So, clarification of the purposes of schooling in science is fundamental and an ideal place to continue our seminar on science education.

Chapter 5 provides an important historical perspective on the changing priorities in science education. History uncovers the roots of tensions among contemporary goals such as college and career readiness, a 21st-century workforce, understanding science and engineering concepts and practices, and applying those concepts and practices to real-life situations.

The premise of Chapter 6 is that there have been a limited number of goals in science education. The terms used to identify the goals have changed, the relative priority of the goals has varied, and the importance of some goals has increased or decreased across the history of science education.

SUGGESTED READINGS

Bybee, R., and G. DeBoer. 1994. Research on the goals for science curriculum. In *Handbook of research on science teaching and learning,* ed. D. Gabel, 357–387. New York: Macmillan Publishing Company. A description of goals in science education and historical changes of the goals.

National Research Council (NRC). 2012. *A framework for K–12 science education: Practices, crosscutting concepts, and core ideas.* Washington, DC: National Academies Press. A fundamental report influencing all dimensions of contemporary science education.

Roberts, R., and R. Bybee. 2014. Scientific literacy, science literacy, and science education. In *Handbook of research on science education: Volume II,* ed. N. Lederman and S. Abell, 549–558. New York: Routledge.

CHAPTER 5

THE PURPOSES OF EDUCATION

Identifying the purpose of education sounds like a major philosophical task. It should not come as a surprise that long before No Child Left Behind (NCLB), the Every Student Succeeds Act (ESSA), and other national, state, and local requirements, individual philosophers and education groups have described various purposes and goals of education. In *Republic*, for example, Plato argued that the purpose of education was to develop a just state. In the 20th century, in *Democracy and Education* (Dewey 1916), philosopher John Dewey argued that the purposes of education included both individual development and the development of a progressive democratic society. This chapter continues with examples from the mid- to late 20th century.

HISTORICAL PERSPECTIVES

Progressive Education

John Dewey and fellow philosopher George Counts proposed different purposes for schooling. In *Experience and Education* (Dewey 1938), Dewey analyzed traditional and progressive education. For Dewey, "traditional" education centered on subjects that had a history and culture in the schools. For progressive education, Dewey argued, the purpose of education was a pragmatic response to experiences in students' immediate environments. In Dewey's perspective, both immediate experiences and current problems in a changing society are essential foundations for student learning.

In *Dare the Schools Build a New Social Order?* (1932), Counts offered a none-too-subtle critique of Dewey's ideas. Counts stated, "[T]he weakness of Progressive Education thus lies in the fact that it has elaborated no theory of social welfare, unless it be that of anarchy or extreme individualism" (p. 7). For Counts, the purpose of education was preparation for life as a citizen. Schooling should emphasize the knowledge and skills necessary for participation in one's community and contribution to improving the social order.

To be clear and fair, Dewey responded to Counts's criticism by making the point that schools probably could not build a new social order, even if they wanted to. The reason was relatively simple:

In a modern society, the school could never have enough power to be the main determinant for political, intellectual, and ethical changes in the society. At best, schools could contribute to understanding the reasons for change and clarify the disposition to change.

We will give the final point to Counts. In *Education and American Civilization* (Counts 1952), he states,

> *There is no quick and easy road to a great education. ... There is no simple device or formula for achievement of this goal. ... It can come only from a bold and creative confronting of the nature, the values, the conditions, and the potential abilities of a civilization. An education can rise no higher than the conception of the civilization that provides it, gives it substance, and determine its purposes and direction. (p. 36)*

General Education in a Free Society

In the mid-1940s, a committee composed of Harvard faculty published *General Education in a Free Society* (Harvard University 1945). Early in the report, the committee makes clear the aims of general, as opposed to specialized, education:

> *General education, as education for an informal and responsible life in society, has chiefly to do with . . . the question of common standards and common purposes. Taken as a whole, education seeks to do two things: help young persons fulfill the unique, particular functions in life which it is in them to fulfill, and fit them so far as it can for those common spheres which, as citizens and heirs of a joint culture they will share with others. (p. 4)*

This quotation sets the stage for discussions that evaluate the merits of general versus specialized education. In the end, the committee proposed a basic general education for all students. The general education would include an introduction to our common heritage through nonspecialized study of the natural and social sciences as well as the humanities.

Public Education and Paideia

In *Public Education* (1976), historian Lawrence Cremin described an approach to education that used ecology as a metaphor. This approach is "one that views educational institutions and configurations in relation to one another and to the larger society that sustains them and is in turn affected by them. . . ." (Cremin 1976, p. 36). Cremin includes both formal education (i.e., schooling) and informal education (i.e., media, museums, churches and synagogues). Cremin also underscores the connection between education institutions and larger society.

There are several points worthy of note concerning an ecological view of education. One is that education should be directed to an emerging lifestyle of the individual. In particular, efforts should be made to help individuals increase their striving to develop their unique qualities. A second aspect of education is based on the Greek concept of paideia. Paideia is a sense of social, political, and ethical aspiration. So, for Cremin, education should be directed to development of both individual qualities and social aspirations (Cremin 1976).

In 1982, Mortimer Adler published *The Paideia Proposal: An Educational Manifesto*, in which he addresses the objectives of schooling: personal growth or self-improvement of mental, moral, and spiritual dimensions; citizenship; and preparing for a career. Adler is concerned with a lifelong education as he points to the fact that these objectives should really be achieved in adult life. He summarizes his point:

> *Here then are the three common callings to which all our children are destined: to earn a living in an intelligent and responsible fashion, to function as intelligent and responsible citizens, and to make both of these things serve the purpose of leading intelligent and responsible lives—to enjoy as fully as possible all the goods that make a human life as good as it can be. (Adler 1982, p. 18)*

A Nation at Risk

In April 1983, the National Commission on Excellence in Education (NCEE) called for a return to excellence in education. The report had the provocative title *A Nation at Risk: The Imperative for Educational Reform* (NCEE 1983). The first section of the report reviewed a litany of factors indicating that the United States was a nation at risk: levels of functional illiteracy, low achievement scores, a decline in SAT scores, and a decline in science achievement. The theme of science and technology was prominent in this influential report. The report presented warnings of a generation of scientifically and technologically illiterate Americans, a growing chasm between scientific and technological elites, and a citizenry both ill-informed and uninformed in scientific matters. A quotation from the section on content recommendations in *A Nation at Risk* introduces the purposes for science education:

> *The teaching of science in high school should provide graduates with an introduction to: (a) the concepts, laws, and processes of the physical and biological sciences; (b) the methods of scientific inquiry and reasoning; (c) the application of scientific knowledge to everyday life; and (d) the social and environmental implications of scientific and technological development. Science courses must be revised and updated for both the college bound and those not intending to go to college. (NCEE 1983, p. 25)*

The first two aims, (a) and (b), are longstanding goals of science education, but (c) and (d) represent purposes for science programs that echo the prior discussions by philosophers and historians of education. We note the need to develop goals for science programs that present basic concepts and processes in the context of personal and social applications and the role of science in society. In review, it seems there is rather close agreement among educators such as Cremin and Adler and the cited reports on education goals. Attention is directed in part toward the individual and in part toward the individual as a citizen.

What Schools Are for, and the Public Purpose of Education

In 1979, John Goodlad published a book titled *What Schools Are for*. One section of the book reports 12 goals that emerged from data gathered in "A Study of Schooling," a project Goodlad and colleagues had conducted. The study involved analyzing goals for schooling articulated by state and local boards of education, various commissions, and other reports. The team identified approximately 100 goals, which were synthesized into 12 goals. Goodlad noted that the goal statements have shifted across time to include different emphases, including religious beliefs; vocational responsibility; worthy membership in home, community, state, and nation; concern for justice; appreciation for democratic values; and development of individual talents (Goodlad 1979, p. 45).

Goodlad and his colleagues (Goodlad 1979, pp. 46–52) identified the following as the 12 goals for schooling. The original list included subdivisions under each goal.

1. Mastery of basic skills and fundamental processes
2. Career education—vocational education
3. Intellectual development
4. Enculturation
5. Interpersonal relations
6. Autonomy
7. Citizenship
8. Creativity and aesthetic perception
9. Self-concept
10. Emotional and physical well-being
11. Moral and ethical character
12. Self realization

The Meaning of Public Education

Theodore Sizer contributed a chapter called "The Meaning of Public Education" for *The Public Purpose of Education and Schooling* (Goodlad and McMannon 1997). In his chapter, Sizer succinctly stated what the public wants from public education. According to Sizer, schools should serve four goals. The first is a civic function—preparing students for the needs and workings for a civil society. Second, schools serve an economic function by assuring an adequate and appropriate work force. For this goal, Sizer mentioned language and mathematical skills. Third, there is a goal of cultivation. This goal includes introducing students to the best that culture has to offer. The goal is both personal, (e.g., creative expression) and collective (e.g., understanding the humanities). Finally, Sizer indicates a goal of intellectual strength—the need for individuals to have the knowledge and abilities to make decisions, recognize and avoid false persuasions, and express themselves (Sizer 1997, p. 37).

A NOTE ON CITIZENSHIP AND SCIENCE EDUCATION

When considering the purposes of science education, the goals of scientific knowledge and methods take immediate priority. After the primary place of disciplinary concepts and scientific inquiry, there may be consideration of the history and nature of science. Finally, there may be recognition of science-related personal and social perspectives. In contemporary discussions of science education, we hear about preparation for college and career, so we ask, "What about citizenship?" Here, we present a brief discussion of that purpose.

American public education was born out of the need to maintain and improve the democratic social order. Educated citizens play a crucial role in a democracy. Here, then, is the foundation, the one fundamental purpose of education: to develop the knowledge, values, skills, and sensibilities of citizenship.

Citizenship is defined as the condition of being a member of a city, state, and nation. A result of this allegiance is an expectation of both security and freedom. Citizenship is at best a reciprocal obligation. There are duties, obligations, and privileges of the citizen toward the government, and rights and requirements of the government toward the individual.

One purpose of public education is to promote citizenship. There is a long history of education's intention to develop citizens capable of informed, rational, and responsible participation in the democratic process. As it turns out, this may be a goal at least as basic as "the basics" of reading, writing, and computing. The basics need to be learned so an individual will be better prepared to participate in the role of citizen. Public education has the important aim of helping individuals understand and act on the reciprocal obligation between themselves and the local, national, and global systems of which they are citizens. In the context of this discussion, one purpose of public education could be expressed in two questions: "What are the duties, rights, and privileges of science and technology toward individual citizens and society?" and "What are the duties, rights and privileges of individual citizens and society toward the science and technology enterprise?" Focusing on citizenship establishes a vital link between public education and science and technology.

A purpose of contributing to students' future roles as citizens is a worthy but abstract goal. How do you think this goal translates to the elementary, middle, or high school classroom?

CHAPTER 5

A PERSONAL PERSPECTIVE

SCIENTIFIC LITERACY AND THE PURPOSES OF SCIENCE EDUCATION

Rodger Bybee

Science education is a social institution and, as such, its purposes should be directed toward meeting the needs of both society and individuals in ways unique to the institution. In our society, the most immediate way citizens learn about the products, processes, and effects of science is through formal and informal science instruction. This understanding defines the uniqueness of our discipline and establishes a connection to scientific literacy, an idea I argued for and tried to clarify in *Achieving Scientific Literacy: From Purposes to Practices* (Bybee 1997).

The idea of scientific literacy, or science education for the populace, has been around for some time. For example, in 1847, James Wilkinson published a lecture called "Science for All." I have found the actual term *scientific literacy* was first used in 1952 by James Bryant Conant in the foreword to *General Education in Science*, which was edited by Fletcher Watson and I. Bernard Cohen. Paul DeHart Hurd set the term in the lexicon of science education when he used it in a 1958 article titled "Science Literacy: Its Meaning for American Schools" for *Educational Leadership* (Hurd 1958). Hurd provided a clear perspective when he described scientific literacy as an understanding of science and its applications to an individual's experience as a citizen. Hurd made clear connections to the science curriculum and the selection of instructional materials that provide students with the opportunities to use the methods of science; apply science to social, economic, political, and personal issues; and develop an appreciation of science as a human endeavor and intellectual achievement (Hurd 1958, pp. 13–16). Although there have been variations, Hurd's definition expresses the application of scientific knowledge to the situations individuals will encounter as citizens. So, you see that there have been objectives for students such as to function as an intelligent and responsible citizen, to earn a living in an intelligent and responsible fashion, and to lead an intelligent and responsible life—to enjoy life as fully as possible—throughout our educational history.

Although individuals continue to claim that we do not have a definition of scientific literacy, their claims simply betray a lack of historical review or a failure to accept publicly expressed definitions. In either case, individuals and organizations have defined scientific literacy in considerable detail in the 1960s, 1970s, 1980s, and 1990s. Table 5.1 provides examples of articles or position statements that define scientific literacy. My intention in providing this table is simply to show that we *have* defined scientific literacy. In recent times, *Science for All Americans* (Rutherford and Ahlgren 1989) and the *National Science Education Standards* (NRC 1996) have presented full, complete, and detailed definitions of scientific literacy.

Table 5.1. Characteristics of Scientific Literacy in Educational Literature

The 1960s: "Goals Related to the Social Aspects of Science" (Hurd and Gallagher 1966)	The 1970s: "What Is Unified Science Education?" (Showalter 1974)	The 1980s: "Science-Technology-Society: Science Education for the 1980s" (NSTA 1982)
1. Appreciate the socio/historical development of science	1. Nature of science	1. Scientific and technological processes and inquiry skills
2. Be aware of the ethos of modern science	2. Concepts of science	2. Scientific and technological knowledge
3. Understand and appreciate the social and cultural relationship of science	3. Processes of science	3. Skills and knowledge of science and technology in personal and social decisions
4. Recognize the social responsibility of science	4. Values of science	4. Attitudes, values, and appreciation of science and technology
	5. Science and society	5. Interactions among science, technology, and society in the context of science-related societal issues
	6. Interest in science	
	7. Skills associated with science.	

In the late 1980s, I worked with Hurd on several projects. He is one reason I was originally drawn to the idea of scientific literacy. A second reason had to do with what I identified as a misconception of the idea. Let me explain.

Many individuals hold a view that could be described using the either/or perspective. That is, science literacy has a defined and singular set of characteristics, thus allowing judgments about other individuals' or a population's scientific literacy. So, it is common to hear that one or some population is scientifically illiterate because, for example, they do not know the difference between atoms and molecules, or what DNA is, or the dynamics of plate tectonics. I argue that such a perspective is both inaccurate and educationally inappropriate.

The perspective I propose suggests that for individuals and populations, scientific literacy is a variation of scientific understandings. For example, an individual may know a considerable amount of Earth science, some life science, little physical science, some about inquiry, less history, more about science in selected social problems, and very little about technology. Such a perspective enables us to see that all individuals demonstrate varying degrees of scientific literacy and can continue developing their knowledge, understanding, and abilities over a lifetime. Being able to declare individuals or populations as scientifically illiterate makes great rhetoric and has political power. Thinking and acting on a view that recognizes scientific literacy as variation in understandings and abilities has educational value and provides a basis for policy, programs, and practices.

The general use of scientific literacy has the advantage of unifying the community by centering on what is perceived to be a primary purpose of science education. The disadvantage of general use is the loss of the term's specific meaning, which was an understanding of science and its applications to personal, national, and global perspectives.

In 2000, George DeBoer published a historical review of scientific literacy. There have been many different goals of science education, all related to scientific literacy. DeBoer suggested a broad conceptualization of scientific literacy, one allowing for variations in curriculum and instruction. The broad goal suggested by DeBoer is consistent with earlier definitions—namely, to enhance the public's understanding and appreciation of science. Here are several insights about scientific literacy: Scientific literacy involves an adult population's level of understanding and appreciation of science; this literacy changes over time; and school experiences certainly affect the public's attitudes toward science and the disposition to continue developing an understanding and appreciation of science (DeBoer 2000).

In the decades since the first use of the term *scientific literacy*, there has emerged a critical distinction between an emphasis on education for future *citizens* and education for future *scientists*. In 2007, Douglas Roberts published a significant essay on scientific literacy in *Handbook of Research on Science Education* (Roberts 2007). Roberts identified a continuing political and intellectual tension with a long history in education. The two conflicting perspectives can be stated in a question: Should curriculum emphasize subject matter itself, or should it emphasize the application of knowledge and abilities in life situations? Roberts refers to curriculum designed to answer the former as Vision I and curriculum that addresses the latter as Vision II. Vision I looks within science disciplines; it is internal and foundational. Vision II uses external contexts that students are likely to encounter as citizens. I was honored to join Roberts for an update of his chapter for a new edition of *Handbook of Research on Science Education* (Roberts and Bybee 2014).

A significant contemporary issue for those developing standards, designing curriculum, and providing professional development is recognizing the difference between the two perspectives just described. One perspective centers on disciplines such as biology, chemistry, physics, or the Earth sciences. In this perspective, programs and teaching practices answer questions such as, "What knowledge of science and its methods should students learn?" and "What facts and concepts from science should be the basis for school programs?" In contrast, there is a *contextualist* (Fensham 2009) perspective that emphasizes situations that require an understanding and application of science. When thinking about standards, curriculum, and instruction from a contextualist viewpoint, questions include, "What science should students know, value, and be able to do as future citizens?" and "What contexts could be the basis for science education?" The difference between these two perspectives is significant because it has implications for various emphases within a curriculum, selection of instructional strategies, design of assessments, and professional education of teachers.

The subsequent outcomes—what students learn about science; the attitudes they develop; the skills they acquire; and their ability to competently identify, analyze, assess, and respond to life situations—also differ significantly.

As a result of a thorough analysis, Roberts and I concluded there was evidence of a withdrawal from Vision II in contemporary policies such as *A Framework for K–12 Science Education* (NRC 2012) and PISA 2015 (Roberts and Bybee 2014).

For many individuals with responsibility for national and state standards, district programs, and assessments, the distinction between Vision I and Vision II is blurred, to say the least. For most state standards, the dominant emphasis for content and learning outcomes is Vision I; the principal (sometimes exclusive) emphasis is on discipline-based science knowledge and methods. There is an assumption that if students understand science concepts, they will apply that knowledge to the personal, social, and global problems they encounter as citizens. I certainly question that assumption. For those interested in achieving higher levels of scientific literacy, school science programs should incorporate Vision II clearly, consistently, and continually. Students should have experiences where they confront appropriate socioscientific issues and problems within meaningful contexts.

A productive step toward achieving higher levels of scientific literacy would be engaging in discussions and open dialogue based on a variation of what I have called the Sisyphean question in science education: "What knowledge, values, skills, and sensibilities relative to science and technology are important for citizens in the 21st century?"

You should note several aspects of this question. First, the question is being asked in the early decades of the 21st century. So, the way you answer should be guided by the contexts of this time in history—that is, present social issues, scientific needs, and projected problems for future decades. Second, the question includes both science and technology. Citizens actually encounter and have concerns about technologies in their lives. Next, there is the assumption that not all discipline-based knowledge and skills may be included in school programs and classroom practices. Finally, the question clearly includes developing informed citizens as a purpose of science education.

To conclude, I will restate a theme introduced earlier in this chapter. The goal of achieving higher levels of scientific literacy incorporates the purposes of public education and expresses an essential challenge for 21st-century science educators. Most national and state standards and school programs emphasize fundamental knowledge and processes of the science disciplines, as well they should. These standards and science programs implicitly provide students with a foundation for professional careers as scientists and engineers. However, there is a need to address the applications of science knowledge and the practices of inquiry to personal and social situations. With the centrality of science and technology in contemporary life, full participation in society requires that all adults, including those aspiring to careers as scientists and engineers, obtain higher levels of scientific literacy—that is, that they not only develop understandings of science fundamentals but also learn how to apply that knowledge to real-life situations. The science education community must understand that the level of a society's scientific literacy depends on citizens' knowledge, receptivity, and appreciation of science as a human endeavor and its significant influence on their lives and society.

CONCLUSION

This chapter provided a historical perspective that addressed fundamental aims for education in the sciences. Those aims include fulfilling an individual's unique potential and applying knowledge, values, and skills to real-life situations they will encounter as citizens. The chapter also included a personal perspective on scientific literacy and the purposes of science education.

REFERENCES

Adler, M. J. 1982. *The paideia proposal: An educational manifesto.* New York: Collier Macmillan.

Bybee, R. W. 1997. *Achieving scientific literacy: From purposes to practices.* Portsmouth, NH: Heinemann.

Counts, G. S. 1932. *Dare schools build a new social order?* New York: John Day Company.

Counts, G. S. 1952. *Education and American civilization.* New York: Teachers College Press, Columbia University.

Cremin, L. A. 1976. *Public education.* New York: Basic Books.

DeBoer, G. E. 2000. Scientific literacy: Another look at its historical and contemporary meanings and its relationship to science education reform. *Journal of Research in Science Teaching* 37 (6): 582–601.

Dewey, J. 1916. *Democracy and education.* New York: Macmillan.

Dewey, J. 1938. *Experience and education.* New York: Collier Books, 1966.

Fensham, P. J. 2009. Real world contexts in PISA science: Implications for context-based science education. *Journal of Research in Science Teaching* 46 (8): 884–896.

Goodlad, J. I. 1979. *What schools are for.* Los Angeles: UCLA and Institute for Development of Educational Activities.

Goodlad, J. I., and T. J. McMannon, eds. 1997. *The public purpose of education and schooling.* San Francisco, CA: Jossey-Bass.

Harvard University. 1945. *General education in a free society: Report of the Harvard Committee.* Cambridge, MA: Harvard University Press.

Hurd, P. D. 1958. Science literacy: Its meaning for American schools. *Educational Leadership* 16 (1): 13–16.

Hurd, P. D., and J. G. Gallagher. 1966. Goals related to the social aspects of science. In *Sequential programs in science for a restructured curriculum,* 12–18. Cleveland, OH: Educational Research Council.

National Commission on Excellence in Education (NCEE). 1983. *A nation at risk: The imperative for educational reform.* Washington, DC: U.S. Government Printing Office.

National Research Council (NRC). 1996. *National science education standards.* Washington, DC: National Academies Press.

National Research Council (NRC). 2012. *A framework for K–12 science education: Practices, crosscutting concepts, and core ideas.* Washington, DC: National Academies Press.

National Science Teachers Association (NSTA). 1982. Science-technology-society: Science education for the 1980s. A position statement from NSTA. Arlington, VA.

Roberts, D., and R. Bybee. 2014. Scientific literacy, science literacy and science education. In *Handbook of Research on Science Education: Volume II,* ed. N. Lederman and S. Abell, 549–558. New York: Routledge.

Roberts, D. 2007. Scientific literacy/science literacy. In *Handbook of research on science education,* ed. S. Abell and N. Lederman, 729–780. Mahwah, NJ: Lawrence Erlbaum Associates.

Rutherford, F. J., and A. Ahlgren. 1989. *Science for all Americans: A Project 2061 report on literacy goals in science, mathematics, and technology.* Washington, DC: American Association for the Advancement of Science.

Showalter, W. 1974. What is unified science education? Program objectives and scientific literacy. *Prism II* 10: 1–6.

Sizer, T. 1997. The meanings of "public education." In *The public purpose of education and schooling,* ed. J. Goodlad and T. McMannon, 33–40. San Francisco, CA: Jossey-Bass.

CHAPTER 6

GOALS OF SCIENCE EDUCATION

In the introduction for this section, we stated that asking and answering questions about the purposes and goals of science education are among the more important activities for those in leadership positions. We said this because clear goals give direction for the clarification and emphasis of curriculum, instruction, assessment, and professional development.

As you can observe in the prior sections, educators have long thought about and expressed the goals of education. In this chapter, we turn from the broad statements of educational purpose to a more specific discussion of goals and science education. If you are interested in more background and greater discussion of goals in the history of science education, we recommend *A History of Ideas in Science Education* (DeBoer 1991) and "Research on Goals for the Science Curriculum" (Bybee and DeBoer 1994). The history of science education indicates a limited number of fundamental goals. Although the terms used for the goals have varied with history, the categories of goals are scientific knowledge, scientific methods, societal issues, personal development, and career awareness. To be clear, the terms used to identify the goals have changed, details within the goals have varied, and categories have been combined. Let's examine the goals.

SCIENTIFIC KNOWLEDGE

There is a body of knowledge concerning biological, physical, and Earth systems that educators have deemed important for students to learn. For more than 200 years, our science education programs have aimed toward informing students about these natural systems. This goal has been, and will continue to be, of significant importance for science teachers. Stated formally, this goal is, *Science education should develop fundamental understandings of natural systems.*

School science programs have been, and likely will continue to be, oriented toward knowledge associated with the science disciplines. In history, the knowledge has been referred to as facts, concepts, principles, conceptual schemes, conceptual organizers, structure of the disciplines, and

unifying concepts. In contemporary standards such as the *Next Generation Science Standards* (*NGSS*), science knowledge is referred to as disciplinary core ideas and crosscutting concepts.

SCIENTIFIC METHODS

A second goal has centered on the abilities and understandings of the methods defining science. Descriptions of this goal have changed; for example, terms such as *processes of science, inquiry,* and *discovery* have been used to describe the goal of scientific methods. The goal can be stated as, *Science education should develop a fundamental understanding of, and ability to use, the methods and practices of scientific inquiry.*

The goal of helping students develop both an understanding of and the ability to use scientific inquiry has a long history in science education. The goal has an equally long history of not being fully realized in school science programs. In the *NGSS,* this goal is described as "science and engineering practices" and is explicitly connected to science knowledge—that is, disciplinary core ideas and crosscutting concepts.

Understanding the nature of science as distinct from the methods of science investigation has been a goal related to scientific methods and scientific knowledge. However, this, too, has been an aim not fully developed in school science programs.

SOCIETAL ISSUES

Science education exists in society and should contribute to the maintenance and aspirations of the culture. This goal is especially important when there are social challenges directly related to science. This goal is, *Science education should prepare citizens to make responsible decisions concerning science-related social issues.*

The early decades of the 21st century have witnessed a number of "grand challenges," such as climate change, resource use, environmental quality, emerging and re-emerging infectious diseases, and a variety of natural hazards such as fires, droughts, and severe weather. To be clear, these issues have both a scientific basis and social and political implications. In addition, there are ethical concerns associated with these and other social issues.

PERSONAL DEVELOPMENT

All individuals have needs related to their own biological and psychological systems. Briefly stated, this goal is, *Science education should contribute to an understanding and fulfillment of personal needs, thus contributing to personal development.*

School personnel and parents express their concerns about meeting the personal development of students through science education. This rhetoric takes the form of health-related issues, life-and-work and school-to-work skills related to science, the preparation ethic, and vocational or college and career aims. In response, science courses often emphasize content that is seen as useful in everyday living and fundamental to 21st-century work. It appears, however, that the goal of fulfilling personal development is not generally emphasized. In recent times, a theme to ensure students are ready for college or career represents this aim and the next.

CAREER AWARENESS

Scientific research, development, and application continue through the work of individuals within science and technology and through the support of those not directly involved in scientific work. One important goal has been, *Science education should inform students about careers in the sciences.*

One of the currently important goals of science education is to provide information and training that will be useful in the 21st-century workforce. Recent increased emphasis on this goal is due in part to public opinion and the economic downturn beginning in the early 2000s. The career awareness goal was found consistently across recommendations from various groups, although it was not a primary goal of science education. These goals have been repeated, with continual variations, through the history of science education in the United States. The goals have changed by updating the content—for example, clarifying what is meant by the process of science in the 1960s, scientific inquiry in the 1990s, and science and engineering practices in contemporary standards. Likewise, the goals of knowledge, personal development, social issues, and career awareness have been updated and clarified (see, e. g., DeBoer 1991; Bybee 1993; and Bybee and DeBoer 1994).

FROM PURPOSES TO PROGRAMS AND PRACTICES

Curriculum and instruction, the programs and practices of science education, are influenced by the way these goals are structured. For example, if a school district decides that knowledge in the form of science concepts is the important goal of its science program, then the other goals become subordinate in the design for curriculum and priorities for instruction. The other goals may be included, but not with the same emphasis. A different curriculum would be designed if the district decided that the methods of scientific inquiry would be the major goal of the curriculum and instruction.

Examination of the goals and their structure in science programs and practices or their implied emphasis in policy statements is a useful way to identify and portray differences in present curricula and a way to describe historical changes and contemporary trends in science education.

To the prior point, examining historical or current curricular structures and understanding the emphasis of goals are important activities for leaders: "The design of school science programs presents a *coherent set* of messages to the student *about* science (rather than *within* science). Such messages constitute objectives which go beyond learning the facts, principles, laws, and theories of the subject matter itself. . ." (Roberts 1982, p. 245). Roberts goes on to describe seven curriculum emphases that result from the organization and priority of goals in science education. Here are the curriculum emphases and the associated messages about science:

- **Everyday coping:** Science is an important means for understanding and controlling one's environment.
- **Structure of science:** The message is—how science functions intellectually in its own growth and development.
- **Science, technology, and decisions:** This message concentrates on the *limits* of science in coping with practical affairs. The message distinguishes science from technology and those disciplines from others in personal and social decision making.
- **Scientific skill development:** The primary goal is competence in the use of skills and abilities basic to all sciences. The message is that methods, processes, or inquiry have a priority greater than concepts or development.
- **Correct explanations:** As the title indicates, the curriculum exclusively stresses mastery of science facts and concepts.
- **Self as explainor:** Science is presented as a cultural institution and an expression of human capabilities. History of science (i.e., the growth and change in scientific ideas) is the general message.
- **Solid foundation:** The science curriculum is organized to facilitate the students' understanding of future science courses–for example, elementary school prepares students for high school (Roberts 1982, pp. 246–249).

These categories help leaders think about and understand the form of goals and their functions in school science programs. We return to this discussion later in the seminar and in Section VI of the book.

GOALS AND STANDARDS IN SCIENCE EDUCATION

Beginning in the 1990s, standards and benchmarks as policy statements became important statements that embedded the goals for science education and an implied emphasis for the goals in curriculum, instruction, teachers' development, and assessments.

Because it is important to understand the connection between goals and standards and the changes between the 1996 and 2013 standards, we present brief summaries in Tables 6.1 and 6.2 (p. 78). Chapters 7 and 8 present more details on the 1996 and 2013 standards, respectively.

Table 6.1. Science Education Goals as Represented in the *National Science Education Standards* (NRC 1996)

	Statements of Goals	Categories of Content Standards	Comments
SCIENTIFIC KNOWLEDGE	• Experience the richness and excitement of knowing about and understanding the natural world	• Unifying concepts and processes in science • Physical, life, and Earth and space science • Science and technology • History and nature of science	• Understandings about scientific inquiry were the content of the standards on inquiry.
SCIENTIFIC METHODS	• Use appropriate scientific processes and principles	• Science as inquiry • History and nature of science	• Abilities of inquiry was one component of the content standards.
SOCIETAL ISSUES	• Engage in public discourse and debate about matters of scientific and technological concern	• Science in personal and social perspectives	• Population growth, natural resources, environmental quality, natural and human-induced hazards, and science and technology in local, national, and global challenges were stated as content in the standards.
PERSONAL DEVELOPMENT	• Use appropriate scientific processes and principles in making personal decisions	• Science in personal and social perspectives	• Personal and community health were specifically stated as content in the area of science in personal and social perspectives.
CAREER AWARENESS	• Increase economic productivity through use of the knowledge, understanding, and skills of the scientifically literate person in their careers	• Career awareness had NO content standards.	• Although stated as a goal, the standards did not directly address the goal of career awareness.

Table 6.2. Science Education Goals as Represented in *A Framework for K–12 Science Education* (NRC 2012) for the *Next Generation Science Standards* (NGSS Lead States 2013)

	Statements of Goals	Categories of Content Standards	Comments
SCIENTIFIC KNOWLEDGE	• Educating all students on science and engineering	• Crosscutting concepts (CCCs) • Disciplinary core ideas (DCIs)	• In general, the CCCs and DCIs align with prior standards and other documents, such as the College Board standards.
SCIENTIFIC METHODS	• Actively engage in scientific and engineering practices and apply crosscutting concepts to deepen understanding of core ideas in these fields	• Science and engineering practices (SEPs)	• The SEPs have been strategically connected to the knowledge goals by connecting them in the standards—the performance expectations. • The SEPs also include both science and engineering. • The SEPs emphasize the importance of using models, constructing explanations, and engaging in argument based on evidence.
SOCIETAL ISSUES	• Insights from studying and engaging in the practices of science and engineering ... should help students see how science and engineering are instrumental in addressing challenges that confront society.	• Influence of engineering, technology, and science on society and the natural world	• Some standards, especially in Earth science, emphasize natural hazards, natural resources, human impact on Earth systems, and global climate change.
PERSONAL DEVELOPMENT	• See above.	• No specific content standards	• The *NGSS* have very little emphasis on personal development.
CAREER AWARENESS	• Providing the foundational knowledge for those who will become the scientists, engineers, technologists, and technicians of the future	• No specific content standards	• Connections to 21st-century careers are lacking.

Science education has a limited number of fundamental goals. We have used the following goals, which are expressed in general terms to provide clarity in discussions of the way goals are described at different times in our history. As you can see in Tables 6.1 and 6.2, the goals involve scientific knowledge, scientific methods, social issues, personal development, and career awareness. These figures provide examples of updating the content of goals, with greater priority given to some goals and less to others.

As we mentioned, the goals are embedded in the standards and the standards have to be translated from policies to programs of curriculum, instruction, assessment, and professional development. We underscore a basic point made in both standards' documents: The standards are not a curriculum. They do, however, have implications for the content and emphasis of school science programs. This discussion continues in the chapters on standards and science education programs.

A PERSONAL PERSPECTIVE

STANDARDS: A VOICE FOR CHANGE

Stephen Pruitt

This chapter makes an excellent case for the goals of science education and the role of standards for the presentation and emphasis of those goals. I will not belabor the point. The historical and current goals of science standards have been discussed thoroughly in the preceding sections. I take the initial perspective that standards present the contemporary content and emphasis of goals or purposes for science education. The representation of goals in standards is important, but understanding the importance of standards as an educational entity also has a purpose. From my perspective, I think there are three additional purposes that center on standards that should be explored. The first purpose for standards is, as it should be, to guarantee a quality science education for *all* students. The second purpose is to provide an opportunity for the scientific community to speak to the education community to ensure a new generation of scientists. The third purpose of standards is to force change in a system whose equilibrium is very difficult to disrupt.

I recognize that there are recurring, and perhaps even repetitive, themes in my perspectives. As such, I will keep this portion brief. However, the purpose and importance of standards really do hinge on the fact that every child has a right to a full and rigorous science education. A quality education is an equalizer that can change our entire society and economy. In the 21st century, a quality science education for all students is paramount. Teachers do not set out to lower their expectations for struggling students or those in underserved populations, but human nature makes it likely that teachers empathize with those who struggle. Despite good intentions, without standards, the gaps

between underserved populations and their more traditional classmates will just continue to expand. Obviously, standards alone will not solve inequities in our education system, but they are a key first step to set the expectation that *all* students can, and should, learn science.

To my second point, it has always been critical for the scientific community to play an active role in the setting of standards at the national or state level. In fact, it has been more critical in science than in other areas such as English language arts and mathematics. Of course, mathematics has had the "math wars" based on placing priority on process or content, which played out in all levels of education, including higher education, but science has had a particular need to have scientists involved early and often. Science has had a similar but less intense debate over process and content, but I say that it was more important to have the scientific voice in science than the other subjects, for two reasons. First, the depth of knowledge needed for conceptual science learning (the concept needed to understand phenomena or topics such as chemical bonding) is at a deeper level than we sometimes acknowledge. Second, science continues to change and evolve at incredible rates so the scientific voice is needed to ensure the concepts in the standards are current and deep enough to endure. While I will discuss the role of national standards in the development of state standards in another chapter, I must bring them up here, as it helps make this second point. As previously discussed, the American Association for Advancement of Science (AAAS) released *Science for All Americans* in the late 1980s, and this led to *Benchmarks for Science Literacy* in the early 1990s. About that time, the National Research Council (NRC) released the *National Science Education Standards.* Both sets of standards intended to illustrate what students should know and be able to do to be considered scientifically literate. Both the AAAS and NRC working groups included individuals that represented many different stakeholders, but what was most critical was the scientific voice expressed in both sets of standards.

Science, by its nature, is complex in that so many concepts are interrelated. As such, even placing a concept in a physical science "chapter" can be misleading, as that concept could easily have been placed in a life or Earth science chapter. I bring this up because for both AAAS and NRC, a report or narrative on what students should know to be scientifically literate was not enough to guide the development of state standards. After all, much of what ended up in *Benchmarks* also was in the *National Science Education Standards*. The education community needed the scientific community to be clear with the content students needed to know. There was another reason the scientific community needed to be involved. Whereas the controversy in the math community was over instructional practices, which led to the math wars, science has always faced a different issue—the divide between scientific understanding and public understanding of science. With issues such as evolution and climate change, scientists must make clear how essential it is that students learn such material. If suggested by a state or even teacher groups such as NSTA, those concepts could have easily been pushed off as unnecessary for scientific literacy. Let's face it— even with the scientific backing, these two issues alone provide plenty of discomfort with, and often rejection of, quality science. Still, the role of the scientific community having a voice in K–12 education cannot be understated. In fact, this involvement is probably the most prominent vehicle for that community to influence K–12 education. Standards affect all of the different components of the education system. Having the most informed individuals in their respective fields speak together with one voice—as they did in the *Framework* and *NGSS*—gives these standards the needed credibility to shape educational change.

As to final consideration of the importance of standards, we must think about standards as the fulcrum of a lever system. As previously discussed, standards affect curriculum, instruction, assessment, and professional development. The system really revolves around the standards. As such, change in the system can only be accomplished by changing the position of the fulcrum. I have said that the *NGSS* were meant to perturb the system. They have done that and continue to churn the waters of change; there is a full range of feelings and emotions around them. While most classroom teachers seem supportive of the new standards, there are still some who do not. For the latter, I suggest the change caused by standards is analogous to how Thomas Kuhn described scientific revolution. In essence, Kuhn describes the history of science as a series of revolutions (changes) and adjustments to new knowledge. Education, like science, is an ever-evolving process. There is never a final answer for either, so understanding the scientific process allows us insight into education reform and its ever-changing process. Whether it was the change from alchemy to chemistry, from the Bohr model of the atom to Schrödinger's model, or from Newton's laws of motion to Einstein's theory of relativity, science has experienced major changes followed by years of adjustment to the new understanding, only to be disrupted again by change. Kuhn coined the now-overused phrase paradigm shift when describing how science works. Kuhn's description of science was different than the more traditional view of scientific philosophers. While we know that new theories arise from old ones, it is not an incremental change, as is insinuated by past philosophers. Kuhn described the change as a period of discovery based on new evidence, a struggle to accept the new theory, and acceptance with a period of redefining practice before the cycle begins again. Standards are a similar cycle. Left alone, the education system would continue as is. Standards act as the paradigm shift education needs to make the necessary changes to direct reform. Based on available research, the guidance of practicing scientists, and the knowledge of education experts, standards bring about a new conversation about what is essential to achieve scientific literacy. As with scientific revolution, new standards are built on the knowledge and experience of previous standards, but with new research in a new context. Some embrace the change and others resist it. The hope with new standards is that the change does not require the old generation to "die off," as Kuhn suggests. Following new standards, there is a time of learning and implementation. The new standards drive new conversations about instruction, curriculum, assessment, and policy. Make no mistake, the standards from the 1990s had a significant effect on all state standards. The *Framework* and *NGSS* were based on those documents, but with new cognitive evidence and a 21st-century context.

The *NGSS* represent the latest paradigm shift in education, so we stand at the beginning of the next cycle of change in education. Some will say it's "just another change," and it is true that education has seen its fair share of new things. However, one thing that does not change often are the standards. We may have a better understanding of them and we may find new ways of implementing them, but the use of standards has stayed firm for more than two decades. A significant difference in *NGSS*, however, is that the standards were developed to be readily adopted by states and districts. *Benchmarks* and *NSES* were meant to inform development of state-specific standards. So, the *NGSS* offer states and districts the opportunity to share resources and drive change by having larger numbers of students with the same expectations. But let's also be clear about a few more key shifts. Yes, as a practical matter, the *NGSS* were developed to be adopted by states, but there are other factors that actually will make the difference for our students if we can pull it off. In developing the *NGSS*, it became clear that there were several characteristics that illustrate the change, including requiring students to operate at the nexus of the three dimensions, as opposed to separately. In short, this means that students should be able to use the three dimensions to explain phenomena; whereas in the past, state standards required focus on each dimension separately, students must now use the three dimensions together. Another way to say this is that *what* students know is not as important as *what they can do* with what they know. Another shift in the standards involves the coherence from kindergarten through twelfth grade. The *NGSS* were designed with learning progressions in mind to ensure a coherent story line throughout school without meandering or having one off-topic discussion. As a final example of the paradigm shift, I must address engineering. The engineering processes are expected outcomes in the science standards. This provides a great opportunity for our students, as science and engineering are so complementary in practice and in the workforce. This is a shift almost to the level of the three-dimensional learning, as this may be the shift that created some of the most concern. Although many teachers have used engineering, even if they did not call it that, they did not engage in full processes, such as dealing with resources, failing (a key component of the process), and revising. Standards are the voice of change, but we have work to do if we are to give that voice the necessary influence to carry out the vision of science education.

OUR COMMON PERSPECTIVE AND LEADERSHIP OPPORTUNITIES

Although it may seem strange or unusual, too often the purposes of science education are given little consideration in the formulation of frameworks, curricula, and assessments. In this seminar, we encourage you to think about and consider the fundamental purposes of science education, the critical role they play in the formulation of standards, and, subsequently, the design of school programs.

Review and consider the different purposes of education as they apply to science education. As you have seen, the purposes of science education range from specific knowledge of concepts to general obligations of citizenship. Between these purposes, there are goal areas that include the methods and practices of science, awareness of careers, and skills and abilities needed by the 21st-century workforce. We encourage you to explore the range of goals in any reform.

Reforms inevitably begin with the revising and restructuring of goals. There are reasons and justifications for both the revising (e.g., knowledge becomes disciplinary core ideas, inquiry becomes practices, and unifying concepts become crosscutting concepts) and restructuring of sets of goals (e.g., placing more emphasis on social perspectives and less emphasis on careers). Pay attention to influences on the goals and consider the implications on implementation for curriculum, instruction, and assessments.

Take national and state standards seriously, as they describe the orientation and emphasis expressed by the scientific and education communities. Standards for science education have made use of thoughtful input from both scientists and educators. Changing, reorganizing, or eliminating standards should only be done with review and justification.

Both equity and excellence are continuing themes of science education. How will your goal statements and translation to programs support both of these themes?

Exercise both caution and clarity when using the term *scientific literacy* to express the goals of science education. The first caution is to guard against definitive negative statements of scientific illiteracy. Such generalized statements likely belie an either/or perspective and are based on single categories of science (e.g., a single concept, process, or application). Our idea of scientific literacy includes a continuum of scientific understandings, abilities, and competencies that develop over one's lifetime.

ISSUES AND QUESTIONS FOR DISCUSSION

1. The term *purpose* refers to various goal statements of what science education should achieve, such as scientific literacy for all learners. Science educators express many and diverse aims and goals in various documents such as national and state standards, curriculum frameworks, school syllabi, and teaching lessons. What do you perceive as the strengths and weaknesses of these purpose statements?
2. There is a long history of tensions between advocates for liberal education and supporters of science programs as preparation for future careers. Should K–12 science education emphasize one perspective, a balance between the two, or neither?
3. In many cultures and throughout most of history, education of youth was the responsibility of institutions other than schools. The family, church, and community are examples of institutions that have educated youth and continue to do so. What do you perceive as the special role of schools in general and education in science in particular?
4. In *Public Education* (1976), historian Lawrence Cremin made this statement (a variation of which was included in Rodger's perspective statement):

   > *[T]he questions we need to ask about education are among the most important questions that can be raised in our society. … What knowledge should "we the people" hold in common? What values? What skills? What sensibilities? When we ask such questions, we are getting to the heart of the kind of society we want to live in and the kind of society we want our children to live in. (pp. 74–75)*

   From a science perspective, how would you answer Cremin's questions? What is your rationale—how do you justify the emphasis in your answers?

5. Statements on citizenship are an important purpose of science education. How would you propose responding to this aim? What would you include in your school science program that addresses this goal of citizenship?

6. Write a short editorial on the question, "Why educate students about science?"

REFERENCES

Bybee, R. W. 1993. *Reforming science education: Social perspectives and personal reflections.* New York: Teachers College Press, Columbia University.

Bybee, R. W., and G. E. DeBoer. 1994. Research on goals for the science curriculum. In *Handbook of research in science teaching and learning*, ed. D. Gable, 357–387. New York: MacMillan.

Cremin, L. A. 1976. *Public education.* New York: Basic Books.

DeBoer, G. E. 1991. *A history of ideas in science education.* New York: Teachers College Press, Columbia University.

National Research Council (NRC). 1996. *National science education standards.* Washington, DC: National Academies Press.

National Research Council (NRC). 2012. *A framework for K–12 science education: Practices, crosscutting concepts, and core ideas.* Washington, DC: National Academies Press.

NGSS Lead States. 2013. Next Generation Science Standards: For states, by states. Washington, DC: National Academies Press. *www.nextgenscience.org/next-generation-science-standards.*

Roberts, D. 1982. Developing the concept of "curriculum emphasis" in science education. *Science Education* 66 (2): 243–260.

SECTION IV

NATIONAL STANDARDS

POLICIES FOR SCIENCE EDUCATION

SECTION IV

Standards are specific policy statements and action plans based on the purposes of science education. Policies are concrete translations of the purpose and apply to specific components such as teacher education, K–12 curricula, and assessments.

National standards have become useful maps that provide purpose and direction in American education by answering questions about what students should know and be able to do after 13 years of school. At the same time, discussions of national standards and the implied reforms have raised questions about the purposes of education, the standards' impact on equity and excellence, who decides the content students should learn, and how society knows if students have learned the content and abilities the standards describe.

National standards identify the purposes and goals for education and—based on those aims—describe clear, consistent, and challenging learning outcomes. Who could be critical of this? After all, common sense and reasonable judgment suggest that educational quality and teaching are better if goals are clear and teachers' knowledge and skills, instructional materials, and assessments are all coherent.

The first generation of standards, the *National Science Education Standards* (*NSES*; NRC 1996), influenced state and district standards until the *Next Generation Science Standards* (*NGSS*; NGSS Lead States 2013) were released. Table IV.1 summarizes the states that have adopted *NGSS* and those that have been influenced by *A Framework for K–12 Science Education* (NRC 2012) and the *NGSS* (NGSS Lead States 2013) in the development of their standards.

REFERENCES

National Research Council (NRC). 1996. *National science education standards.* Washington, DC: National Academies Press. The "first generation" volume with the title *National Standards.*

National Research Council (NRC). 2012. *A framework for K–12 science education: Practices, crosscutting concepts, and core ideas.* Washington, DC: National Academies Press. A fundamental report influencing all dimensions of contemporary science education.

NGSS Lead States. 2013. *Next Generation Science Standards: For states, by states.* Washington, DC: National Academies Press. Both Volume 1 (the standards) and Volume 2 (the appendixes) are the current national standards for science education.

SUGGESTED READINGS

American Association for the Advancement of Science (AAAS). 1993. *Benchmarks for science literacy.* New York: Oxford Press. A product of Project 2061, these were the initial standards for American science education.

Ravitch, R. 1995. *National standards in American education.* Washington, DC: Brookings Institution. An excellent introduction to and history of education standards.

Rutherford, F. J., and A. Ahlgren. 1990. *Science for all Americans.* New York: Oxford University Press. This volume set the stage for an era of national standards in science education.

Table IV.1. States That Have Adopted and Adapted the *NGSS*

States That Have Adopted the *NGSS*	States That Have Adapted the *Framework* and *NGSS*
• Arkansas • California • Connecticut • Delaware • District of Columbia • Hawaii • Illinois • Iowa • Kansas • Kentucky • Maryland • Michigan • Nevada • New Jersey • Oregon • Rhode Island • Vermont • Washington	• Alabama • Georgia • Idaho • Indiana • Massachusetts • Missouri • Montana • Oklahoma • South Carolina • South Dakota • Utah • West Virginia • Wyoming
States With Standards in Development or Not Formally Adopted	**States That Have Not Revised Their Science Standards**
• Colorado • Louisiana • Minnesota • Nebraska • New Hampshire • New Mexico • New York • North Dakota • Pennsylvania • Tennessee • Virginia	• Alaska • Arizona • Florida • Maine • Mississippi • North Carolina • Ohio • Texas • Wisconsin

Note: Updated September 23, 2016.

CHAPTER 7

NATIONAL STANDARDS AND SCIENCE EDUCATION

HISTORICAL PERSPECTIVES

This chapter provides background on the idea of standards, the context for contemporary national standards, and perspectives on the *National Science Education Standards* (*NSES*). Chapter 8 is an introduction to the *Next Generation Science Standards* (NGSS).

NATIONAL STANDARDS ARE A NEW IDEA—RIGHT?

National standards may seem like something new in American education, but they are not.

As you have seen, American education has a long history of committees, reports, and groups defining content and required courses that amount to standards. If the basic idea of standards is to provide clear and consistent statements of what students should know and be able to do, then standards, even if they were not called standards, have been a part of American education since Harvard established admission requirements in 1643, followed by Yale (1745) and Columbia (1778). With time, the admission requirements broadened from, for example, reading classical Latin (Harvard) to include the rules of arithmetic (Columbia; Ravitch 1995).

In 1892, the National Education Association (NEA) established the Committee of Ten, a panel of experts charged with making recommendations to improve the nation's high school curricula. As a national panel, the Committee of Ten had no precedent to make recommendations to thousands of school districts. The report was relatively effective, primarily due to the stature of the panel members as national leaders. The report recommended physical science (physics, astronomy, and chemistry), "natural history" (biology, which included botany, zoology, and physiology), and geography (physical geography, geology, and meteorology).

The standards described in these reports were clearly directed toward college preparation. Because of this orientation, many educators objected to the standards. As a result, the NEA established a Commission on the Reorganization of Secondary Education (CRSE), whose 1918 report (standards) was distinctly different from prior college preparatory standards. The committee included academic subjects and industrial arts, household arts, vocational guidance, agriculture, and other areas not generally considered academic. This committee report identified seven cardinal principles as the main objectives of education. Those principles included health, citizenship, worthy use of leisure, and ethical character. The individual academic subjects needed to be shown as making contributions to achieving these objectives. The emphasis was clearly on utility (What was useful for the student?) and social efficiency (How could school programs serve the needs of society?). In this justification of courses, in terms of educational objectives, geography was part of social studies (Ravitch 1995).

With the CRSE example, we point out the two different purposes of education, one with knowledge of academic disciplines and the second with a liberal arts orientation. The Committee of Ten example stressed an education primarily for college-bound students. In contrast, the CRSE underscored the purpose of social efficiency, an education for non-college-bound students. The CRSE report resulted in vocational and general tracks for some students and academic and college tracks for others. The contemporary perspective of college and career preparation is a possible resolution of the conflicting purposes expressed by the two committees and their respective national "standards."

Tests and Textbooks as Standards

We cannot leave this discussion of implied or suggested national standards without mentioning tests and texts. To be specific, we are referring to standardized tests and commercial textbooks. In the early decades of the 20th century, standardized achievement tests were introduced, as were college entrance examinations (Ravitch 1995). Both types of tests served as implicit academic standards for states and school districts. The American College Testing (ACT) and Scholastic Aptitude Test (SAT) examinations serve similar purposes today.

Textbooks, such as those used in science, also serve as de facto national standards in education. An estimated 75% (or more) of instructional time in classrooms is structured by textbook programs (Woodward, Elliott, and Nagel 1988). Although the report on this is several decades old, we have little reason to suggest significantly different percentages; however, this could change as states set new frameworks and adoption requirements based on contemporary national standards.

So, the term *national standards* may be a new addition to American education, but the idea of clarifying purposes and describing the content for curricula and assessments is by no means new. We continue with contemporary national standards for science education, beginning with their origins.

The Origins of Contemporary National Standards

In 1983, the landmark report *A Nation at Risk* (NCEE 1983) stimulated concerns and reforms among states. The report warned that the American education system was far behind its international competitors and that there were eminent threats to the country's economic future. The report recommended high expectations in academic subjects including science and a nationwide system of assessments. In time, it became clear that 50 states and thousands of school districts working independently could not meet the challenges and reduce the risks America faced. There was a need for national leadership.

On September 27 and 28, 1989, President George H. W. Bush gathered the country's governors in Charlottesville, Virginia, to discuss a single issue—education. This historic meeting resulted in the proposed America 2000 legislation (1991), which called for voluntary national standards. Congress did not pass the legislation. However, the idea of national goals and standards had risen to prominence. In 1989, the National Council of Teachers of Mathematics (NCTM) published *Curriculum and Evaluation Standards for School Mathematics*.

Given the increasing corporate attention to total quality management based on raising performance to meet higher standards of quality, it is not surprising that the National Educational Goals Panel (NEGP) found the idea of standards in different subjects and performance-based assessments attractive. When the National Council on Education Standards and Testing (NCEST) reported on the merits and feasibility of national standards and assessments, the NCTM standards had already provided the proof that NCEST needed—the mathematics standards. The standards set focus and direction, not a national curriculum; they were national, not federal; they were voluntary, not mandatory; and they were dynamic, not static.

NATIONAL SCIENCE EDUCATION STANDARDS: THE FIRST GENERATION

As you can see, support for national standards in science formalized in 1989, when the nation's governors and President George H. W. Bush established six national education goals, which were adopted by Congress and later expanded to a total of eight goals. In 1994, Congress enacted Goals 2000: Educate America Act and formed the National Education Goals Panel (NEGP) to support and monitor progress toward the goals. (See Table 7.1 [p. 94] for historical highlights of the *NSES*.)

Developing National Standards for Science

In science, two important publications preceded initial work on national standards. In 1989, the American Association for the Advancement of Science (AAAS), through its Project 2061 led by F. James Rutherford, published *Science for All Americans* (Rutherford and Ahlgren 1989). This publication defined science literacy for all high school graduates and provided the foundation for *Benchmarks for Science Literacy* (AAAS 1993), which had a significant influence on the development of national standards for science. Three years later, the National Science Teachers Association (NSTA), through its Scope, Sequence, and Coordination Project, published *The Content Core* (1992).

In 1991, the National Research Council (NRC) was formally asked by Dr. Bonnie Brunkhorst, then president of NSTA, to assume a leading role in developing national standards for science education. The NRC was encouraged by leaders of several other science and science education associations, the U.S. Department of Education, the National Science Foundation (NSF), and the NEGP. The effort—funded by the U.S. Department of Education, NSF, and the National Aeronautics and Space Administration (NASA)—was led by the National Committee on Science Education Standards and Assessment (NCSESA), advised by the chair's advisory committee that consisted of representatives from major science education organizations, and carried out by three working groups (i.e., content, teaching, and assessment) composed of science teachers, educators, scientists, and others involved in science education.

Preparations for work on the intellectual substance of the standards began in the fall of 1991. NRC staff were assigned to produce summaries of the proposed standards, based on the work of NSTA's Scope, Sequence, and Coordination; AAAS Project 2061; and other projects, as well as state science frameworks and science standards from other countries.

One early decision was to develop standards for content, teaching, and assessment all displayed in mutually re-enforcing ways. Another decision committed the working group chairs to function as a team throughout the project. A third decision was to take the critique and consensus process seriously, issuing frequent updates on the project and materials suitable for intense critique by teachers, subject matter experts, and others. Discussion and working papers were released in October 1992, December 1992, and February 1993. The first draft of content, teaching, assessment, professional development of teachers of science programs, and system standards appeared late in 1993.

Early drafts of the NRC standards were subsequently reviewed by groups of experts and large numbers of educators across the country. More than 40,000 copies of a complete draft were distributed in December 1994 to approximately 18,000 individuals and 250 groups for review. The comments and recommendations received from these reviewers were used to prepare the final document, which was formally released in December 1995 as the *National Science Education Standards* (NRC 1996).

National Science Education Standards: An Overview

In early 1996, the NRC consolidated its education activities into the Center for Science, Mathematics, and Engineering Education (CSMEE). CSMEE took on support for the new *National Science Education Standards* as an important priority, and Rodger Bybee was hired as the executive director. The first initiative of CSMEE was the preparation of an introduction to *NSES* (NRC 1997).

The *NSES* defined the science content that all students should know and the practices they should be able to do and provided guidelines for assessing the degree to which students have learned that content. The *NSES* detailed the teaching strategies, professional development, and support necessary to deliver high-quality science education to all students. The *NSES* also described policies needed to bring coordination, consistency, and coherence to science education programs. You can see from this summary that the *NSES* were a comprehensive set of standards for science education. Specifically, the *NSES* included standards for science content, teaching, assessment, professional development, school science programs, and the education system's support of *NSES*.

In *NSES*, the content standards included the following:

- Unifying concepts and processes

- Science as inquiry
- Physical science
- Life science
- Earth and space science
- Science and technology
- Science in personal and social perspectives
- History and nature of science

The first category of the content standards, unifying concepts and processes, identified powerful ideas that are basic to science disciplines. These standards included both conceptual and procedural content (e.g., systems, order, and organization; evidence, models, and explanation). The other content categories included knowledge and abilities in inquiry, which ground students' learning of subject matter in physical, life, and Earth and space sciences. Science and technology standards introduced the similarities and differences between the natural and designed worlds and questions and problems. The personal and social perspectives standards introduced students to science in life situations and helped them develop decision-making skills. The history and nature of science standards helped students see science as a human experience that is both ongoing and ever-changing (NRC 1996).

Benchmarks and Standards

The *Benchmarks for Science Literacy* (AAAS 1993) were also statements of standards and caused some confusion within the science education community. Which should be used, the *Benchmarks* or *NSES*? There were differences. For example, the *Benchmarks* included components for different grade levels and included more content in social and behavioral sciences and mathematics. The *NSES* gave greater emphasis to inquiry both as science content and as a teaching strategy. Finally, as mentioned above, the *NSES* addressed a broader range of standards. There was, however, an estimated 90% consistency of content between the *Benchmarks* and *NSES*. Use of either document by states or local school districts would improve science education.

Finally, the *NSES* content clarified scientific literacy. Here is an answer to the question, "What is scientific literacy?" Scientific literacy is the knowledge and understanding of scientific concepts and processes required for personal decision making, participation in civic and cultural affairs, and economic productivity. People who are scientifically literate can ask, find, or determine answers to questions about everyday experiences. They are able to describe, explain, and predict natural phenomena.

Scientific literacy has different degrees and forms; it expands and deepens over a lifetime, not just during the years in school. The *NSES* outline a broad base of knowledge and skills for a lifetime of continued development in scientific literacy for every citizen and provide a foundation for those aspiring to scientific careers (NRC 1996).

Table 7.1. *National Science Education Standards:* Historical Highlights

Year	Highlights
1983	*A Nation at Risk* is released by NCEE.
1989	*Curriculum and Evaluation Standards for School Mathematics* is released by NCTM. National Governors Association releases national educations goals. President George H. W. Bush forms the NEGP. *Science for All Americans* is released by AAAS.
1991 **1992** **1993**	NSTA's president and executive director request the NRC coordinate development of national standards for science education. The NRC establishes the National Committee on Science Education Standards and Assessment (NCSESA). The first meeting of NCSESA takes place.
1994	*Benchmarks for Science Literacy* are released by AAAS-Project 2061. The first complete draft of standards for science education are developed and released.
1996 **1996–2013**	Professional organizations, focus groups, and the NRC report review teams evaluate the first draft of standards. The second draft of the standards is released, and 40,000 copies are distributed for review
	The *National Science Education Standards* are released (NRC 1996). The *NSES* continue to influence components of the science education system until 2013, when the *Next Generation Science Standards* are released.

A PERSONAL PERSPECTIVE

THE *NATIONAL SCIENCE EDUCATION STANDARDS*: REFLECTIONS AFTER A DECADE

Rodger Bybee

"Nationwide Standards Eyed Anew." This headline appeared in the December 7, 2005, issue of *Education Week*. The story highlighted the diversity of demands by states and the resurgence of national standards. The article (Bybee 2006) quoted Diane Ravitch: "Americans must recognize that we need national standards, national tests, and a national curriculum" (p. 1). This article appeared 10 years, almost to the day, after the release of the *National Science Education Standards (NSES)*. The article and quote expressed views generally consistent with my own. The United States needs national standards for science education and for technology and mathematics as well. National standards provide the means for improving student achievement while maintaining the authority for states and local school districts to determine their science programs. In principle, this is possible. In practice, it is far from reality. But that is not a reason to reject national standards. Indeed, quite the opposite is the case. We should embrace the national standards for science education. Although originally written in 2006, I think these ideas still hold in 2016.

It is worth noting several things about standards at the beginning of this essay. National, state, and local standards are primarily a reflection of values and priorities of those individuals, organizations, and agencies responsible for developing the standards. They are not research reviews or based on research. Questions about the influence of the *NSES* on education are important for research, albeit they are very complex issues to investigate. Second, the *NSES* provide policies for curriculum, instruction, assessment, and professional development. They must be interpreted by those responsible for designing and implementing programs, facilitating changes in instructional practices, and instituting new assessments and accountability measures. Finally, any significant influences of national standards on the education system will take time, likely more than a decade; a reasonable estimate would be two decades. The time it takes for national standards to influence the system is the reason the headline captured my attention and influenced my comments about the standards for science education.

This essay describes my reflections and opinions based on more than a decade's experience with the *NSES*. My work on the *NSES* began in 1992 as a member (and later chair) of the Content Working Group. In 1995, I became executive director of the Center for Science, Mathematics, and Engineering Education (CSMEE) at the National Academies, where my work completing and disseminating the *Standards* continued until 1999, when I returned to Biological Sciences Curriculum Study (BSCS). At BSCS, we used the *NSES* as the content and pedagogical foundation for curriculum materials and professional development. So, my experiences with the *NSES* have been quite varied and include the perspectives of policy, program, and practice. This essay does not include a discussion of the project that produced the *NSES*, but Angelo Collins has provided an excellent history (Collins 1995). Also worth noting is the October 1997 issue of *School Science and Mathematics*, a theme issue for which my colleague Joan Ferrini-Mundy and I served as guest editors. With this as context, I continue with my reflections on the 1996 *National Science Education Standards*.

WHY ARE NATIONAL STANDARDS IMPORTANT?

The power of national standards lies in their potential capacity to change the fundamental components of the education system at a scale that makes a difference. Very few things have the capacity to change curriculum, instruction, assessment, and the professional education of science teachers. National standards must be on the short list of things with such power. The changes also are systemwide and thus at a significant scale. To the degree that various agencies, organizations, institutions, and districts embrace the standards, there is potential to bring increased coherence and national unity among state frameworks, criteria for adoption of instructional materials, and other resources for science education.

How Do the Standards Change Components of the Education System?

Early in my work on the standards, I realized there were several ways they may affect the system. The importance of teaching biological evolution provides excellent examples for this discussion. First, including content such as biological evolution in standards in turn affects the content in state and local standards. A review by *Education Week* (November 9, 2005) found that a majority of states (39) included some description of biological evolution and 35 states described natural selection. In short, national standards influence the priorities for content in state and local standards.

My second point centers on feedback within the systems. Using the *NSES* as the basis for their review, *Education Week* provided insights about which states did *not* mention evolution—Florida, Illinois, Kentucky, and Oklahoma. It also indicated the significant variation in the presentation of evolution among other states. The latter was a major finding in the review.

The *NSES* also can be used to define the limits of acceptable content. This is my third point. When Kansas again planned to adopt state standards that would promote non-scientific alternatives to evolution and liberally borrowed from the *NSES* and the National Science Teachers Association's (NSTA) *Pathways to Science Standards* (2005), both organizations denied Kansas the right to incorporate any of their material into its new standards (*Science* 2005).

Briefly, the *NSES* indicate what should be included in state standards, school science programs, textbooks, and assessments. The standards provide the basis for feedback about content of other standards and programs. Finally, they can be used as defense against efforts to include non-scientific content. These are three important ways the *NSES* influence the science education system. This was true for the 1996 standards and is still true today.

Contrast the potential influence of standards on the instructional core and education system with the possibility of improving student achievement at national levels using other contemporary education ideas and priorities, such as vouchers, charter schools, and site-based management. To be very clear, I am not opposed to such ideas; they may embrace important goals and result in some improvement of student achievement. But they do not necessarily result in fundamental changes at the place where students and teachers meet. Furthermore, the changes are usually local and thus at a scale that lacks significance. They may represent high political priorities, but they have low value when viewed nationally.

What About Equity?

There is a second feature of the standards that demonstrates their importance: They present policies for all students. By their very nature, national standards are policies that embrace equity. When the *NSES* answered the question, "What should all students know and be able to do?" the standards became clear statements of equity. In the decade since the release of the *NSES,* I have had many individuals ask if we really meant *all.* The answer is yes. Of course, there are exceptions that prove the rule; severely developmentally disabled students would be an example. But the standards are still clear statements of equity. While developing the NSES, we were quite clear about the fact that many aspects of the education system would need to change to accommodate the changes the standards implied. The need for changes such as the reallocation of resources to increase achievement of those students most in need was clearly understood by those most closely associated with the *NSES.*

Have the *NSES* changed the fundamental components system-wide and achieved equity? No. But you will notice that I indicated they had the *potential* to do so, not that they actually *did* do so. I would note for readers that this nation has not achieved equal justice for all, but we hold this as an important goal, one that we do not plan to change simply because it has not been achieved.

How Have National Standards Influenced Science Education?

Using national standards places emphasis on outputs of the education system. The *NSES* clearly define the goals for 13 years of science education and assume the various inputs to the system will change to accommodate the goals. For example, textbooks, tests, teaching, and technologies would change to achieve the stated goals. Ultimately, we could assume that national standards would influence student achievement. Of course, educational change does not work as planned. The rubber of education innovation always meets the road of reality.

While directing the Center for Science, Mathematics, and Engineering Education at the National Academies, I initiated a report intended for researchers interested in answering questions about national standards, *Investigating the Influence of Standards: A Framework for Research in Mathematics, Science, and Technology Education* (NRC 2002), directed by Dr. Iris Weiss. Although the goal of standards is student achievement, the influence of national standards is proximate and often compromised by countervailing forces and conditions in the education system. The NRC committee identified three primary channels of influence: curriculum, teacher development, and assessment and accountability. This said, the channels of influence are complex and interactive; significant time is needed for national standards to influence components of the system. The standards may be altered or ignored at the interface of system components such as the design of instructional materials, development of state standards, requirements for teacher certification, and national, state, and local assessments. Several questions from the NRC report (2002) present the basis for the following discussion.

How Have the *NSES* Been Received and Interpreted?

The answer should not surprise any reader—it depends. Release of the *NSES* signaled change, and this by its nature resulted in resistance from some individuals and groups. Interestingly, the resistance primarily was about the *idea* of standards, not the actual content of the standards. On the other hand, I think it is safe to say that the standards have been positively received within the science education community. Science educators recognized the importance and potential value of the standards on the education system. Unfortunately, at the national level, policy makers did not embrace the *NSES.* I attribute this to politics and the need for a Republican Congress to set new policies and reject many aspects of the prior Democratic administrations. The *NSES* and *Before It's Too Late: A Report to the Nation from the National Commission on Mathematics and Science Teaching for the 21st Century* (U.S. Department of Education 2000) suffered this fate.

Textbook publishers did not receive the *NSES* well. We held a meeting for publishers at the National Academy to introduce and review the standards and help publishers interpret the various features for their textbooks and programs. The reception of the *NSES* by the representatives of major publishing houses was cold and largely dismissive. When I asked several individuals about their responses, I was told they were very upset because they had an excellent gauge of the current market for school science programs. Their "gauge" was well calibrated because they had influenced the market using a variety of strategies. The *NSES,* however, would cause the school district priorities to change, and publishers would need to change marketing strategies and publish new programs. Depending on whether you have economic or educational priorities, the publishers' views were seen as good or bad. Of course, I had a positive view of changes based on the *NSES.* I still do.

Interpretations of the *NSES* have varied. Initially, individuals had to make sense of the *NSES* in terms of background, potential use, and priorities. For example, some interpreted national standards in terms of the *Benchmarks for Science Literacy* (AAAS 1993), which had been released earlier. Other interpretations included equating standards on scientific inquiry with the traditional processes of science, equating the *NSES* with a curriculum framework, and confusing the statements of the *NSES* with other aspects of the narrative. Although the *NSES* included discussions of their use and function, it seemed that many individuals did not read the discussions. For example, the *NSES* state:

> *The content standards are not a science curriculum. Curriculum is the way content is delivered: It includes the structure, organization, balance, and presentation of the content in the classroom.*

> *The content standards are not science lessons, classes, courses of study, or school science programs. The components of the science content described can be organized with a variety of emphases and perspectives into many different curricula. The organizational schemes of the content standards are not intended to be used as curricula; instead, the scope, sequence, and coordination of concepts, processes, and topics are left to those who design and implement curricula in science programs. (NRC 1996, pp. 22–23)*

Still, the *NSES* were interpreted as a curriculum framework. Even now, a decade later, one hears that the *NSES,* for example, recommend an integrated approach to science curriculum or a particular scope and sequence for curricula. I will state again that the *NSES* do not represent a science curriculum. They present science content and abilities that all students should learn or develop, respectively. How curriculum developers, states, and local school districts organize the content can and should vary.

What Actions Have Been Taken in Response to the *NSES*?

The first point is that numerous and varied actions have been taken. States have used *NSES* as a basis for science standards, so the influences and actions are wide, but the variations in state standards are significant. *The State of State Science Standards,* a report from the Thomas B. Fordham Institute (Finn and Gross 2005), bears witness to the variation. I suspect the variation is even greater among the standards developed at the district level. Requests for proposals (RFPs) from federal agencies such as the National Science Foundation (NSF), the National Institutes of Health (NIH), and the National Aeronautics and Space Administration (NASA) have required alignment of proposed projects with the *NSES.* A review of journal citations reveals recognition of the *NSES* in articles that range from policy to practice.

BSCS, for example, paid very close attention to the *NSES* when designing new NSF-supported programs, such as *BSCS Biology: A Human Approach* and *BSCS Science: An Inquiry Approach,* and the revision of the elementary program *TRACKS.* Content from the *NSES* has been central to our professional development programs and research. Other developers—such as Lawrence Hall of Science (LHS), Education Development Center (EDC), and Technical Education Research Center (TERC)—used the *NSES* in the development of new programs.

A close review of national and international assessment frameworks for NAEP (2009), PISA (2006), and TIMSS (2003) also reveals the influence of the *NSES.* The actions have been national and even international and have bridged policies to practices.

ISSUES, INSIGHTS, AND IDEAS CONCERNING THE STANDARDS

During more than a decade of involvement with the *NSES,* I have read, heard, and seen many things, some of which are worthy of comment. Following are some of those issues.

Why Do We Have Both *NSES* and *Benchmarks*?

From the beginning of the work on *NSES*, we heard this question and associated questions, such as, "What are the differences between *NSES* and *Benchmarks?*" and "Which document should be used?" To the lead question, I have to answer that it is probably a function of timing and politics. Certainly, *Science for All Americans* (AAAS 1989) set the stage for national standards. The publication of *Benchmarks for Science Literacy* (AAAS 1993) presented the major ideas from *Science for All Americans* as practical outcomes for the science education community. From the beginning of my work on the *NSES*, I paid very close attention to the *Benchmarks*. Although many had questions and complained about the two documents, for some time I thought that this situation had the positive benefit of facilitating review, thought, and discussion about the fundamentals of science education and the importance, role, and function of pivotal documents such as *Science for All Americans*, the *Benchmarks*, and the *NSES*. I still believe this.

The *NSES* and *Benchmarks* are comparable sets of policies. In 1995, Project 2061 completed an analysis of the two documents and concluded there was a "consensus on content." There is an estimated 90% agreement on content associated with the traditional disciplines of life, Earth, and physical sciences. The congruence should not surprise anyone (Rutherford 1996). Indeed, we acknowledged the *Benchmarks* in the introduction to the *NSES* (NRC 1996, p. 15). When asked which document I recommend, my response has been "either": Pick either the *NSES* or *Benchmarks* and use it consistently. *Consistency* is the operational term here. I, for obvious reasons, prefer the *NSES* but have supported use of the *Benchmarks* (Bybee 1997).

There Is a Persistent Confusion of Policy, Program, and Practice

The *NSES* is a policy document. It is not a school science program or instructional materials. It is not a document to be used in actual classroom practice or science teaching. Confusion about the purpose and function of the *NSES* centers on a fundamental lack of understanding and misconceptions about standards in general. Primary audiences for the *NSES* included state coordinators, curriculum developers, preservice and in-service teacher educators, and those responsible for assessments and accountability. These individuals, by nature of their jobs, have the responsibility of translating the policies of *NSES* to programs of curriculum, instruction, assessment, and teacher education and facilitating the effective implementation of those programs in classrooms. One challenge associated with the translation of policies to programs and eventually to classroom practices is understanding the time involved in developing and implementing new instructional materials and assessments (i.e., programs) and then providing professional development that results in changes in classroom teaching practices. One has to ask how long it takes to develop and implement standards-based curricula, instructional strategies, teacher professional development, and assessments. Furthermore, one might wonder how soon after those changes have been implemented we can reasonably expect changes in teacher practices and student achievement. Based on my experience at BSCS with curriculum development, my answer to questions such as these is that one can expect achievement changes between three and six years after funding for new curriculum programs and between seven and ten years after new instructional practices have been adopted (Bybee 1997). A 2002 headline in *Education Week*—"Science Standards Have Yet to Seep Into Class"—should not have surprised anyone. Yet, the media characterized the situation as a failure of the *NSES*. This is the kind of report to which I have become quite sensitive. The *NSES* should not be deemed a failure to change instructional practice because they must be translated into materials, assessments, and professional development. These processes take time and money. Indeed, the report on which this article was based did have a more positive, albeit preliminary, evaluation. This *Education Week* article was based on the release of an NRC report (2003) called *What Is the Influence of the* National Science Education Standards*?* This report commissioned authors to review more than 200 studies related to the *NSES*. The authors reported on the following areas: curriculum, teacher professional development, assessment and accountability, and student learning. Although most authors reported that research was inconclusive, it did tend to support the positive influence of the *NSES*. Given the short time between release of the *Standards* and the report, I would consider the results somewhere between very good and excellent.

Shouldn't the *NSES* Include Contemporary Issues and Specific Courses?

Personally, my position on contemporary issues, particularly those related to the environment, is that science education programs should address such issues. But standards are not and should not reflect personal biases that conflict with federal policy, such as the Constitution's Tenth Amendment (that is, states' rights). Through all of our work on the *NSES*, we had to avoid anything that would suggest, or even hint at, a national curriculum or set of policies that would reduce the states' rights to select content. Why, for example, did we not provide grade-level-specific standards instead of standards for the grade-level ranges of K–4, 5–8, and 9–12? Why didn't we indicate that Earth and space science should be a ninth-grade course, thus assuring a place in school programs? Why didn't we include specific problems such as global warming or other contemporary issues? The answer centers on the potential for any of these to reduce the potential influence of the *NSES* due to the politically controversial nature of these positions. The potential controversy has two components: the issue of an organization such as the National Academies suggesting a national curriculum and the social-political acceptance of topics such as global warming and stem-cell research, among others.

We did respond appropriately to some issues by including standards for science in personal and social perspectives. We included concepts in the *NSES* that lend themselves to understanding environmental issues, the nature of science, and the relationships among science, technology, and society. However, these standards have, for the most part, been ignored. In these standards, for example, we introduced fundamental conceptual understandings of population growth, natural resources, and environmental quality. Educationally, these can be defended on the basis that they are fundamental to many contemporary environmental issues; they present the conceptual basis for understanding topics such as climate change. Students should understand scientific concepts fundamental to an array of science-related issues they may confront now and in the future.

Scientific Inquiry Includes Both Content and Teaching Strategies.

A decade later, confusion continues about what is meant by *scientific inquiry*. For some, scientific inquiry is the same as skills, and for others, scientific inquiry is associated with a variety of teaching strategies. In efforts to criticize the theme of scientific inquiry as expressed in the *NSES*, Chester E. Finn Jr., recently stated, "Science education in America is under assault with 'discovery learning' attacking on one flank and the Discovery Institute on the other. That's the core finding of the first comprehensive review of state science standards since 2000" (Finn and Gross 2005). This statement is in a report from the Thomas B. Fordham Institute, a conservative Washington, DC, think tank. Finn later stated that "'discovery learning' is getting more weight than it can support in science. This is largely due to states' over-eager, over-simplified, and misguided application of some pedagogical advice enshrined in the so-called 'national standards'" (p. 10). To show what it is like to take a reasonable idea and reduce it to the ridiculous, I cite the final conclusion of Finn's discussion. He stated, "American students run a grave risk of being expected to replicate for themselves the work of Newton, Einstein, Watson, and Crick. That's both absurd and dysfunctional" (p. 10). Inflated rhetoric such as this from one person may appeal to colleagues with similar views, but it does not diminish the potential of national standards, either the *Benchmarks* or *NSES*, especially since it is politically motivated, is not grounded in an accurate view of the presentation of science as inquiry in the content standards, and fails to recognize that the majority of instructional materials and teaching strategies currently in schools can only be characterized as old-fashioned traditional instruction for which we have evidence of their lack of effectiveness. The evidence for my statements can be found in reports on the status of science education including curriculum, textbooks, and teaching strategies by Horizon Research, Inc., on the one hand, and the results from NAEP, TIMSS, and PISA on the other hand. I do not think America is under assault with discovery learning attacking on one flank; there is little or no evidence for this. It well may be under attack by the Discovery Institute. There is ample evidence for this!

The *NSES* Can Resolve the Paradox of International Comparisons and States' Rights.

For some time, I have been intrigued by the paradox of our education system and the role of national standards. International assessments such as TIMSS and PISA present a situation where we view results as one nation. We ask, "How does the United States compare to other countries?" Yet, we maintain the right of each state to set its own standards and assessments. To magnify the situation, each of 14,000 school districts makes decisions on what science to teach, when to teach it, and how to teach it. This is a situation designed for incoherence. What is the role of national standards in the paradox of results for one nation versus 15,000 school districts? The *NSES* can facilitate increased coherence by establishing agreement on fundamental concepts that all students should learn while maintaining the freedom of states and school districts to select instructional materials, implement assessments, and provide professional development. It is not a perfect system, but one that may resolve the paradox.

I concluded the 2006 editorial on national standards with an answer to these questions: "Should the standards be revised? If so, how?" My answer was yes, and I described changes that should be made to the 1996 standards.

Note: In late 2005, Dr. Norman Lederman, then editor of *School Science and Mathematics*, asked me to prepare a guest editorial in which I reviewed my experiences developing and implementing the *National Science Education Standards* (1996) and reflected on the importance of standards for science education. That editorial was published in February 2006, a decade after the standards were released. The editorial is included here with minor editorial changes to describe my reflections after two decades and experience working on *A Framework for K–12 Science Education* (NRC 2012) and the *Next Generation Science Standards* (NGSS Lead States 2013).

CONCLUDING REFLECTIONS ON THE 1996 AND 2013 NATIONAL STANDARDS

For almost two decades, the 1996 standards had a positive influence on fundamental components of the science education system. The same can be said for the 2013 standards, even after the brief period since their release. Yes, both standards have caused debates, agitated critics, and resulted in political issues for states and districts. That said, both sets of national standards have maintained the integrity of science, the aims of science education, and the highest aspirations of the United States. Given the complexity of our education system, one could hardly ask for, or expect, more from national standards for science education.

REFERENCES

American Association for the Advancement of Science (AAAS). 1989. *Science for all Americans: A Project 2061 report on goals in science, mathematics, and technology*. Washington, DC: AAAS.

American Association for the Advancement of Science (AAAS). 1993. *Benchmarks for science literacy*. New York: Oxford University Press.

Bybee, R. W. 2006. The *National Science Education Standards*: Personal reflections. *School Science and Mathematics* 106 (2): 57–63.

Bybee, R.W. 1997. *Achieving scientific literacy: From purposes to practices.* Portsmouth, NH: Heinemann.

Bybee, R.W., and J. Ferrini-Mundy, eds. 1997. Special issue, *School Science and Mathematics* 97 (6).

Collins, A. 1995. *National Science Education Standards* in the United States: A process and a product. *Studies in Science Education* 26: 7–37.

Finn, C., and P. R. Gross. 2005. *The state of state science standards.* Washington, DC: Thomas B. Fordham Institute.

National Commission on Excellence in Education (NCEE). 1983. *A nation at risk: The imperative for educational reform.* Washington, DC: U.S. Department of Education.

National Council of Teachers of Mathematics (NCTM). 1989. *Curriculum and evaluation standards for school mathematics.* Reston, VA: NCTM.

National Education Goals Panel (NEGP). 1994. *Goals 2000: Educate America Act.* Washington, DC: U.S. Government Printing Office.

National Research Council (NRC). 1996. *National science education standards.* Washington, DC: National Academies Press.

National Research Council (NRC). 1997. *Improving student learning in mathematics and science: The role of national standards in state policy.* Washington, DC: National Academies Press.

National Research Council (NRC). 2002. *Investigating the influence of standards: A framework for research in mathematics, science, and technology education.* Washington, DC: National Academies Press.

National Research Council (NRC). 2003. *What is the influence of the* National Science Education Standards*? Reviewing the evidence, a workshop summary.* Washington, DC: National Academies Press.

National Research Council (NRC). 2012. *A framework for K–12 science education: Practices, crosscutting concepts, and core ideas.* Washington, DC: National Academies Press.

National Science Teachers Association (NSTA). 1992. *Scope, sequence, and coordination of secondary school science. Volume 1: The content core: A guide for curriculum designers.* Arlington, VA: NSTA.

National Science Teachers Association (NSTA). *Science.* 2005. Pathways to Science Standards. November 4.

NGSS Lead States. 2013. *Next Generation Science Standards: For states, by states,* Washington, DC: National Academies Press. *www.nextgenscience.org/next-generation-science-standards.*

Ravitch, D. 1995. *National standards in American education.* Washington, DC: Brookings Institution.

Rutherford, F. J., and A. Ahlgren. 1989. *Science for all Americans.* New York: Oxford University Press.

U.S. Department of Education. 2000. *Before it's too late: A report to the nation from the National Commission on Mathematics and Science Teaching for the 21st Century.* Washington, DC: U.S. Department of Education.

Woodward, A., D. Elliott, and K. Nagel. 1988. *Textbooks in school and society.* New York: Garland Publishing Company.

CHAPTER 8

NEXT GENERATION SCIENCE STANDARDS

CONTEMPORARY PERSPECTIVE

This chapter introduces contemporary standards that are having an effect on science education at the national, state, and local levels.

The Foundation for the *NGSS*

The *Next Generation Science Standards* (NGSS; NGSS Lead States 2013) began with the development of *A Framework for K–12 Science Education* (NRC 2012). However, we must go back even further. In 2009, the Carnegie Corporation of New York and the Institute for Advanced Study established a commission that released a report, *The Opportunity Equation*, that recommended development of a common set of standards for science education (Carnegie Corporation 2009). The following introduction is adapted from the *Framework.*

The *Framework* is based on a body of research on teaching and learning in science, as well as on nearly two decades of efforts to define foundational knowledge and skills for K–12 science and engineering education. From this work, the *Framework* committee concluded that K–12 science and engineering education should focus on a limited number of disciplinary core ideas and crosscutting concepts, be designed so that students continually build on and revise their knowledge and abilities over multiple years, and support the integration of their knowledge and abilities with the practices needed to engage in scientific inquiry and engineering design (NRC 2012).

The committee recommends that science education in grades K–12 be built around three major dimensions (see Figure 8.1, p. 107):

- Scientific and engineering practices
- Crosscutting concepts that unify the study of science and engineering through their common application across fields

- Core ideas in four disciplinary areas: physical sciences; life sciences; Earth and space sciences; and engineering, technology, and the applications of science

All three dimensions must be integrated into standards, curriculum, instruction, and assessment. Engineering and technology are featured alongside the natural sciences (physical sciences, life sciences, and Earth and space sciences) for two critical reasons: to reflect the importance of understanding the human-built world and to recognize the value of better integrating the teaching and learning of science, engineering, and technology.

The broad set of content in the *Framework* guided development of new standards that, in turn, will guide reforms of curriculum, instruction, assessment, and professional development for educators. A coherent and consistent approach throughout grades K–12 is key to realizing the vision for science and engineering education embodied in the *Framework*—that students, over multiple years of school, actively engage in science and engineering practices and apply crosscutting concepts to deepen their understanding of each field's disciplinary core ideas.

Before publication, a draft of the *Framework* was sent out for review. The Council of State Science Supervisors (CSSS) played an important role in this review by organizing focus groups and providing feedback to the National Research Council (NRC).

The *Framework* represented the first step in a process that informed state-level decisions and provided a research-grounded basis for improving science teaching and learning across the country. The *Framework* guided standards developers, curriculum designers, assessment developers, state and district science administrators, professionals responsible for science teacher education, and science educators working in informal settings.

The NRC *Framework* provides guidance for the development of standards. The following list summarizes key points from the NRC recommendations. Standards for K–12 science education should

- set rigorous goals for all students;
- be scientifically accurate;
- be limited in number;
- emphasize all three dimensions;
- include performance expectations that integrate all three dimensions;
- be informed by research on learning and teaching;
- meet the diverse needs of students and states;
- have a coherent progression across grades and within grades;
- be explicit about resources, time, and teacher expertise;
- align with the *Common Core State Standards*; and
- account for diversity and equity (NRC 2012, pp. 297–307).

Figure 8.1. The Three Dimensions of the *Framework*

1. **Scientific and Engineering Practices**
    1. Asking questions (for science) and defining problems (for engineering)
    2. Developing and using models
    3. Planning and carrying out investigations
    4. Analyzing and interpreting data
    5. Using mathematics and computational thinking
    6. Constructing explanations (for science) and designing solutions (for engineering)
    7. Engaging in argument from evidence
    8. Obtaining, evaluating, and communicating information
2. **Crosscutting Concepts**
    1. Patterns
    2. Cause and effect: Mechanism and explanation
    3. Scale, proportion, and quantity
    4. Systems and system models
    5. Energy and matter: Flows, cycles, and conservation
    6. Structure and function
    7. Stability and change
3. **Disciplinary Core Ideas**

    Physical Sciences

    PS1: Matter and its interactions

    PS2: Motion and stability: Forces and interactions

    PS3: Energy

    PS4: Waves and their applications in technologies for information transfer

    Life Sciences

    LS1: From molecules to organisms: Structures and processes

    LS2: Ecosystems: Interactions, energy, and dynamics

    LS3: Heredity: Inheritance and variation of traits

    LS4: Biological evolution: Unity and diversity

    Earth and Space Sciences

    ESS1: Earth's place in the universe

    ESS2: Earth's systems

    ESS3: Earth and human activity

    Engineering, Technology, and Applications of Science

    ETS1: Engineering design

    ETS2: Links among engineering, technology, science, and society

CHAPTER 8

Development of the *NGSS*

Development of the *NGSS* began after the NRC released *A Framework for K–12 Science Education: Practices, Crosscutting Concepts, and Core Ideas* (2012). The report identified the key content and practices all students should learn by the time they graduate from high school. The *Framework* served as a vision for K–12 science education and the foundation for new science education standards. The prior national standards were released in the mid-1990s and influenced science education for nearly two decades.

As the *Framework*'s subtitle suggests, science and engineering practices, crosscutting concepts, and core ideas from the physical, life, and Earth and space sciences were defined. Figure 8.1 summarizes the three content dimensions from the *Framework.* These dimensions became the basis for the *NGSS.*

This figure presents the content of the *NGSS.* The core ideas for science disciplines are similar to prior standards (see, for example, the *National Science Educations Standards* [NRC 1996]) that have influenced most state standards. The crosscutting concepts are updated statements of several unifying themes from prior standards, and the science and engineering practices also are elaborated statements of prior science practices and scientific inquiry.

The *NGSS* were developed using the following foundational ideas. The science standards

- present standards as performance expectations;
- describe policies for school programs and classroom practices, not a curriculum;
- clarify equity and excellence;
- integrate engineering with science; and
- define college and career readiness.

The genesis and support for both the *Framework* and *NGSS* came from the Carnegie Corporation of New York and was based on the report *The Opportunity Equation: Transforming Mathematics and Science Education for Citizenship and the Global Economy* (Carnegie Corporation 2009). It is important to note that development of neither the *Framework* nor the *NGSS* received financial support from the federal government.

Achieve, Inc., an independent, bipartisan, nonprofit education organization, managed the development of the *NGSS.* Leadership for the *NGSS* initiative came from 26 states. The *NGSS* were released in April 2013 after several years of development and thorough review by the scientific and education communities, as well as by key stakeholders and the public (see NGSS Lead States 2013, Volume 2, Appendix B).

Innovations in *NGSS*

Although there are similarities between the *NGSS* and the prior standards, such as the *National Science Education Standards* (*NSES*; NRC 1996), there also are significant differences. Those differences present innovations that must be accommodated by corresponding changes in instructional materials, assessments, and teachers' knowledge and skills.

The following innovations established in the *NGSS* are hallmarks of current thinking on how students learn science and set a vision for science education. These innovations will not only cause a shift in state standards but also must influence and refocus state assessments, the development of comprehensive school science programs, and the preparation and professional development of K–12 teachers.

Innovation 1: The *NGSS* reflect three dimensions of science and their interconnectedness. In the *NGSS,* science is presented as three distinct dimensions, each of which describes equally important learning outcomes: science and engineering practices, crosscutting concepts, and disciplinary core ideas. The *NGSS* provide for connections among all three dimensions. Students gain an understanding of what is known about the natural world and how that body of scientific knowledge came to be known. Students develop the skills and abilities expressed by the practices and how they are applied to gain a better understanding of the phenomena of the natural and designed worlds.

Innovation 2: The *NGSS* incorporate engineering and the nature of science as practices or crosscutting concepts. The *NGSS* includes engineering design and the nature of science as significant innovations. The unique aspects of engineering (e.g., identification of and designing solutions for problems), as well as aspects essential to science (e.g., designing investigations and developing evidence-based explanations), are incorporated within practices and crosscutting concepts. In addition, unique aspects of the nature of science (e.g., scientific investigations use a variety of methods; scientific knowledge is based on empirical evidence; science is a way of knowing; and science is a human endeavor) also are included as practices and crosscutting concepts.

Innovation 3: The *NGSS* describe performance expectations in which students study natural phenomena. The *NGSS* provides clear expectations for students studying natural phenomena as the basis of what they should learn (i.e., what they should know and be able to do) at the end of a grade or grade band. Past standards provided the isolated content and inquiry abilities but did not provide for the full integration of the science practices with the content.

Innovation 4: The *NGSS* present coherent learning progressions for K–12 science instruction that are structured into science and engineering concepts and practices. The *NGSS* provide for sustained opportunities from elementary through high school for students to engage in and develop a deeper understanding of the three dimensions of science. Students require a coherent learning progression or story line to fully understand the content of science. These coherent learning progressions must be built both within the grade level and across grade levels. Through the building of the cohesive story line, students have multiple opportunities to revisit and expand their understanding of the science and engineering practices, disciplinary core ideas, and crosscutting concepts by twelfth grade.

Innovation 5: The *NGSS* make connections to *Common Core State Standards* for English language arts and mathematics. The *NGSS* not only provide for coherence in science teaching and learning but also unite science with the basics—*Common Core State Standards* for English language arts and mathematics. The skills of *Common Core* subjects, both linguistic and mathematical, are applied and enhanced in the science classroom and ensure coordinated learning in all content areas.

This meaningful and substantive overlapping of skills and knowledge affords all students equitable access to the learning standards.

Table 8.1 summarizes the five innovations in a "from/to" form and locates a component of the education system where the innovations will be implemented.

SCIENCE EDUCATION GOALS FOR THE 21ST CENTURY: THE *NEXT GENERATION SCIENCE STANDARDS*

The *NGSS* define the essential science concepts and practices for contemporary reform of science education. This is especially true for those states that have adopted the *NGSS* and is also relevant for states and school districts that may not have adopted the *NGSS* but use standards based on *A Framework for K–12 Science Education*. We think it is reasonable to review the goals for science education discussed in Chapter 6 (see Bybee and DeBoer 1993) and assess the degree to which the *NGSS* as policies accommodate the five goals. We are clear that neither the *Framework* nor the *NGSS* are the curriculum materials. They do, however, indicate priorities and emphasis for the goals and, by extension, for school science programs.

Scientific Knowledge

In the *NGSS*, scientific knowledge has a primary emphasis in disciplinary core ideas (DCIs) and crosscutting concepts (CCCs). While nature of science and engineering design are included, they do not have an equivalent level of emphasis.

Scientific Methods

Science and engineering practices (SEPs) represent the goal of learning scientific methods in *NGSS*. The eight practices contribute a thorough list of abilities and knowledge necessary to achieve this goal. The detail and emphasis for "signature" practices (e.g., developing and using models, constructing explanations, engaging in argument from evidence) are significant dimensions with added value for the scientific method's goal in the *NGSS*. Additionally, the fact that the practices are integral to the statements of standards—the performance expectations—increases the probability that these strategies will be included as teaching strategies and learning outcomes in school programs.

Social Issues

One fundamental purpose of science education is to provide students with knowledge about and the abilities to act on various issues they may confront as individuals and citizens. The *NGSS* recognize this goal through the general emphasis on disciplinary core ideas, crosscutting concepts, and science and engineering practices. Compared to the *National Science Education Standards* (*NSES*), the 1996 standards, there is reduced emphasis on science in personal and social perspectives in the *NGSS*. In

Table 8.1. A Summary and Implications of *NGSS* Innovations

From	To	Reform of System Components
Single concepts in science disciplines	Integration of three dimensions (science and engineering practices, disciplinary core ideas, crosscutting concepts)	Instructional approach
Engineering/nature of science as supplemental	Engineering and nature of science incorporated as practices or crosscutting concepts	Lessons, units, and programs
Standards as description of content	Standards as performance expectations and basis for studying natural phenomena and design problems	Context for student experiences and basis for assessments
Grade level or course emphasis	K–12 learning progressions	School science program—the curriculum
Few connections to other disciplines	Explicit connections to Common Core State Standards for ELA and math	Within the sequence of lessons

the *NGSS*, the Earth science standards present content emphasizing societal issues such as climate change.

Personal Development

Similar to what we just noted in the prior statement, the *Framework* and *NGSS* have the general aim of addressing the personal development of students; however, this is not a goal with particular emphasis.

Career Awareness

The *NGSS* were reviewed for the effect on college and career readiness and certainly passed muster. The various practices, connections to *Common Core* literacy and math goals, and primary emphasis on scientific knowledge and application of that knowledge all address this goal of science education.

The *NGSS* present a reasonable and fair emphasis on the five goals of science education, with societal issues and personal development as exceptions. The content standards form a thorough set of outcomes representing physical, life, and Earth and space sciences. The practices are an excellent contemporary statement for the historical goal of scientific methods.

CHAPTER 8

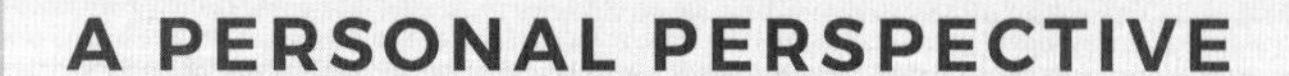

THE *NEXT GENERATION SCIENCE STANDARDS*: PERSONAL REFLECTIONS AFTER LESS THAN A DECADE

Stephen Pruitt

Similarly to Rodger, I am including an edited version of an article I have done for the National Science Teachers Association (NSTA) journals, which was published in June 2015.

The *Next Generation Science Standards* (*NGSS*; NGSS Lead States 2013) were released several years ago. Work tied to the *NGSS*, their adoption, and their implementation continues to move forward around the country. I am most frequently asked about the pace of adoption by states, the implementation of the standards, and how the *NGSS* will be assessed. In this personal perspective, I discuss where we are at the time of this publication and what I have learned during the process so far. As we implement the *NGSS*, it is important to remember that education is a journey, not a destination.

WHERE ARE WE NOW?

As of September 2016, 17 states and the District of Columbia—encompassing approximately 40% of the nation's public school population—have adopted the *NGSS*. Other states and districts continue to consider adoption. Additionally, a growing number of districts in non-adopting states are embracing the *NGSS* as the best way to move scientific literacy forward. Many of these are large districts that see the need to significantly change how they approach science education, regardless of the state-level politics. As a result, the *NGSS* are significantly influencing science education throughout the country. The excitement around the *NGSS* that I see at the NSTA national conferences is palpable.

From the beginning, adoption needed to proceed at a pace befitting each state, occurring if and when it made sense. Each adopting state, even those that were not lead states due to their undertaking of long review and public comment periods, can lay claim to owning the *NGSS*. As such, they can and should choose their own timing. A host of issues face states beyond adopting and implementing new science standards. These issues include developing timelines for adopting instructional materials, revising science standards statutes, and building the will within a state's education community to make the changes called for in *A Framework for K–12 Science Education* (NRC 2012) and the *NGSS*.

Any teacher will tell you that adopting and implementing the *NGSS* cannot be done without a way to assess the outcomes. Given the political climate around assessments, the conversation can be harrowing. As a key first step, the *NGSS* adopter states are committed to building classroom capacity. The focus has been, and must continue to be, on classrooms first rather than on building a test. The more we focus on educators and how to make the *NGSS* real in classrooms before developing an assessment, the better. Assessments that support classroom practice will come as we learn more from classroom experience. The *NGSS* and the *Framework* were developed to identify a more effective way to engage students in science. To do this, instruction must change, the planning of instruction must change, and the expectations of what happens in science classrooms must change. The type of change called for in the *NGSS* will not happen just because there is a new test. In fact, the change is significant enough that we should learn from the classroom first before a statewide, large-scale assessment is developed and administered.

It's time to move from valuing what we measure to measuring what we value. In Kentucky, for instance, the state department of education hired a "thought partner" before awarding assessment contracts to ensure that any new assessment fully evaluates the *NGSS*. California is using a similar structure with two different groups as they consider new science assessments. So, I am encouraged with the direction and pace of implementation. A thoughtful and deliberate approach has always made the most sense. It is tough to have the courage to be patient, but it is a necessity—not for the adults, but for the students.

WHAT HAVE WE LEARNED?

We have learned much in the first two years of the *NGSS*. Implementation, as expected, is far more complicated than was development of the standards themselves. The way the *NGSS* outline how students show proficiency makes sense, so teachers are embracing it. That does not mean everyone is an expert, at least not on the *NGSS* and not right away. Research from various places, including *The Cambridge Handbook of Expertise and Expert Performance* (Ericsson et al. 2006), shows that it takes many hours of practice before expert thinking is acquired. As such, teachers will need hours of thinking about *NGSS* and instructional strategies to become experts. Teachers are among the brightest and most innovative individuals on the planet, in my opinion, which does not equate to them being perfect at instruction right off the bat. As research about expert thinking points out, the move from novice to expert will require practicing all the elements of the *NGSS* with reflection and feedback and practicing quality science instruction through this new lens that allows them to develop a conceptual model of their own instruction. Finally, just like the *NGSS* require students to operate at the nexus of the three dimensions, Ericsson's research found that experts recognize knowledge is only meaningful if it is integrated with practice. That is to say, teachers could quote the three dimensions, use the language, and even quote the performance expectations from the *NGSS*, but all of that is irrelevant if they never put it into practice. The reaction to the *NGSS* has been incredible, but that alone does not translate into an automatic change in our science education system. It does mean, however, that

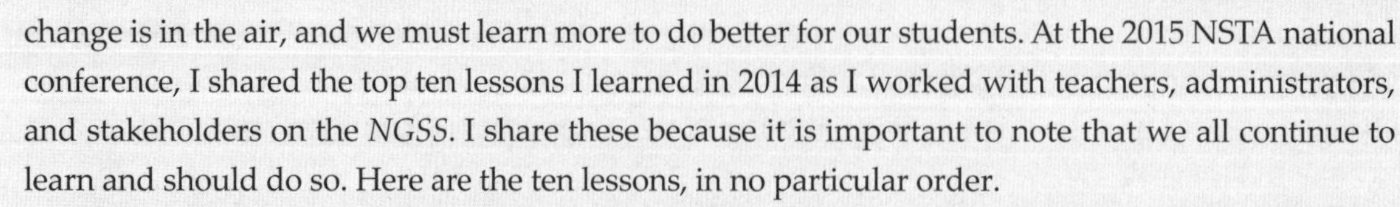

change is in the air, and we must learn more to do better for our students. At the 2015 NSTA national conference, I shared the top ten lessons I learned in 2014 as I worked with teachers, administrators, and stakeholders on the *NGSS*. I share these because it is important to note that we all continue to learn and should do so. Here are the ten lessons, in no particular order.

Three-dimensional learning is hard. We do not help teachers or students by pretending it's not. If anyone claims to know everything about the three-dimensional learning embodied in the *NGSS*, be skeptical. This is hard. But, like other professions that deal with hard changes, we will surmount these challenges, too. Learning how to create a three-dimensional culture in our classrooms takes time and effort. Why is this so difficult? First, I believe it is hard because the three dimensions in and of themselves are not new. The scientific and engineering practices involve a more expansive view of scientific processes but have similarities to inquiry. The disciplinary core ideas are similar to content standards. There are fewer but, with the possible exception of waves and their applications being an actual core idea and not a subsection, they are not new. The crosscutting concepts are similar to unifying themes from the 1990s standards documents. As such, some educators rationalize that they already incorporate these components. A good friend and excellent science leader, Sean Elkins, identified what I refer to as the Elkins Principle. He says, "There is an inversely proportional relationship between the number of times a person says, 'I already do that' and the number of times they actually do." Creating a culture of three-dimensional learning is hard because we were not taught to use the practices to gain deeper understanding of core ideas and apply to new or unique phenomena by understanding the crosscutting concepts. I think an error we made early was talking about the three dimensions, not focusing on three-dimensional learning itself. For the *NGSS* to be successful and for us to make a difference in students' lives, we have to give teachers room to get comfortable with three-dimensional learning. For this to happen, one must acknowledge this process is hard and realize that that is okay. No other profession backs down from a hard procedure if it is good for their patients, clients, or products, and neither should educators.

Eliminating the black box is tough. A black box is created when current science learning is predicated on future science learning. This means that when you say to your students, "You will not understand this until next year," you create a mystery rather than understanding. The *NGSS* provide an opportunity to look at science instruction coherently by connecting the different disciplines to better understand a phenomenon, removing the black box. Understanding the role of photosynthesis in the cycling of matter, for example, means you must understand a little about physical sciences in terms of matter and Earth science in terms of distribution of matter. I believe this to be one of the biggest issues facing science education. It has forced us, due to our siloing of concepts, to push memorization on students. This leaves students with a disconnected view of science and the world around them. In particular, it leaves students with a "Why does this matter to me?" attitude. To be clear, I am not pushing for integrated science across K–12; I am simply saying we must take full advantage of what the disciplinary core ideas afford us. If we do this, we no longer have to discuss "high-energy phosphorus bonds" in adenosine triphosphate (ATP) because students would understand that bond is the first to break and release its energy due to its position in relation to the larger molecules, the forces holding the bond, and the stability of energy. This is a more difficult concept to grasp, but it is a far better learning experience than memorizing that one phosphorus bond has more energy than the others.

Rather than teaching topics, educators should help students understand phenomena. Teaching science is about helping students understand the world around them, both natural and designed. Teaching topics such as gas laws, volcanoes, and photosynthesis without connecting them to core ideas that help students explain the world provides no reason for students to learn or retain that information. Gas laws describe part of the structure and properties of matter. The deeper understandings of gas laws are found in the *NGSS*, but they are couched in explaining the bigger picture of the structure of matter. The understanding needed for gas laws is spread throughout the years and across three core ideas in high school physical science. Understanding forces, energy, distribution of energy, and interactions of particles are far more powerful in explaining the world than simply calculating Charles' law.

Simply reading the *NGSS* does not lead to *NGSS* expertise. We have a history in the United States that when new standards are developed, we construct professional development designed to "teach" the new standards to teachers. This simply does not work. In our work with the Educators Evaluating the Quality of Instructional Products (EQuIP) rubric, we have seen that professional development that dwells only on the *NGSS* does not help educators see the innovations required in the *NGSS*. A key feature of quality *NGSS* professional development is putting teachers into a position to really see how the *NGSS* are different from their existing standards and practices. So, having educators engage in EQuIP, curriculum design, task design, or even an intense discussion about standards that preceded the *NGSS* stimulates greater understanding. Since the *NGSS* are developed based on learning progressions, professional development should also push educators to think outside their grade band and discipline when considering the *NGSS*. This involves looking not only at the core ideas but also at the practices and crosscutting concepts.

If you can eat it, it's probably not a model. Understanding the science and engineering practices takes time. There are "models" in elementary classrooms across the country; I imagine about 80% are edible. Models that students will construct and use for the *NGSS* classroom are quite different. Students will need to use models to explain, use evidence, or predict phenomena. Most "edible" models do not allow for that experience. There are a few components of the scientific and engineering practices that need to be understood before students can use them effectively. First, one must understand the practices are what students *do*; they are not teaching strategies. Students should be able, for example, to identify the components of a model, articulate the relationship of those components, and explain or predict future phenomena based on the model. The same can be said of all the practices. (For more information, see the appendix of the evidence statement at *www.nextgenscience.org/ngss-high-school-evidence-statements*.)

Crosscutting concepts are still the third dimension. The *NGSS* have three dimensions: scientific and engineering practices, crosscutting concepts, and disciplinary core ideas. The crosscutting concepts dimension is still the most difficult one to implement but also is incredibly powerful. This dimension helps students connect what they learn to the world around them in a meaningful way. The crosscutting concepts are implicit to many of us who have studied the sciences. What we know is that if the concepts are implicit in our instruction, they will be hidden from students. This dimension is challenging, but clear instruction about how crosscutting concepts fit with the other dimensions will change science education.

Phenomena are underplayed and underappreciated. The *Framework* and the *NGSS* are very focused on phenomena. We need to bring the wonder back to science classrooms, which can be done through studies of phenomena. We have found that this is tough to do because of our conditioning, but doing so is essential to making science real to students. Phenomena are observable actions or events that naturally happen in a student's life. Phenomena can look different for different ages of students, but teaching phenomena is a key feature if we are to help students pursue, or even be interested in, the world in which they live. Common phenomena can be condensation on the side of a pitcher of ice water, flags waving in the wind, rainbows, weather, or even someone having "brain freeze" after eating ice cream too quickly. Engaging students in instruction about phenomena gives them a reason to learn the content, perpetuates curiosity, and helps them retain that knowledge for years to come.

Bundling is not easy. Bundling performance expectations in the study of phenomena is critical to painting a coherent science picture for students. The idea of bundling is not as easy as it sounds. Bundling involves assembling a set of performance expectations that represent the understanding students need to address an essential question or explain a phenomena. There is no single correct way to bundle; rather, it must make sense to the teacher. So, pick a phenomenon and look at all of the standards to find a way to better explain the world. Discuss your thoughts with colleagues. Bundling will only get easier with discussion and practice. Many teachers start with the disciplinary core ideas as the driving force behind bundling, and this is an acceptable way to go about the process. One could also use crosscutting concepts as the driving force. The key is to remember that performance expectations should be understood deeply so teachers will recognize how they can be arranged and bundled to leverage the concept most needed to explain or answer questions.

Communicate, communicate, communicate, and then communicate some more. The *NGSS* represent a lot of what we want science classrooms to be, but they also depart from how most of our parents were taught. We must make every attempt to be clear about purposes, development processes, and how the *NGSS* will better prepare our students for the world. Teachers are a significant voice in a community; as such, they must be given time to understand the vision of the *NGSS*.

Leadership makes the difference. Educators, and specifically teachers, make the difference in classrooms. It is time we realize that our profession also makes a difference in society. Teachers are leading the way to our future. What we see in states and districts that are effectively implementing the *NGSS* is that teachers and administrators are assuming greater leadership roles. Yes, there is more to learn, and, yes, it is not easy, but the early implementers have shown us that quality leaders make the difference.

As was mentioned earlier, achieving expertise (thinking like an expert) takes many hours. Teachers should, as engineers do, give ourselves time to learn and room to grow. We will not get it right the first time, and that is okay. We will get better at *NGSS* instruction, but we must first acknowledge that it will take time and we will have varying degrees of initial success. The *NGSS* represent a great opportunity for students and science education. To me, they also represent a great opportunity for teachers to teach science the way we know we should and to be real leaders as we prepare our students for the future.

As one final thought for this essay, I want to speak to the standards as a sitting commissioner of education. Implementation is hard. In fact, I have come to say often that no great education initiative ever died in the vision phase; it dies in implementation. We have much work ahead of us. I have seen it at the national and state levels. Every time we think we have *NGSS* down, it moves away again, showing us bigger and better things we can do in our science classes. I am reminded of an Advanced Placement chemistry student who once told me that my class was like trying to catch a lizard: Every time you think you have it, you realize you just grabbed the tail and it broke off in your hand while the lizard escaped. Working with the *NGSS* can feel that way. I know in my state we continue to work to try to "catch it," but it keeps us moving. I know this— education is an ever-changing organism that will not stop being that way. As teachers and, more important, as leaders, we cannot let what is hard get in the way of what is right. I have many things on my plate as a commissioner, but first and foremost I must ensure our students get a first-rate education. I also need to remember, however, that it takes time, effort, and support of all of the people who touch science education. So, implementation is tough; leaders have to be tougher.

OUR COMMON PERSPECTIVE AND LEADERSHIP OPPORTUNITIES

Not surprisingly, our perspective centers on the *NGSS*. As states and districts adopt or adapt the *Framework* for *NGSS*, the need for leadership is clear. Our perspective is based not only on experience with the development and implementation of national standards but also, it is important to note, on their use as the basis for state standards and translation to curricula, instruction, assessments, and professional development.

Adopting, or even adapting, national standards for states and school districts will involve politics. Our experiences have borne witness to the reality of politics as an integral part of the process of adopting standards for science education. The leadership opportunities must include informing the decision makers about the new standards and addressing any potential problems. Those in leadership positions must be ready because, in time, the politics will emerge.

Implementing new standards requires change. By their very nature, new standards do not represent the status quo. So, the majority of teachers, for example, are not already implementing the innovations. Leaders should be prepared with examples of what the standards look like for curriculum, instruction, and assessments.

Reform based on new standards is complex and takes time. Leaders are encouraged to provide time, make plans, and proceed slowly. Extended professional development for teachers is required. This requirement is not a "one and done" workshop.

Pay attention to the concerns of teachers. The ultimate step in implementing new standards involves teachers changing their curriculum, instruction, and assessments. They reasonably will express concerns about the process and will require concrete responses to their concerns. This is both a challenge and an opportunity for leaders.

ISSUES AND QUESTIONS FOR DISCUSSION

1. What do you perceive as the appropriate statements for, and functions of, national standards in the contemporary reform of science education?
2. What does implementation of standards mean in various settings (state frameworks, teacher education, curriculum, classrooms)? What do you think is "acceptable"—wholesale but superficial consistency? Deep consistency with a few ideas? People mention the concept of "fidelity" to standards. What does this mean?
3. Is there a paradox in that standards documents are umbrella-like, general, and non-prescriptive, yet we need to be able to measure and describe levels of implementation as these levels relate to teaching and student achievement?
4. People generally seem to acknowledge that there are many ways of implementing or interpreting standards and "successful implementation" can look quite different in different places. How would you identify successful implementation of national standards for science education?
5. We propose that you explore the idea of models for standards-based reform—model programs, model practices, and model instructional units. In the context of standards-based reform, what are models for? Helping people envision "reformed" practice in some way? So they can imitate it? So they can make choices for themselves? Is it possible, in providing models, to offer them as rich examples, with enough contextual description to provide choices for individuals, schools, and so on? What problems might arise from thinking about "exemplary models"?

REFERENCES

Bybee, R., and G. DeBoer. 1993. Research on goals for the science curriculum. In *Handbook of research on science teaching and learning*, ed. D. Gabel, 357–387. New York: Macmillan.

Carnegie Corporation. 2009. *The opportunity equation: Transforming mathematics and science education for citizenship and the global economy.* New York: Carnegie Corporation of New York.

Ericsson, K. A., N. Charness, P. J. Feltovich, and R. R. Hoffman, eds. 2006. *The Cambridge handbook of expertise and expert performance.* Cambridge, UK: Cambridge University Press.

National Commission on Excellence in Education (NCEE). 1983. *A nation at risk.* Washington, DC: U.S. Department of Education.

National Research Council (NRC). 1996. *National science education standards.* Washington, DC: National Academies Press.

National Research Council (NRC). 2012. *A framework for K–12 science education: Practices, crosscutting concepts, and core ideas.* Washington, DC: National Academies Press.

NGSS Lead States. 2013. *Next Generation Science Standards: For states, by states,* Washington, DC: National Academies Press. *www.nextgenscience.org/next-generation-science-standards.*

SECTION V

STATE STANDARDS AND DISTRICT LEADERSHIP

POLICIES AND POLITICS IN SCIENCE EDUCATION

SECTION V

State policies for science education and associated politics have a significant influence on assessments and accountability, teacher education and professional development, and K–12 curriculum and classroom instruction. District leaders have the important task of working with classroom teachers to implement programs and change teaching practices based on state and local policies.

Fifty states have standards for science education from kindergarten through grade 12. Some states adopted national standards, some states adapted national standards, some standards were created by independent organizations, and some were developed by local teams of scientists, educators, and teachers. Developing science education standards at any level—national, state, or district—is a complex and difficult task. Once state standards are in place, other challenges present themselves. What about states' capacity and resources to support implementation of the standards? Are there strategies to identify and promote the selection and use of standards–based instructional materials? What about high school graduation requirements—do they accommodate the new standards? How are teacher certification and reform of undergraduate programs addressed? In addition, there is the important issue of aligning assessments with the state standards. No small task, these.

We would be remiss if we did not extend this discussion to district leadership. There are more than 14,000 public school districts, about 130,000 schools, and approximately 4 million teachers in the United States. While the initiatives vary in districts and schools, there remains the need for someone or some group to provide leadership and support the changes in state and local policies. These individuals may be deputy superintendents for curriculum, science coordinators, department heads, or lead teachers. Regardless of the title, these individuals have some responsibility for addressing the challenges of improving science teaching and student learning.

SUGGESTED READINGS

DeBoer, G. 2011. *The role of public policy in K –12 science education.* Charlotte, NC: Information Age Publishing. This book of readings provides an excellent introduction to the many aspects of policy in science education.

Dow, P. 1991. *Schoolhouse politics: Lessons from the Sputnik era.* Cambridge, MA: Harvard University Press. An insightful history of politics relative to science education.

Sunal, D., and E. Wright, eds. 2006. *The impact of state and national standards on K –12 science teaching.* Greenwich, CT: Information Age Publishing.

Thomas B. Fordham Institute. 2012. *The state of state science standards.* Washington, DC: Thomas B. Fordham Institute. Periodically the Fordham Institute reviews state standards. The reviews reveal the variability, poor quality, and omissions in science standards. Examination of recent review is recommended.

CHAPTER 9

STATE STANDARDS

CONTEMPORARY PERSPECTIVE

This chapter will discuss the potential of state standards and provide some initial answers to the introductory questions. This discussion continues by describing the influence of state standards on science education and challenges for district leaders.

SCIENCE EDUCATION STANDARDS AS STATE POLICIES

This section presents various ways that state standards may influence the education system.

Use of the word *policies* signals the fact that improving science education relies on the translation of standards and then their implementation into curriculum programs that ultimately influence what happens in science classrooms. It is important to recall that standards are policies. To be clear, policies are plans, blueprints, directions, or maps; they are not instructional materials, tests, or programs for undergraduate education, for example. Standards are necessary but not solely sufficient for the process of improving many aspects of science education. Let us examine the potential of state standards, whether they are based on the *Next Generation Science Standards* (*NGSS*) or developed by committees of teachers, scientists, and educators.

State standards have the capacity to change fundamental components of the education system. The standards, as a set of state policies, provide a comprehensive approach to changing the infrastructure to improve science education. By their design, standards inform decisions about various components of the education system. To the degree that various agencies, organizations, institutions, and school districts embrace the standards, they have the potential to bring greater coherence and unity to diverse components such as state curriculum frameworks, assessments, teacher education, certification requirements, professional development, textbook adoptions, and resources that support science education.

Contrast the potential influence of standards on the instructional core (e.g. curriculum, professional development, assessments, and achievement) with the possibility of improving student achievement through education reforms that do not address the fundamental components of education systems. Such initiatives may embrace important goals, but they often are political and displaced from those factors that have the highest probability of enhancing student learning. In fact, the instructional core is central to the work of district leaders.

State standards for science education provide a perspective on education improvement that emphasizes learning outcomes. Using standards shifts the aim of improving education from "inputs" to the system—for example, changes that we assume will enhance achievement such as time in school, homework, and use of instructional technologies—to "outputs" such as defining the goals for 13 years of school science education. State standards should clearly define the goals and content of science education and then influence changes in various means of achieving those ends. Of course, education change never works out as planned, in part due to politics. But standards provide perspective, a different way of thinking about reform and achieving higher levels of scientific achievement for all students.

Let's look at a school district as an example, and particularly the leadership of a district science coordinator. The district coordinator can use state standards as the basis for identifying goals for the K–12 science program—what all students should know and be able to do. The district coordinator could then establish a process for either selecting or designing assessments aligned with those outcomes, then do the same for selecting textbooks and curriculum materials and improving instructional strategies through professional development. Such approaches would result in greater coherence for the school science program.

Implementing state standards for science education facilitates greater coherence among components of the education system. The assumption behind this position is that greater coherence among goals, curriculum, instruction, assessments, teacher education, and professional development will be embraced by teachers and, ultimately, will enhance students' achievement. By some reports, such as the Trends in International Mathematics and Science Study (TIMSS), the United States has an education system with low levels of coherence. Goals are only tangential to instructional materials that are not true to assessments, which are not aligned with professional development, and the list goes on. Using a basic definition, coherence occurs when a small number of basic components are defined in a system, and other components are based on or derived from those basic components. There is an orderly and logical relationship of education components that affords greater comprehension of the whole system. How will state standards bring about greater coherence within science education, especially at the district level? Over time, standards for science education will develop coherence by

- defining the conceptual understandings and procedural abilities of science that all students, without regard to background, future aspirations, or prior interests, should develop;

- presenting criteria for making decisions about science content and school curriculum at different grade levels, including learning goals, design features, instructional approaches, and assessment characteristics;
- providing local leaders the criteria they can use to select instructional materials and learning experiences developed by national projects, state agencies, other districts, schools, or teachers; and
- including standards for the undergraduate preparation and continuing professional development of science teachers.

State standards should emphasize fundamental science concepts, basic practices of scientific inquiry, and applications of science to life situations. Although it may seem that state standards are at the center of science education, curriculum, teaching, and assessments matter most. In the end, students learn the content they are taught. This view directs attention to the content of the curriculum. Indeed, the standards help define that content. This said, district leaders set processes for making decisions about the textbooks and materials used in the school science program, and teachers decide the particular emphasis and activities that students will experience. It is worth noting that national standards such as the *NGSS* and assessment frameworks recommend the development of fundamental concepts of the scientific disciplines and abilities and practices of science as opposed to the current emphasis on facts, information, and topics.

One insight from high-achieving countries (e.g., on TIMSS and PISA) is that there is a need to reduce the number of science topics students encounter in a school year and focus efforts on fundamental science concepts and the practices of inquiry. State standards should do this. However, there is an assumption that those with responsibility for curricular reform would reduce topics and reform programs so the topics in school programs present opportunities to learn fundamental science concepts. Instead, in some cases, school personnel have viewed standards as topics that must be added to current programs. This should not be the intention of the state standards and efforts to reform the science curriculum.

The state standards should present a view of scientific inquiry. Based on the historical importance of scientific inquiry or the practices of science in the *NGSS*, the standards can extend historical views from process skills such as observing, inferring, hypothesizing, and the like to the development of cognitive abilities such as reasoning, critical thinking, and using evidence and logic to form explanations of natural phenomena. In addition, the standards should recommend that students develop some understanding of inquiry and abilities of science practices. Science as inquiry and the practices thus become a part of content, not just teaching strategies.

Finally, the standards can recommend a context within which fundamental concepts can be presented. For example, technology and engineering, personal and social perspectives, and history and nature of science can all provide appropriate contexts for use in curriculum reform.

State standards provide the basis for curriculum materials, instructional approaches, assessment strategies, and professional development programs that are educationally coherent,

developmentally appropriate, and scientifically accurate. If district leaders use standards as the basis for deciding the content of a curriculum, then it is important to make decisions about how the content should be organized. Many school science programs lack coherence—that is, topics do not represent an organized and coordinated K–12 progression for content. The parts should make a whole. Rather, many programs present a grade–level course orientation, especially at the secondary level. Curricular coherence requires a strong vertical perspective that is complemented by the traditional courses or horizontal view.

Developmentally appropriate refers to a curricular perspective that includes the number, duration, repetition, sequencing, specificity, and difficulty of science concepts and inquiry practices in the curriculum. Decisions relative to these issues should be based on students' developmental and learning capacities (NRC 1999) and the fundamental concepts and intellectual abilities as presented in the standards.

Science concepts must be accurate given students' ages and developmental levels. In general, the standards should recognize this criterion in the grade–level orientations for content. Another essential feature of *scientifically accurate* is expressed by the requirement of some topics at the heart of science but may cause some groups concern and political criticism. Biological evolution and climate change are contemporary examples of this point. Omitting concepts, as opposed to topics, that are central to science disciplines results in curriculum that is not scientifically accurate. Holding the line on socially and politically controversial concepts is a part of our educational responsibility to science. In a very real sense, representatives of states, school districts, and science teachers have responsibilities to maintain the integrity of science as presented in school science programs.

State standards should facilitate alignment of state, district, and classroom assessment practices with curriculum goals and instructional approaches. Teachers often lament, "I have to teach to the test." Why, one might ask, is this a problem? The complaint betrays the observation that the tests do not contain the content valued by teachers and parents. A second complementary point can be identified within this problem. There exists a basic lack of coherence between what is taught and what is tested. It seems a simple educational task is required: Align assessments with the curriculum goals and instructional approaches. State standards should contribute to this alignment.

What would "tests worth teaching to" be like? In simplest form, they would be consistent with the content of the curriculum. They would emerge from what teachers, school districts, and states value. The content would be clearly defined in the state standards, and curriculum and instruction would be aligned with those standards.

THE INFLUENCE OF STATE STANDARDS ON THE EDUCATION SYSTEM

The National Research Council (NRC) report *Investigating the Influence of Standards* (NRC 2002) discusses various influences on policy—for example, resource allocation at the national level, graduation requirements at the state level, and textbook adoptions at the district level. The report states what one would expect, which is that if standards are to influence the curriculum, for example, then

the design, development, and selection of materials should reflect this priority. School districts and states would have policies in place to support the priority of standards-based instructional materials. The following quotation summarizes the NRC's position.

> *People who understand the standards would develop instructional materials and textbooks and that understanding would be reflected both in the content they include and the nature of the tasks they use to develop student knowledge of that content. Textbook adoption processes would be carried out by selection committees knowledgeable about standards-based materials. Textbook adoption criteria would be based on features congruent with the standards, such as inquiry-based learning, an emphasis on problem solving, and an emphasis on conceptual understanding as well as skill development. Teachers would have appropriate resources for teaching standards-based curricula, including laboratory equipment and supplies, and support for learning to use them effectively.* (NRC 2002, p. 45)

The NRC report also recognized the fundamental influence of teacher development, assessment, and contextual forces in the education system. See Figure 9.1 (p. 128) for the framework that forms the basis of this NRC report. Figure 9.1 has been adapted to represent the influence of state standards on the education system.

Figure 9.1. Understanding the Influence of State Standards for Science Education Leaders

How has the system responded to the introduction of standards in science education?

What are the consequences for student learning?

STATE STANDARDS FOR SCIENCE EDUCATION →

RESPONSES

- Politicians and policy makers
- Public
- Business and industry
- Professional organizations
- District leaders

Among those who have responsibility for policies and political support for state standards—

- *How are state standards being received and interpreted?*
- *What are proposed changes within the educational system?*
- *What political issues and concerns have emerged?*

CHANNELS OF DIRECT INFLUENCE WITHIN THE EDUCATIONAL SYSTEM

Curriculum

- State and district policy decisions
- Instructional materials adaptation or development
- Text and materials selection

Teacher Development

- Initial preparation
- Certification
- Professional development

Assessment and Accountability

- Accountability systems
- Classroom assessment
- State and district assessment
- College entrance and placement practices

→ **Teachers and Teaching Practice** in classrooms and school contexts →

Among teachers who have been exposed to state standards—

How have they received and interpreted those standards?

What actions have they taken?

What about their classroom practice has changed?

Who has been affected and how?

Student Learning

Among students who have been exposed to standards-based practice—

How have learning and achievement changed?

Who has been affected and how?

Source: Adapted from NRC 2002.

THE POLITICS AND POLICIES OF STATE STANDARDS

Stephen Pruitt

The development of standards for science education is as complex as it is difficult. As such, the development of the *Next Generation Science Standards* (*NGSS*; NGSS Lead States 2013) had to take into account everything from the "big-*P* politics" to the "little-*p* politics." Both types have an effect on adoption, and both can also have a great influence on implementation. However, the latter may prove to be even more difficult in implementation if these issues are not addressed in development. *Big*-P *politics* refers to the very public politics that exist in the actual political system, which includes governors and legislators. *Little*-p *politics* refers to the ideas, issues, and concerns at a more local level and includes the people and organizations that need to be at the table to move the vision forward. So, little-*p* politics includes teachers, the right people from the partner organizations to ensure proper representation, and those who understand the research around current practice. I make this point about little-*p* politics because this area often is overlooked in the process of reform.

POLITICS AND DEVELOPING THE *NGSS*

Having managed a standards development process before the *NGSS*, I believed there needed to be an approach to the process that focused on states yet engaged all areas of science, education, and business. At the end of the day, ownership of the standards needed to belong to the states. This was not for any political reason or even because it makes for a good sound bite; rather, it was important because the implementation needed to be driven at the state and local levels. Given this fact, the first step in the development process was to recruit states to lead the process. From a leadership standpoint, the recruitment of states to lead meant having the right people at the table at the state level. As critical as they are to implementation, the state science supervisors could not be the people to commit to the process. That commitment had to come from the chief state school officers and the state board chairs. As part of that commitment, states were asked to put together state-level teams that included all the various players, such as teachers from all grade bands, disciplines, and specialties (i.e., students with disabilities, English language learners, gifted and talented students, and so on), as well as administrators, postsecondary faculty, governors' policy advisors, and practicing scientists and engineers. These teams helped guide the development and also provided invaluable expertise that led to a quality set of standards.

A set of writers was then selected to translate *A Framework for K–12 Science Education* (NRC 2012) to standards by following the direction of the states. The writers were selected based on an application process and how well each writer addressed the needs of the total group. That is to say, the makeup of the writing team needed to resemble the state-level groups. In particular, a committee was required to address the access and equity issues that needed to be at the forefront of all conversations from the beginning.

It would be easy at this point to ask, "What does all of this detail on the process have to do with politics and policies of state standards?" It has everything to do with both of them. When considering an education initiative, leaders should consider two things: (1) developing and executing a common vision with all those involved, and (2) making sure the individuals involved have the capacity to implement the vision. It is shortsighted to think that the goal of standards development is adoption. The goal is better instruction in the classroom. To achieve this goal, you need the expertise for development, but you also need people to understand the decisions made to better explain and advocate for the standards. If the initiative does not have transparency, lots of communication, and influence from many key stakeholders, the initiative likely will fail. So, decisions and plans must be made early as to how these tasks will be accomplished. The process cannot be approached as a content exercise; it must be approached as a communication and engagement process. As I have said before, you have to choose when you prefer to make life easier, at the beginning or end. Personally, I would rather be uncomfortable during the process to try to ensure understanding and support than to work in a bubble that eventually will lead to mistrust.

From my perspective, the political issues associated with education often play a bigger role than whether or not standards are the right thing for students. With regard to the *NGSS*, there is a significant change in our approach to science and teaching. To the general public, science is about facts, not concepts, and especially not the practices of inquiry. One may ask, "What does this have to do with the politics of standards?" It has everything to do with it. We know now how students best learn science, and it is not by using lists of facts that could become obsolete at any time.

A key first hurdle to the adoption of new standards is building the common vision as we have discussed. That is particularly difficult, in my opinion, because you are actually trying to change a mindset. The idea of three-dimensional teaching and learning is a tough goal to accomplish. In fact, it is incredibly tough. The first attempts at developing curriculum have taken typical approaches. Rather than connecting the three dimensions as the *NGSS* require, there tended to be a lesson on each dimension separately or, at best, an activity that blended two of the three dimensions. As we have seen with the little resistance that has been out there to the standards themselves, it has always come down to the critics wanting to see the things they remembered from their school days that either lived in the world of curriculum or were simply no longer valid measures of whether a student understands science. As those who have heard me speak can attest, I use mitosis most often as an example. At no time have I or any state or member of the writing team stated mitosis should never be taught; rather, students should learn the concept of how the process works to aid in the survival of an organism. That is very different if you believe that one simply should be able to state or even identify the phases. Another great example comes from chemistry. As you can tell from my biography, I am very much a chemistry guy. These standards were meant to bring about scientific literacy for all students. As such, the standards focus on core ideas. I have been asked many times, "Where is all the chemistry?" My answer is simple: The concepts to understand even some of the more complex ideas of chemistry are there. I think people wanted to see the requirement for students to know how to name and write formulas, gas laws, and maybe even more advanced equilibrium. The reality is that understanding how charged objects act at a distance, the effect that particles have on one another, and the basics of the periodic table actually provide a much deeper understanding of the concepts than simply reciting or calculating the ideal gas law. So, a key political piece is building these understandings. The involvement of states and initiatives such as the Building Capacity for State Science Education (BCSSE), managed by the Council of State Science Supervisors (CSSS), was critical to the standards' success because these groups focused on building an understanding of the direction of the standards and, perhaps more important, helped build an understanding of the research that supported the new direction.

Another key political issue was the effect that the *Common Core State Standards* (*CCSS*) were having on states. I will not spend a lot of time discussing this initiative, but there are a few key aspects that should be known, as they place the political challenges of the *NGSS* in a better perspective. The *CCSS* were also a state–led initiative that was managed by the National Governors Association (NGA) and the Council of Chief State School Officers (CCSSO). The *CCSS* were the standards for English language arts (ELA) and mathematics and were released in spring 2010. The standards were funded privately and completely without federal funding. When they were first released, there was a very rapid adoption that saw 45 states adopt the standards fully and one state adopt only the English language arts standards. As adoption was ongoing, *NGSS* development began to ramp up. Some were concerned that the science standards would hurt the *CCSS* because of the more controversial aspects, but that has not proven to be the case. Aiding in this adoption was the release of the federal Race to the Top grant that stated simply that states need to adopt college- and career-ready standards *like* the *CCSS* to be eligible.

There were several lessons learned from the *CCSS* development process—frankly, many more than I have the space to discuss here, but there were a few key ones that we should discuss due to the politics they inspired. The biggest problem was the appearance that the U.S. Department of Education (USED) not only endorsed the standards but had pushed their adoption, thereby overreaching into state-controlled systems of education. The irony is that the *CCSS* were the first set of math and ELA standards to have not been funded, at least in part, by the Department of Education. The impact of this for the *NGSS* was that we realized we needed a different process. The process had to start with the scientific community and could not have an affiliation with the federal government. For the most part, that goal was accomplished, even though there are many who still try to conflate the *NGSS* with the U.S. Department of Education. Another key lesson learned was that capacity had to be built in

the states if the initiatives were to live on. This was a key reason each state had to have large teams reviewing the standards. Besides their primary role of giving direction to the writers, the idea was that these teams would be able to aid in the explanation of and advocacy for the standards. Finally, and perhaps most important, states needed to adopt the standards when and if they were ready. Adoption should be slow and allow states to go through their usual processes to adopt new standards. This has gone well and, I believe, has helped states hold strong to the standards. States needed to own the standards and be able to make the adjustments they needed to their system (consider policy here) to be successful. The lessons learned from *CCSS* were very important to the *NGSS*. My hope would be that just as much attention to the details of *NGSS* development will help with future state standards. The politics can simply never be overlooked. It is important to realize that no matter how much you love your content or how much you believe it is needed for future success, I would contend that those arguments will lose every time if the content is not shown in the bigger context of the education system. I am proud of the fact that states pushed for the *NGSS* to include math and ELA; I am also proud that they paid attention to the other things going on in their states and did not overburden the system with even more weight that the *CCSS* and the new assessments were placing on the system.

As a bit of a side note, there was an important phenomenon that began to take place in all of this. That phenomenon was that teachers, schools, and districts began to adopt and implement ahead of the states. This type of grassroots acceptance was in part, I believe, a byproduct of all of the state politics. However, I believe strongly that it was actually due in a larger part to the engagement teachers and districts had in the process, a better understanding of the research and what is best for students, and the involvement of great national partners such as the National Science Teacher Association (NSTA) and CSSS playing a large role in pushing information out to the field. In short, the fact that the politics were part of the considerations from the beginning has helped more students receive *NGSS*-supported instruction than if a set of standards simply had been developed and states were left to their own devices.

POLICIES AND STATE STANDARDS

Let's turn our attention to policy. The implementation of new standards has a large impact on policy. It is interesting that this often goes unnoticed. With new standards, I think it is often extremely easy for state or district staff to go immediately to the professional development of teachers. Of course, this is a sound assumption and should absolutely take place. However, in the absence of proper policies, or at least recognition of those policies, the standards will most likely not have the effect they should. Earlier in this piece, I noted that states should adopt the standards in their own time. In that context, I discussed the importance of buy–in and political will. Policy must be part of that discussion as well. First, policy must be considered in light of the stress on the entire system. A key part to any education reform is change management. For instance, if a state was in the throes of implementing ELA and mathematics standards in K–12, adding science immediately may not be a great idea. In 2004, when I was at the Georgia Department of Education, we undertook the huge task of revising mathematics, ELA, science, and social studies. The revision took place all at once and for the first time in more than a decade. The change required by the new standards in all areas was significant. I remember sitting in a meeting with all of my colleagues and the state superintendent discussing the implementation of the new standards. For a brief moment, it seemed time slowed down and it hit all of us at once that we could not expect our elementary teachers who were teaching all four of these areas to implement this level of change all at once. As such, we chose to do a phased –in approach to implementing the standards. When considering changes to the system, one should also consider other aspects that could stand in the way, such as when standards were last updated, when instructional materials will be available or how recently materials were adopted, and what funding or partners will help with the implementation.

No big surprise here, but for me policy begins and ends with equity. The purpose of the standards is to begin that movement, but policy is also what puts the movement in gear. This part is also where the movement becomes quite the challenge. Changing policy is every bit as large a task as developing the standards themselves. Just because new standards are adopted does not mean that everyone sees the need to change policies such as graduation requirements or instructional materials adoption. For instance, when the *CCSS* were being adopted by the states, everyone knew that mathematics concepts through Algebra II were included in the standards. However, a significant number of states have, to this day, not changed their high school graduation requirements to include Algebra II. So, it is great that states have these standards, but without requiring all students to have that course or the requirement to have all standards, students still do not have equity in their educational opportunities as a result. To be fair, even though this step seems logical, it is very difficult to implement. To change graduation policy, a state must have all of its shareholders at the table, including postsecondary faculty and state and campus leadership. Creating a policy takes many months of discussion, negotiation, public hearings, and explanation to get to approval. Changing this policy is particularly tough because increasing the expectations leads to a concern that it will hurt the all–important graduation rate. If done well, this does not have to be the case. In Kentucky, the graduation rate continues to climb and is now in the top 10 in the country. The particularly important part to know about the rate is that Kentucky ranks first in states that increased the graduation requirement to include four years of ELA and math courses through Algebra II (Kentucky Department of Education 2016. This runs counter to what some would assert. However, it is well supported in research that an increase in expectations actually leads to students rising to meet the new expectations. I would love to say that changing graduation requirements is easy, but it is not, for good reason, as it impacts all areas of education. This is why changing these requirements should not happen often and should never happen for only one discipline or course. The requirements should reflect a full education experience, and innovation and creativity often are needed to ensure students have access to all types of experiences. Again, this is a key feature for future science leaders; while you should advocate for what is needed for a quality science education, it must be done in context of the overall system. The necessary changes cannot be implemented in an "us vs. them" environment.

There are other aspects of policy that must be addressed. Living and working in a state where science standards are in the process of being implemented brings a full complement of issues to consider. Instructional materials adoption is a big one. Criteria for adoption often can be adjusted to align with new standards, but the policy is often forgotten and not considered when considering implementation. The whole system really needs to be evaluated. Policies around assessment must be considered, especially as overtesting has become a part of the conversation. Other policies such as class size, instructional time for K–8 students, safety, and policies affecting exceptional children (including students with disabilities, English language learners, alternative school students, and gifted and talented students) should all be reviewed to ensure the system is efficient and serving all students while not limiting opportunities.

Finally, states and science leaders must consider federal policies as well. They do not typically have actual input into these policies, but they are part of the education landscape. Currently, science is required to be assessed once in elementary school, once in middle school, and once in high school. The scores must be reported, but they do not have to be part of the accountability formula for schools. In late 2015, the Elementary and Secondary Education Act (ESEA) was reauthorized as part of the Every Student Succeeds Act (ESSA). This act did not change how science was assessed or reported but gives states the ability to develop their accountability systems from scratch. There is much for chief state school officers to consider in this area. Again, one must be very familiar with what is needed for science, as well as with what is best for all subject areas and performances. It is an exciting time, but one in which science leaders need to be involved. I would argue that science leaders need to be problem solvers and have a willingness to be innovative in figuring out how science fits into accountability. The opportunity is there to do so, but science leaders need to step up to be involved and make the case for science in the context of teaching the whole child and not just one portion.

CONCLUSION

At this point, you may be wondering why we have not discussed standards in a while. As I stated earlier, standards are simply the first step. We cannot think of them as the end, or that implementation in classrooms represents "mission accomplished." Standards will not lead to equity if the policies for successful implementation are not in place. These policies will not be in place if science leaders stand on the sidelines; policy makers need guidance from the field. I would argue that it must be in the spirit of leadership, not pleading or especially not a "victim's mentality." Just as with the development of the standards, there must be drive on the part of the science community to change policy and to provide support to find the ways to make those changes. Finally, as I have said repeatedly, timing is everything. Science leaders cannot be tone–deaf to politics or the stress on the field. If time is spent learning about that and the right levers are identified, the result will be a quality science education for all students.

REFERENCES

Kentucky Department of Education. 2016. *The state of K–12 education in the Commonwealth of Kentucky.* Frankfort, KY: Kentucky Department of Education. *http://education.ky.gov/comm/news/Documents/SoE%20 Web.pdf*

National Research Council (NRC). 1999. *How people learn: Brain, mind, experience, and school.* Washington, DC: National Academies Press.

National Research Council (NRC). 2002. *Investigating the influence of standards.* Washington, DC: National Academies Press

National Research Council (NRC). 2012. *A framework for K–12 science education: Practices, crosscutting concepts, and core ideas.* Washington, DC: National Academies Press.

NGSS Lead States. 2013. *Next Generation Science Standards: For states, by states.* Washington, DC: National Academies Press. *www.nextgenscience.org/next-generation-science-standards.*

CHAPTER 10

DISTRICT LEADERSHIP

CHALLENGES

We need only mention the obvious: A district leader faces numerous challenges. In the more than 14,000 school districts in the United States, we cannot guess at the various challenges each one faces. Still, someone will have to make decisions about where to spend time and energy. You can probably identify several challenges. If so, what are they? Here are several we would place on the list. First, in this post–No Child Left Behind (NCLB) era, with its emphasis on English language arts and mathematics, we would recommend providing elementary students opportunities to learn science. The NCLB emphasis on English language arts and mathematics had consequences for science—it was reduced or eliminated from the curriculum. It is time to provide opportunities for students to learn science, and new state standards present an excellent rationale for district leaders. Second, there is the continuing need to improve science teaching. New state standards likely have implications for classroom instruction, so professional development is in order. Third, district leaders have to attend to budgets, regulations, proposals, and other small but routine tasks. Fourth, there are requests from community members for new science-related topics, career-oriented presentations, and additions and changes to the curriculum. Finally, there are issues we could not identify. These, too, will present challenges that take time and energy. In the next section, we describe general strategies for district leaders.

Implementing State Standards: Strategies for District Leaders

What can district leaders do to ensure the standards will constructively influence the education system? Although no individual or organization can guarantee success, district leaders can establish a process that will increase the probability of fulfilling the promise of state standards.

Improving Student Learning in Mathematics and Science: The Role of National Standards in State Policy (NRC 1997) presents a strategic framework for standards-based reform. Such a framework helps

district leaders anticipate problems so they can realize the potential of standards to improve science education. Table 10.1 summarizes that framework.

Table 10.1. A Strategic Framework for Standards-Based Reform

Phase	Goal	Theme
Dissemination	Develop awareness	"Getting the word out"
Interpretation	Increase understanding and political support	"Getting the idea"
Implementation	Change policies, school programs, and instructional practices	"Getting the job done"
Evaluation	Monitor and adjust policies, programs, and practices	"Getting it right"
Revision	Improve the efficacy and influence of state standards	"Doing it all again"

Source: Adapted from NRC 1997.

Actions by many individuals and organizations are needed if meaningful and lasting changes are to occur in science education. And the larger the system (e.g., the state vs. a school), the more extensive and coordinated the effort must be. The framework provided in this section is intended as an organizing tool for science education leaders.

Similar to many models for change and improvement, the Strategic Framework for Standards-Based Reform has several different dimensions, each with particular goals. In the framework, the developer of the standards plays a role, as do other participants in the education system. State organizations—for example, the state science teachers associations—play a major part in initial dissemination of the national standards, but they do not implement the standards. The framework helps organize thinking about what strategies are needed and clarifies where responsibility and authority lie for making changes in the various components of the education system. Although the framework is designed as a means of thinking about state standards, it is equally appropriate as a means of thinking about standards at district and school levels.

Dissemination involves developing a general awareness of the existence of the standards among those responsible for making policy, developing programs, teaching, and providing political support and encouragement for the changes that will be required. Dissemination includes addressing these questions: "What are the standards?" "Why are they needed?" and "How could they be used to shape policy and practice?" Especially during dissemination, it is important to be clear about what the standards can do (and cannot do) and why they are worth supporting. In addition, leaders should be aware of content or political issues that may be controversial, such as biological evolution and climate change in science standards. Being clear in the dissemination phase will help neutralize some criticisms and build support for the changes implied by the standards. A final note on dissemination:

District leaders will need support from the education community, district administrators, the board of education, and the general public.

Interpretation is about increasing understanding of and support for standards. This strategy involves careful analysis, dialogue, and the difficult educational task of challenging current conceptions and establishing a knowledge base that helps respond to potential critics. Deeper and richer understanding of standards is the goal, and this goal is accomplished through professional development.

Implementation involves changing policies, programs, and practices to be consistent with the new standards. Providing guidance to modify district and school science curriculum, revising criteria for the selection of instructional materials, changing teacher credentialing and recertification, and developing new assessments are all critical to the implementation process. Enacting new policies, programs, and practices builds understandings that can feed back into interpretation and lay the foundation for evaluation.

In the evaluation dimension, information gathered about the impact of standards can contribute directly to improvement. Monitoring of and feedback to various parts of the system result in modification and adjustment of policies, programs, and practices.

At some point, as a planned element of the process, revision of standards occurs, incorporating the new knowledge developed through implementation and evaluation and drawing heavily on input and discussion generated in the field by the original documents. There is some logical sequence to the phases. For example, district leaders need to become aware of standards before they deepen their understanding through interpretation activities. Likewise, implementation without understanding can lead to change that is mechanical and superficial and that can—in the extreme—imperil reform with the dismissal that "it doesn't work." Effective implementation requires interpretation and understanding. Revision without adequate evaluation will not reflect what is learned from the original effort. Note, however, that while this discussion may seem linear, its dimensions are intertwined. For example, because practice informs understanding, implementation can lead to a new or deeper interpretation of the standards or their elements. Evaluation and reflection pervade all other dimensions.

The different dimensions we describe are played out with various stakeholders, as shown in Table 10.2 (p. 140). These audiences are organized into four categories that reflect each audience's primary role in the state system: policy, program, practice, and political and public support.

Although the state developers of standards for science education likely will have major responsibility for dissemination, they can be assisted by state agencies, science education associations such as NSTA, special coalitions, and district leaders. Responsibility and authority for implementation do not necessarily lie with the organizations that developed standards. The organizations or agencies can provide support and expertise, as well as help in networking various implementers, but they are not always positioned to change policies and practices at the state level. State supervisors, curriculum developers, teacher educators, science coordinators, and classroom teachers may assume major responsibility for leadership and implementation.

Table 10.2. Participants in Standards-Based Reform

POLICY	• Governors and state legislators • State education departments • State and district school boards • School districts' administrators • School personnel
PROGRAMS	• Colleges and universities • Publishers • Curriculum and assessment developers • School districts' leaders • Business and industry members • Informal educators • Professional organizations
PRACTICES	• Teachers • Students
POLITICAL SUPPORT	• Scientists and engineers • Business and industry members • Federal, state, and local governments • Parents • General public • Teacher associations

Source: Adapted from NRC 1997.

A PERSONAL PERSPECTIVE

SOME THOUGHTS ON STATE STANDARDS AND DISTRICT LEADERSHIP

Rodger Bybee

Early in the era of standards-based reform, the *National Science Education Standards* (*NSES*; NRC 1996) were released, with little attention to their effect on state policies, programs, and politics. Sure, all associated with those standards were aware of the state standards, frameworks, adoption criteria, and assessments. We recognized the need for support beyond content for K–12 by including program and system standards. Standards in these chapters were largely ignored. So, I will remind readers of ideas we stated and standards we proposed in the 1996 documents.

The power of state government to influence science education has two sources: (1) constitutional, legislative, or judicial authority; and (2) political and economic actions. Education is not mentioned as a federal function in the U.S. Constitution. The authority for education resides in states or local jurisdictions. In the end, state agencies have more direct power over science education activities than federal agencies do (NRC 1996, p. 228).

At this point, I should clarify the fact that the National Research Council (NRC) is *not* a federal agency, so the *NSES* were not a federal mandate. This fact also applies to the development of *A Framework for K–12 Science Education* (*Framework*; NRC 2012) and the *Next Generation Science Standards* (*NGSS*; NGSS Lead States 2013). All of this said, it is also safe to say that in time, the *NSES* and *Benchmarks for Science Literacy* (AAAS 1993) influenced state standards as they became primary resources for the design and content of those standards.

REFLECTIONS FROM THE FIRST GENERATION

Development of the *NGSS* had a different approach, one that recognized the direct influence that state agencies have on classrooms. From the beginning, the form and function of standards, their review and revision, and the proposed use and implementation were informed by leadership from 26 states and an Achieve team led by Stephen Pruitt. This attention to states likely will result in greater use of the *NGSS* as the basis for state policies. To date, 17 states and the District of Columbia have adopted the *NGSS* as their state standards for science education, and 11 states have based their standards on the *Framework* and *NGSS*, with some adaptations. Finally, 13 states are reviewing the *Framework* and *NGSS* and considering their adoption (see Figure IV.1 in the prior section for a summary). Because

the states do have a critically important role, the advice embedded in the *NSES* seems important, so I have listed that advice in Table 10.3 and Figure 10.1.

Table 10.3. Science Education Standards: Changing Emphasis in School Districts

Less Emphasis on	More Emphasis on
Independent initiatives to reform components of science education	Partnerships and coordination of reform efforts
Funds for workshops and programs having little connection to the standards	Funds to improve curriculum and instruction based on the standards
Frameworks, textbooks, and materials based on activities marginally related to the standards	Frameworks, textbooks, and materials adoption criteria aligned with state standards
Assessments aligned with the traditional content of science education	Assessments aligned with the standards and the expanded view of science content and practices
Current approaches to teacher education	Professional development of teachers including science-specific pedagogy aligned with the standards
Teacher certification based on formal, historically based requirements	Teacher placement that is based on certification (i.e., knowledge and abilities in science and science teaching)

Figure 10.1. Advice From Past Reforms to District Leaders of Contemporary Reform

- Policies for science education must be congruent with the program, teaching, professional development, assessment, and content standards and allow for adaptation to district priorities.
- Policies should be coordinated within and across state agencies and school districts.
- Policies need to be sustained for sufficient time to assure continuity necessary to implement the changes required by the state standards.
- Policies must be supported with resources.
- Policies must be equitable.
- Policy instruments must be reviewed for possible unintended effects on classroom practices.

REFLECTIONS ON DISTRICT LEADERSHIP

In our use of "purposes, policies, programs, and practices" as a map for reform and categories within which to locate challenges, I would point out the need, for example, to translate policies to programs and practices. The individuals serving as district leaders provide a critical link between state standards, new school programs, and classroom practices. Using the translation metaphor, these leaders are the *translators*. They introduce standards, describe the implied changes, field questions from teachers, inform administrators, talk to parents, and likely address school boards. So, I would say the role of district leaders is not only critical but essential.

One thing I have observed that will be helpful to district leaders relates to questions about assessment. Teachers may ask if assessments will change. Be prepared to describe anticipated changes in state and local tests. But, more important, discuss any accountability measures implied by the new standards. A discussion of accountability may be at least as important as the actual assessments.

Recently, I have seen politics as increasingly important and often overlooked or ignored by leaders at all levels. As a district leader, pay attention to and even anticipate the concerns of administrators, school board members, and the community. Explaining the political, as well as educational, changes implied by new state standards is very important.

I will direct my reflections on state standards and district leadership to three areas: policy, pedagogy, and politics. The first, and obvious, is to learn the details and implications of the new state policies. Make sure you understand the fine points of the standards and the translation of these standards to your district. Using the translation theme, a key point would entail understanding the language of new standards and being able to translate the terms to the language of programs and practices. A good example is translating *performance expectations* and *three dimensions* to assessment and lessons, respectively.

For my second topic, pedagogy, I say, "Pay attention to the instructional core. As a leader, you should be able to address the new content, instructional materials that will engage learners, and the required knowledge of skills of teaching and testing.

Finally, my third topic, politics. In general, district leaders have not attended to the politics of teachers, unions, administrators, school boards, and the public. More and more, I realize the important and often disturbing place of politics in education. Taking the time to meet and explain proposed changes based on new state standards is worth the time and effort.

CONCLUSION

A question that should be asked as state and district leaders consider new standards is, "As citizens, what science knowledge, values, skills, and sensibilities should our students develop?" Past generations have asked this question, held conversations that considered their answers, and then implemented a new generation of standards. For each state, a new generation of science education standards holds great promises and presents major challenges for state and district leaders. The potential of implementing the changes implied by new standards resides with the science education leadership. With a common bond of high aspirations for students, the past generation should be ready to support the current generation as we get ready for the next generation of science education standards.

CHAPTER 10

OUR COMMON PERSPECTIVE AND LEADERSHIP OPPORTUNITIES

In this section, we turn to the challenges faced by state and district leaders as they initiate changes based on the *NGSS*, adaptations of the *NGSS*, or standards they developed. In one sense, it does not matter the origin or development of state standards—if they are new, then they imply change for components of the state and district systems. We note, however, that the quality of standards may make a difference in that more time, expertise, and review have gone into standards such as the *NGSS* compared to locally developed standards. The point here is the issue of changes implied by new standards, not what it takes to develop high-quality standards.

The challenges and our perspective represent the insights and lessons we have learned during two decades of work on standards, numerous reviews of state standards, presentations on standards, and listening to state and district leaders and classroom teachers express their concerns about new standards for science education.

Initiating Change and Maintaining Continuity in the Education System

With the announcement of new standards for science education, there will emerge some who immediately embrace the change and some who protect the status quo. Here is one apparent contradiction and a paradox of change.

In this situation, state and district leaders are advised to recognize the need for continuity in the system and address those components of the system that will continue—for example, "We are still teaching the basic concepts of life, Earth, and physical science." With this, someone should clarify the changes—for example, "We are adding the practices of science and crosscutting concepts." To this statement it is advisable to add the means for achieving the change *vis* à *vis* how to adapt current lessons or implement new instructional materials.

We cannot overemphasize the need to recognize the concerns of teachers and the need to build their understanding of the innovations that are part of new state standards. The research on concerns of teachers and levels of use for innovations is especially helpful (Hall and Loucks-Horsley 1978; Hall and Hord 1987; Hall and Hord 2010). Briefly, when teachers are confronted with innovations such as those in the *NGSS* or new state standards, addressing their concerns required creating awareness, understanding personal demands, managing resources, and identifying the positive outcomes of changing. These stages of concern may be broadly expressed by teachers in questions such as "What does this mean for me?" "What is the task and how do I achieve it?" and "What is the result for me and my students?"

While it is relatively easy to support teachers who are motivated to change and accept the reform, encouraging teachers who are reluctant to change is facilitated by answering their questions, providing information, and expanding their knowledge and skills through professional development.

Understanding Education Reform as a Long-Term Process and Working With Short-Term Events

Most individuals understand the time required for education change. Fully implementing the changes required for standards-based reform literally takes years, if not a decade. In contrast, the realities of making time for professional development involve increments of hours, a day, or at best a week during the summer. It would be nice if these opportunities were longer, but they are not.

One is advised to envision the events as opportunities in a sequence that may eventually contribute to the process of reform. Consider, for example, where the group is in the reform process. If they are near the beginning, awareness likely will be appropriate, or they may require examples of materials and assessments that apply to classrooms.

Reducing Complexity and Maintaining Integrity

Most standards present an array of features such as developing science concepts, integrating science practices, and incorporating the history and nature of science while also addressing contemporary social issues. To this list, one can add understanding technology and engineering and applying mathematics where appropriate. To any reasonable person, this list presents a complex set of teaching challenges. "How can I do this? It is just too much" is a common response when teachers realize what they are being asked to accomplish.

Reducing the complexity can be accomplished by addressing the elements individually or in sets. This is quite common. Maintaining the integrity or overall purposes of the innovations usually is the challenge. The *NGSS* innovations present an excellent example. Curriculum developers, assessment specialists, and ultimately teachers are encouraged to integrate the dimensions of science and engineering practices, crosscutting concepts, and disciplinary core ideas. The underlying idea is the need to present science as it is actually practiced and not as separate and discrete elements. The latter was an unintended consequence of the way content was presented in the first generation of national standards. While the abilities and understanding of inquiry were included as science content, inquiry did not receive the emphasis intended. The standards for inquiry were separate from those for physical, life, and Earth and space science content. In the context of contemporary innovations, the point is this: It may be important to introduce the science and engineering practices and crosscutting concepts separately, but it is essential to provide experiences where these dimensions are integrated with the disciplinary core ideas.

Accepting Political Reality and Pursuing Improvements Through Education Change

The majority of state standards include biological evolution, climate change, and other contemporary issues. With these and other areas such as inquiry and practices of science, one can expect criticism and political movements to eliminate these standards, even though they are well established as having empirical evidence and, in the case of inquiry, have long-standing status as important content. In 2014, for example, bills that would undermine the teaching of evolution were initiated in Missouri,

Ohio, Oklahoma, South Carolina, South Dakota, and Virginia. Similarly, climate change met with legislative resistance in five states. In other states, some individuals and groups have expressed concerns or lack of support for this fundamental science content. Fortunately, after early legislative resistance, the West Virginia and Wyoming legislatures subsequently voted in favor of the *NGSS*, thus recognizing the overwhelming support for evolution and climate change by the scientific community.

Although some teachers avoid politically (not scientifically) controversial topics, the avoidance erodes the integrity of science and does not represent the strength of empirical evidence in light of political and economic concerns. Science teachers have a responsibility to maintain the integrity of science by presenting the content and processes of science as a way of knowing about the natural world.

A subtle report that had political consequences is the Thomas B. Fordham Institute's report *The State of State Science Standards* (2012). The grades for states vary from A to F and change during the period from 2005 to 2012, for example (Fordham Institute 2012). This is a small, self-appointed review group that carries political weight with generally right-leaning groups. We indicate the latter based on the observation that the reviewers lack support for inquiry, processes, or practices of science based on their misconception that they do not represent legitimate content and are perceived to be instructional strategies that encourage the discovery of scientific ideas. The Fordham Institute did, for example, criticize the NRC's *Framework* for paying too much attention to engineering and technology as well as "science process" skills (Fordham 2012, p. 6).

One other political reality is to question the research base for standards. Individuals or groups ask, "Where is the research on effectiveness for these standards?" When there is no answer, the response is used to suggest the status quo is good (even if there is not a research base for those standards) or to recommend different standards that also lack a research base.

An excellent discussion of this issue was provided by James Hiebert regarding mathematics, specifically the NCTM standards. His points, however, also apply to national or state standards for science education. Hiebert clearly makes his point: Standards are not determined by research but are statements about priorities and goals (Hiebert 1999). We can expect research to provide data, for example, about means to implement standards, strategies for effective professional development, and the effectiveness of standards-based programs.

ISSUES AND QUESTIONS FOR DISCUSSION

1. Do you think states' adoption of the *NGSS* would increase coherence among the components of science education and, thus, our national level of science achievement? Why or why not?
2. Should science educators at the state level be concerned with the United States' ranking on international assessments such as TIMSS and PISA? Why or why not?
3. What do you perceive as the greatest strengths and weaknesses of state standards for science education?
4. Why is "teaching to the state test" a problem?

5. How would you improve the general strategies for standards-based reform (see Table 10.3, p. 142)?
6. Rodger describes the role of the *NSES* in the reform of the 1990s. He also points out several aspects of the standards that were generally ignored. How do you explain the latter? Why did those responsible for reform ignore standards for programs and systems, for example?
7. Stephen describes the politics and policies of state standards. How would you add to the points he makes in his essay?

REFERENCES

American Association for the Advancement of Science (AAAS). 1993. *Benchmarks for science literacy.* New York: Oxford University Press.

Hall, G., and S. Hord. 1987. *Change in schools: Facilitating the process.* Albany: State University of New York Press.

Hall, G., and S. Hord. 2010. *Implementing change: Patterns, principles, and pot holes.* New York: Pearson.

Hall, G., and S. Loucks-Horsley. 1978. Teacher concerns as a basis for facilitating and personalizing staff development. *Teachers College Record* 80 (1): 36–53.

Hiebert, J. 1999. Relationships between research and the NCTM standards. *Journal for Research in Mathematics Education* 30 (1): 3–19.

National Research Council (NRC). 1996. *National science education standards.* Washington, DC: National Academies Press.

National Research Council (NRC). 1997. *Improving student learning in mathematics and science: The role of national standards in state policy.* Washington, DC: National Academies Press.

National Research Council (NRC). 2012. *A framework for K–12 science education: Practices, crosscutting concepts, and core ideas.* Washington, DC: National Academies Press.

NGSS Lead States. 2013. *Next Generation Science Standards: For states, by states.* Washington, DC: National Academies Press. *www.nextgenscience.org/next-generation-science-standards*

Thomas B. Fordham Institute. 2012. *The state of state standards.* Washington, DC: Fordham Institute.

SECTION VI

CURRICULUM PROGRAMS FOR SCIENCE EDUCATION

Previous chapters addressed goals and policies for science education. In this section, we turn to curriculum programs, a critical and practical component in the work of science teaching.

Curriculum programs are the actual materials, equipment, activities, books, and software used in classrooms. Programs are unique to disciplines, K–12 grades, and levels in the education system. Programs are a translation of policies and are designed to the unique requirements of teaching and learning science.

As you will see, curriculum has a long history of definitional ambiguity, if not controversy. Definitions of *curriculum* include "whatever is advocated for teaching and learning" (Schubert 1993) and "the whole body of courses offered by an educational institution" for a particular course of study in a special field, such as science.

Standards such as the *Next Generation Science Standards* (*NGSS*) describe the intended science concepts and practices for activities, textbooks, trade books, computer simulations, charts, and models. The organization of these components within a course and across grade levels represents the enacted curriculum. There remains the learned curriculum that is revealed by formative and summative assessments.

We use curriculum programs in an attempt to add clarity. While the term *curriculum* may have some ambiguity, the term *curriculum programs* suggests the organization, order, or sequence of experiences; the balance of content and practices; and the K–12 learning progressions. The curriculum program also can include instruction, or the way science content and processes are delivered, especially in the classroom environment. Chapter 11 covers historical perspectives illustrating shifts in emphasis for curriculum programs. In Chapter 12, we discuss the design and development of school science curricula.

We conclude this introduction with a list of grand challenges for contemporary curriculum programs (Figure VI.1; AAAS 2013). You may note that these challenges did not include the critical role of curriculum materials and the implementation of state and national standards. We do address this challenge in the following chapters.

REFERENCES

American Association for the Advancement of Science (AAAS). 2013. Special issue, *Science* 340 (6130).

Schubert, W. H. 1993. Curriculum reform. In *Challenges and achievements of American education: The 1993 ASCD yearbook*, ed. G. Cawelti, 80–115. Alexandria, VA: Association for Supervision and Curriculum Development (ASCD).

Figure VI.1. Grand Challenges for Curriculum Programs

- **Help students explore the personal relevance of science and integrate scientific knowledge into complex practical solutions.** Teaching science in this way requires a focus on authentic problems that often cannot be defined in purely scientific terms.
- **Develop students' understanding of the social and institutional basis of scientific credibility.** Science education should empower students to make reasonable judgments about the trustworthiness and local validity of scientific claims, even when they don't have deep background knowledge or access to expertise.
- **Enable students to build on their own enduring, science-related interests.** Schooling that fosters the development of idiosyncratic interests, habitual curiosity, and lifelong science-related hobbies will strengthen students' motivation and confidence in future learning experiences.
- **Create online environments that use stored data from individual students to guide them to virtual experiments appropriate for their stage of understanding.** Online environments can provide students with personalized guidance to maximize outcomes.
- **Determine the ideal balance between virtual and physical investigations for courses in different subjects.** The best combination may vary based on circumstances. Combining both virtual and physical investigation may be optimal.
- **Identify the skills and strategies teachers need to implement a science curriculum featuring virtual and physical laboratories.** The aim is to create a professional development program that enables teachers to revise their lessons based on information obtained from student online work.

Source: Adapted from AAAS 2013.

SUGGESTED READINGS

American Association for the Advancement of Science (AAAS). 2001. *Designs for science literacy.* New York: Oxford University Press. This is a seldom-cited but very insightful book on variations for the design of school science programs.

Bruner, J. S. 1960. *The process of education.* New York: Vintage Books. The book that informed curriculum development in the Sputnik era of science education.

DeBoer, G. E. 2014. The history of science curriculum reform in the United States. In *Handbook of research in science education.Volume II.*, ed. N. Lederman and S. Abell, 559–578. New York: Routledge. An excellent summary of curriculum reform.

Kali, Y., M. C. Linn, and J. Roseman, eds. *Designing coherent science education: Implications for curriculum, instruction, and policy.* New York: Teachers College Press, Columbia University. A well-integrated discussion that calls for increasing coherence in science education by using a systemic approach to curriculum materials.

Roberts, R. 1982. Developing the concept of curriculum emphasis in science education. *Science Education* 66 (2): 243–260. This article should be required reading for science educators engaging in the processes of designing and developing science curriculum programs.

Tyler, R. W. 1949. *Basic principles of curriculum and instruction.* Chicago: The University of Chicago Press. A classic on planning curricula.

CHAPTER 11

REFORMS OF THE SCIENCE CURRICULUM

HISTORICAL PERSPECTIVE

Several excellent histories address this theme (see, e.g., Atkin and Black 2003, 2007; Black and Atkin 1996; DeBoer 1991, 2010, 2014). The following discussion is not intended as a thorough and complete history; rather, we provide a series of examples from history to illustrate the reality of curriculum reform, the factors influencing reform, and the changes that have been implemented, especially in curriculum programs. In organizing this discussion, we asked, "What did teachers use as curriculum materials in response to reforms?"

Social Forces and Curriculum Reform

As societies change, so must science education. Schools, and by extension science curriculum programs, are expected to change so they transmit the knowledge, values, and skills that are priorities in a nation during particular periods. We also would note a significant paradox: At the same time that education systems are asked to address and help resolve contemporary problems, they also are cited as the blame for many of those problems.

As contemporary issues such as economic competitiveness and recessions emerge, we hear the call for financial literacy. This is but one example. The reforms implied by social changes are often represented in the aims for the science curriculum.

Learners, Learning, and Curriculum Reform

A second factor influencing curriculum reform is research on learning and the perception of the learner. For example, learners can be perceived as passive recipients of science content and processes. A second perspective shows the learner as interested in phenomena and actively making sense of what he or she experiences. A third perspective holds that both of these views have some grounding

in research and thus may have merit, so both students and teachers need a variety of resources and experiences within a school science program. A contemporary perspective recognizes the cognitive sciences and themes such as misconceptions and conceptual change.

Subject Matter and Curriculum Reform

Different perspectives on learning and social forces imply different emphasis and orientation for science content and process (Roberts 1982). For example, there may be greater or less emphasis on vocabulary, concepts, memorization, and hands-on experiences.

One way to view science subject matter is by asking, "As a result of science education, *what* should students learn?" We review five aims (introduced in Chapter 4) that have dominated science education history: (1) to acquire scientific knowledge, (2) to learn the methodologies of the sciences, (3) to apply scientific knowledge and abilities to personal and (4) social contexts, and (5) to develop an awareness of scientific and technical careers.

With reference to curriculum, science *knowledge* is often presented as facts, concepts, laws, principles, and theories. In contemporary *Next Generation Science Standards* (*NGSS*) terms, knowledge primarily includes disciplinary core ideas and crosscutting concepts. Science *methods*, referred to as inquiry or scientific practices, involve a manner of acting, a predisposition to behave or perform in a scientific manner. *Application* is an attempt to help students learn science in personal and social perspectives. Finally, the term *careers* seems self-evident. In today's goals, this term refers to the emphasis on college and career readiness.

In reviewing the following historical examples of curriculum reform in science, one should note the changes in terms used for the goals of knowledge, methods, applications, and careers. In addition, note which of the goals is given priority. A priority for the goals often is expressed in discussions of *why* students should learn science. Should students learn science for future careers? To be good citizens? To solve problems like a scientist? Leaders must understand these changes as they work at national, state, or local levels (Bybee and DeBoer 1994).

In the following examples from the history of science education, we point out the social influences, conceptions of learning, and changes to the aims of science education. These themes will be important in later discussions of planning curriculum programs.

Object Study

The first example is that of object study (lessons). The mid-19th century included a period of emphasis on common objects and events within the experiences of elementary students. The aim was to increase the intellectual rigor of science study by having students reason through the patterns and cause-and-effect relationships they observed. This approach to curriculum was introduced as an emphasis quite different from earlier curricula that centered on reading and rote memorization (DeBoer 2014). J. M. Atkin and Paul Black stated:

> *In the history of science teaching, the Object Study curriculum may be one of the most pervasively influenced by a particular theory of learning, namely faculty psychology. (Atkin and Black 2007, p. 785)*

In object lessons, one recognized a goal of personal intellectual development, an aim that centered on acquiring scientific knowledge—and the use of actual objects as the basis for study. Although significantly influenced by faculty psychology (i.e., a perspective that conceived of the human mind as consisting of separate faculties applied to different tasks), object lessons never had an actual impact on the science curriculum. Why? The lack of a specific methodology, the dominance of textbooks, the exclusion of textbooks by those advocating object study, and the fact that development of the lessons relied on the abilities and knowledge of classroom teachers culminated in a reduced influence on the science curriculum. Developing science curricula based on the orientation of object study was a demand that could not be accommodated by many teachers, most likely due to a complexity that required understanding science, faculty psychology, and the use of varied instructional strategies.

In 1893, the Committee of Ten presented an influential report for high schools. The report made specific recommendations about the age at which subjects were to be introduced to students, the number of years the subjects should be taught, the number of hours of instruction each week, the topics to be included in the secondary curriculum, and the differences in subject matter for those going to college versus those not going to college (i.e., those going into the late 19th-century workforce). The Committee of Ten report represented the culmination of efforts by scientists to increase the intellectual rigor of school science by introducing students to natural phenomena. The approach contrasted with the earlier emphasis on reading and memorizing science in textbooks.

In this period during the mid- to late 19th century, scientists became important advocates for curriculum reform. In short, they were the social forces. They argued for the importance of learning the logic of scientific thinking through investigations of natural phenomena. This approach contrasted with earlier emphasis on lectures and reading textbooks. Teaching the subject matter of chemistry, physics, and biology and associated areas and using investigations and activities were popular approaches to the content of science teaching.

The curriculum programs focused on students' intellectual development without connections to personal usefulness or applications to society. The science disciplines provided the organization of the structure of knowledge and the logic required to make sense of direct observations of phenomena and the direction of knowledgeable teachers.

In these historical examples, one can recognize features of contemporary priorities such as the *NGSS*. Cautions based on this example include the lack of practical values, based on a narrowly focused and uniquely specialized approach to science.

Nature Study

By the late 1800s, curricular emphasis was shifting from utility, narrowly presented, to one with much broader social perspectives of community life and citizenship. The nature study movement serves as an example of a perspective that had a clearly defined purpose, one that reflected both a

social movement and curriculum reform, especially for elementary schools. As America entered the 20th century, the country experienced significant social problems associated with increasing urbanization. One solution was nature study, the key purpose of which was an appreciation of nature and rural life. Here, the goals were personal and social. The curricular orientation was a variation on the aforementioned Object Study.

Nature study originated in 1893 as the vision of Liberty Hyde Bailey. Under Bailey's direction, the materials were written and disseminated by the faculty in the Department of Agriculture at Cornell University. The social forces were clear—slow migration from rural areas to urban centers. The approach to learning involved the study of nature, with an emphasis on appreciation.

In *The Nature-Study Idea* (1903), Bailey contrasted nature study with elementary science:

> *Nature Study is a revolt from teaching mere science in the elementary grades. In teaching practice, the work and methods of the two .... integrate. ... Nature Study is not science. ... It is concerned with the child's outlook on the world. (Bailey 1903, p. 4)*

Here one sees the contrast of two different approaches to the science curriculum. Bailey makes this distinction even clearer:

> *Nature may be studied with either of two objects: to discover new truths for the purpose of increasing the sum of human knowledge; or to put the pupil in a sympathetic attitude toward nature for the purpose of increasing his joy of living. The first object, whether pursued in a technical or elementary way, is a science-teaching movement, and its professed purpose is to make investigators and specialists. The second object is a nature study movement, and its purpose is to enable everyone to live a richer life, whatever his business or profession may be. (Bailey 1903, pp. 4–5)*

This quotation makes clear the different purposes of science education and their importance for curriculum programs. The social forces of an emerging industrial-technological society and the need to maintain a substantial agricultural base for society as well as reduce the unemployment in large cities resulted in different curriculum models, each based in science disciplines and appealing to the direct study of natural phenomena.

Progressive Education

The Progressive Era presents another example of science curriculum reform in the United States. The Progressive reforms were identified with the philosophy of John Dewey and lasted for most of the first half of the 20th century. The Progressive Era underscored the importance of applying science knowledge and methods to situations of social importance and the need to make the curriculum meaningful to students.

In 1918, the National Education Association's (NEA) Commission on the Reorganization of Secondary Education (CRSE) presented the Cardinal Principles that influenced the curriculum, including science. The seven principles included (1) health, (2) command of fundamental processes, (3) worthy home membership, (4) vocation, (5) civic education, (6) worthy use of leisure, and (7)

ethical character (NEA 1918). The Cardinal Principles represent a balance of knowledge and the application of that knowledge in personal and social contexts.

Through the first half of the 20th century, the science curriculum was, in general, the textbook. In the mid-1950s, for example, the high school biology text was *Modern Biology*. The text was organized based on a progression from basic ideas about cells through the phyla to a final chapter on change—the latter being the only hint of biological evolution. Along with the fundamental biological knowledge, the text included applications to health, ecosystems, human interactions, commercial applications, the implied vocational opportunities, and the role of citizens. Here, one can see subject matter connections to social issues and topics of interest to students, but the connections to social issues are minor themes.

The Progressive science curriculum stressed experience in a social context. Science curricula should help students understand the usefulness of science to society and recognize that scientific inquiry had both personal and social utility. Science and its applications (i.e., technology) became a significant aim of science education. It also should be noted that some textbooks written by prominent scientists maintained an emphasis on information with a lack of personal and social applications (DeBoer 2014).

Sputnik-Era Curriculum Reform

After World War II, as scientists actively participated in curriculum reform, Progressive themes of science in everyday life and life adjustment were criticized and their emphasis reduced. The Sputnik era and programs such as the Physical Sciences Study Committee (PSSC), Chemical Bond Approach (CBA), Chemical Education Material Study (CHEM Study), Biological Sciences Curriculum Study (BSCS), and later the Harvard Project Physics (HPP) and Earth Science Curriculum Project (ESCP) led the reform. Jerome Bruner's book *The Process of Education* (Bruner 1960), which was based on a National Academy of Sciences meeting, became an agenda for this reform. The science curriculum should help students learn the conceptual structure of the science disciplines and the underlying processes of science. Students' active involvement in laboratory work was an essential means for them to learn the scientific concepts.

The developmental psychology of Jean Piaget (Piaget 1964) was an important influence on programs in this era. The era lasted but a decade or so until calls emerged for more socially relevant programs that addressed environmental issues and the relationship between science and society. The Sputnik era curricula, however, have had a longer and more lasting effect than the actual period of activity for curriculum development groups such as BSCS, the Education Development Center (EDC), and Lawrence Hall of Science (LHS), which have continued their work of developing curriculum programs. These curriculum development groups represent a major innovation of the era, as they were responsible for the design, development, field testing, and implementation of new science curricula.

The Sputnik era can easily lay claim to the most revolutionary changes in science curriculum witnessed in American education. Although curricula changes were already under way in October 1957 when the Soviet Union launched Sputnik, that event surely provided public support and accelerated

curriculum reform. Societal forces were clearly expressed by President John F. Kennedy, as was a goal and timeline for reform when he indicated we were in a space race and should land a man on the Moon and return him safely by the end of the decade (i.e., by 1969). Also, the reform was directly affected by the introduction of Piaget's cognitive psychology and the emergence of scientists as leaders in the reform of curriculum programs. The influence of Jerome Buner's *The Process of Education* (1960) should not be overlooked. The conference on which the book was based involved leading scientists and psychologists of the day. Many of those scientists went on to lead the reform (e.g., Bentley Glass, Arnold Grobman, and Jerrold Zacharias).

AN ERA OF STANDARDS-BASED REFORM

This final example briefly introduces a new period of science curriculum reform. By the 1980s, policy makers became concerned about America's low achievement on international assessments and high national security risk. The policy makers' response was to establish the National Commission on Excellence in Education (NCEE), which produced *A Nation at Risk* (NCEE 1983). The title signals the urgency of curriculum reform. This report called for all components of the education system to increase achievement levels of American students. The commission's recommendations included a return to basic disciplines, especially science and mathematics. This report and *Educating Americans for the 21st Century* (National Science Board 1983) resulted in a predictable emphasis on academic science.

In the late 1980s, an era of national standards emerged. In science, the era was led by F. James Rutherford and Andrew Ahlgren at the American Association for the Advancement of Science (AAAS). They produced *Science for All Americans* in 1989 (Rutherford and Ahlgren 1989). This influential volume described basic content in science, mathematics, technology, and the social sciences and included recommendations that students develop understandings of the nature of science, mathematics, and technology. The leadership by Rutherford and Ahlgren resulted in *Benchmarks for Science Literacy* (AAAS 1993), a predecessor and significant influence on the *National Science Education Standards* (*NSES*; NRC 1996).

The *Benchmarks* and *NSES* affected state standards and school science curricula and assessments from the mid-1990s until the *Next Generation Science Standards* (*NGSS*; NGSS Lead States 2013) were released in 2013, nearly a quarter century later!

A PERSONAL PERSPECTIVE

THE ROLE OF SCIENCE CURRICULUM DEVELOPMENT AND THE TEACHER

Stephen Pruitt

Each year, hundreds of thousands of teachers go through the school year using a set of curriculum materials that may or may not align to their state's standards. In fact, the variety of curriculum materials is as vast as the students who walk into those courses. When I was still in the classroom, I prided myself on not using the book my district selected. I developed many of my own materials. In reflecting on that time, I see there were things I should have been proud of and some I should not have seen with such pride. As the *NGSS* are being implemented, there are many questions about curriculum materials and when they will catch up. I want to be clear that I believe quality instructional materials are a key tool to an adequate and appropriate implementation. However, I also believe we cannot simply assume that published materials are aligned to the *NGSS* because there is a sticker on the cover or they use the *NGSS* color scheme—orange, blue, and green. I also think we must come to terms with the reality that there currently are materials being used by hundreds of thousands of teachers. I will address those issues, but through the three different lenses I use to see the world of contemporary science education. The first lens will be my experience as a high school teacher, the second as a state agency staffer, and the third as a national leader in science education.

TEACHERS AND CURRICULUM DEVELOPMENT

As I mentioned before, when I was in the classroom, I did not use the materials that had been purchased by the district. The order in which I taught concepts was often in conflict with the materials, and how I chose to couple content rarely aligned with the book. In 1998, five years before I left the classroom, I had a professional, maybe personal, epiphany regarding the structure of my course. As I prepared to teach my Advanced Placement (AP) chemistry course, I lamented how tough equilibrium concepts were for students and how heavily weighted they were on the AP exam. I needed to find a way to teach the concept of equilibrium. After some reading of research and attending a few conferences that summer, I realized that if equilibrium was the toughest concept—and, as I believed, a central focus that helped students understand the chemical world—I needed to teach it all year. So, I decided to teach the basics of equilibrium during the first two weeks of the year, then make it pervasive in my course. No published materials I could find did this. I actually liked our AP book, but the sequence of chapters represented a very traditional view of chemistry. It took all summer to plan, and it was not easy to figure out how equilibrium worked with every topic. For instance, I ended up using the Krebs cycle to teach the relationship of equilibrium and molecular geometry, as the control points in the cycle are equilibrium reactions affected by Le Châtelier's principle. Clearly, there weren't curriculum materials with this order or that made these essential, at least to me, connections.

At a basic level, I created my own curriculum. It was not written or published, but it was implemented. From a teacher's perspective, I do not believe teachers can be scripted and expect good results. They need to understand and be able to make sense of the order and depth of the content and skills they teach. From this, I learned the value of the research in the process of curriculum development. The most important thing was how much I learned about a course in which I actually had a major. I learned the incredible professional development opportunities offered when teachers are engaged in evaluating and developing their own materials. In fact, despite having a degree in chemistry, I may have learned more about the applications of chemistry in that work than I did during my undergraduate study. At the same time, I also realized, or at least I do now, how shortsighted I was. Some of the previously discussed resources and research were lost on me and limited the effectiveness of the instruction. Teachers are not trained as curriculum developers, and they certainly do not have a lot of time, but neither of these issues should be confused with ability. A key component to curriculum alignment with the *NGSS*—to me, at least—is coherence. When I was pulling together my own materials, I am afraid I focused on flow. That is to say my focus was on whether the order made sense, rather than considering the question, "Do all of the lessons actually contribute to the students' being able to assemble information and apply it?" Another way to state this is that I did not pay enough attention to the learning progression. I realize now that designing a quality curriculum involves a deep knowledge of the content. Coherence can only be achieved by helping students make connections from practices and crosscutting concepts to the disciplinary core ideas.

I cannot leave the teacher section without addressing how I think quality materials for the *NGSS* go well beyond the content. Just as students need to learn the knowledge associated with each of the three dimensions, the materials need to support a deep, not superficial, use of practices and crosscutting concepts; that is to say that a diorama is not a model, a recipe lab is not an investigation, pointing out patterns in a side box is not the use of a crosscutting concept, and discussing the scientific method is certainly not covering the nature of science. Teachers can benefit greatly from taking their materials, whether they claim to be aligned or not, and evaluating them to make adjustments to enhance their teaching and their students' learning.

CURRICULUM DEVELOPMENT AND REVIEW AT THE STATE LEVEL

For the record, every state claims local control. Granted, some are more locally controlled than others, but they will all claim to be locally controlled. I oversaw an adoption of science instructional materials while I was at the Georgia Department of Education and lived through a few more while in the classroom. State education agencies have some influence, but the materials are really most influenced by the actual consumer—the districts. A growing number of states such as Georgia are

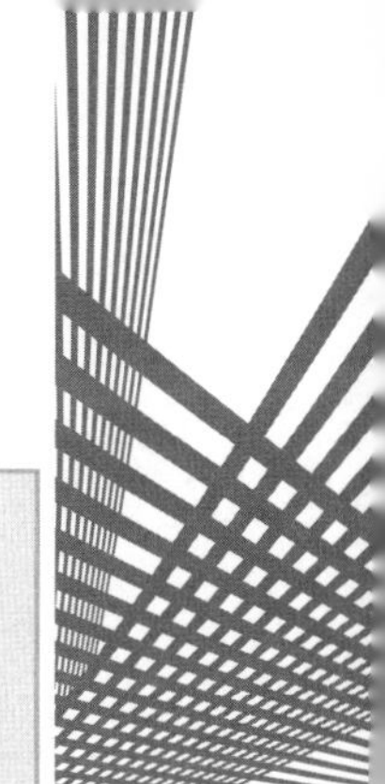

basically involved to ensure districts are not the victims of price gouging. In fact, I used to joke that we guaranteed the ABCs of instructional materials—Availability, Binding, and Cost—but that is a bit of an understatement. States do typically act as an initial screener of materials but can be very limited in what they are able to do with that screen.

Until 2015, the California Department of Education had an incredible amount of control because districts had an economic incentive to buy materials on the official adoption list. But with the move to local control, California's districts find themselves in the position of decision makers. One may wonder why this is important to this conversation. It is a good question. The reason, I am convinced, is that teachers must be involved in the process of selection and development of materials. I realize it is one more thing for teachers to do, but to change the publishing world, the consumer must stand up for what is right. Many state agencies have embarked on developing model units and lessons aligned to the *NGSS*. They do this by working with teachers. So, if given the opportunity, all teachers should experience the development of even one lesson with a set of colleagues to gain a better understanding of the standards and the development process.

A NATIONAL PERSPECTIVE

The *NGSS* were released on April 9, 2013, after four years of work, including the development of *A Framework for K–12 Science Education.* Yet, when walking through the National Science Teachers Association's (NSTA) national conference exhibit hall on April 10, 2013, publishers were already claiming alignment. When I asked one publisher who claimed to be "matched" to the *NGSS* how they could be matched when the standards were only released 24 hours earlier, he said they were really matched to the January 2013 draft of the *NGSS*. He did not know who I was, and he was quite proud of their ingenuity. The January draft was significantly different from the final product. When I asked about that, he complained that "they" had changed a lot, but this matching was good enough for now. I introduced myself and told him I hoped his company would find time to do a real alignment. Two years later, at the same NSTA conference, the publishers had turned everything to *NGSS* colors—orange, blue, and green. So, how was "alignment" attained? Clearly what some publishers did was take existing materials and find the key words and concepts in the *NGSS*. This is not alignment. If one starts with materials and aligns those to the *NGSS,* rather than the other way around, one is really making the materials the standard.

So, from the national perspective, it is clear that if we want change in how curriculum is developed and we actually want to see change, once again teachers and districts come into play. The days of just accepting a crosswalk (a chart that matches curriculum topics to standards) as evidence have to end, and the people who use the materials need to be familiar enough with the standards and, maybe more important, the vision of the *NGSS* to verify claims of alignment. It is neither easy nor quick. It is, however, doable and necessary.

It is also important to acknowledge the reality that instructional materials currently exist in schools and that districts do not have the funds to buy new materials. So, again I make the case that

the teacher must be given the freedom and authority to make changes to existing materials to better support the *NGSS* and, even more important, to better serve their students. There are tools to aid in the transition, such as the most recent version of Educators Evaluating the Quality of Instructional Products (EQuIP; Achieve 2016). There is plenty of research showing the importance of quality instructional materials and their effects on student achievement. While I do not disagree with this research, I also hold strongly to the idea that teachers with a deep understanding of the vision and content of the *NGSS* are key to unlocking the power of curriculum.

CONCLUSION

In this chapter, we presented several examples of curriculum reform and the influences on those changes. Stephen's perspective complemented the historical examples with the perspective on the reality of a science teacher.

The next chapter addresses the design and development of curriculum programs.

REFERENCES

Achieve. 2016. *Educators Evaluating the Quality of Instructional Products (EQuIP) rubric for lessons and units: Science* [Version 3.0, September 2016]. *www.nextgenscience.org/resources/EQuIP*.

American Association for the Advancement of Science (AAAS). 2013. Special issue, *Science* 340 (6130).

American Association for the Advancement of Science (AAAS). 1993. *Benchmarks for science literacy.* New York: Oxford University Press.

Atkin, J. M., and P. Black. 2003. *Inside science education reform: A history of curricular and policy change.* New York: Teachers College Press, Columbia University.

Atkin, J. M., and P. Black. 2007. History of science and curriculum reform in the United States and the United Kingdom. In *Handbook of research on science education,* ed. S. Abell and N. Lederman, 781–806. Mahwah, NJ: Lawrence Erlbaum Associates.

Bailey, L. H. 1903. *The nature study idea.* New York: Doubleday.

Black, P., and J. M. Atkin. 1996. *Changing the subject: Innovations in science, mathematics, and technology education.* New York: Routledge.

Bruner, J. S. 1960. *The process of education.* New York: Vintage Books.

Bybee, R. W., and G. E. DeBoer. 1994. Research on the goals for the science curriculum. In *Handbook of research on science teaching and learning,* ed. D. Gabel, 357–387. New York: MacMillan Publishing Company.

DeBoer, G. E. 1991. *A history of ideas in science education.* New York: Teachers College Press, Columbia University.

DeBoer, G. E. 2010. Leadership for public understanding of science. In *Science education leadership: Best practices for the new century,* ed. J. Rhoton, 277–311. Arlington, VA: NSTA Press.

DeBoer, G. E. 2014. The history of science curriculum reform in the United States. In *Handbook of research in science education: Vol II,* ed. N. Lederman and S. Abell, 559–578. New York: Routledge, Taylor and Francis.

National Commission on Excellence in Education (NCEE). 1983. *A nation at risk: The imperative for educational reform.* Washington, DC: U.S. Government Printing Office.

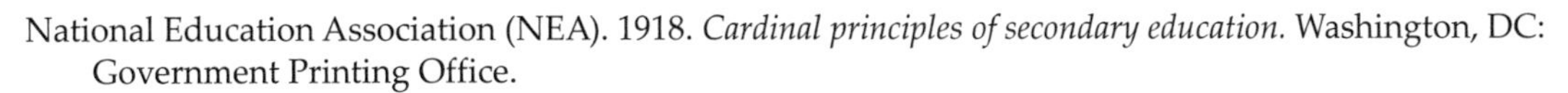

National Education Association (NEA). 1918. *Cardinal principles of secondary education.* Washington, DC: Government Printing Office.

National Research Council (NRC). 1996. *National science education standards.* Washington, DC: National Academies Press.

National Science Board (NSB). 1983. *Educating Americans for the 21st century: A plan of action for improving mathematics, science, and technology education for all American elementary and secondary students.* Washington, DC: National Science Board, National Science Foundation.

CHAPTER 12

DESIGN AND DEVELOPMENT OF CURRICULUM PROGRAMS

Designing a new curriculum presents challenges that may seem daunting or even overwhelming. It need not be so. People with no former training in design engage in the process all the time, such as by planning a trip or a dinner. We will consider the latter—planning a dinner.

- One of the first things to consider is the dinner's purpose. Is the dinner a weekly family gathering? A special holiday meal? A professional affair with business colleagues? An informal barbecue with neighbors? These different situations may imply different plans, such as formal settings of china and silver or an informal buffet approach, using paper plates and plastic utensils.
- What are the constraints? One may consider the time it takes to prepare the meal, the cost of main ingredients, and any dietary restrictions of participants.
- How large a risk do you want to take with the meal? If the participants are good friends, you may take a risk and try something new. On the other hand, if it is a formal dinner, you may wish to lower the risk and prepare something "tried and true."
- How might you consider the meal a success? Participants paying compliments on the salad, main course, and dessert would certainly count. They may ask for the recipes.
- What changes may be made during the preparation? You may add spices to enhance flavors, replace canned fruits with fresh fruit, and rethink the dessert after realizing that a different dessert would be a better balance for the meal.

These few examples illustrate our point: Without realizing it, people engage in the processes of design all the time. They may not articulate formal design terms such as *purpose, specifications, constraints, criteria,* and *costs and benefit*. Cooks realize the best recipes, ingredients, and processes can fail. Such is also the case with the design of bridges, communications, electrical systems, and many objects we encounter on a daily basis. Cooks and those who engage in the design process will also talk about creativity, compromise, and trade-offs in the process.

CHALLENGES

Basic Principles for Designing Curriculum

With this example and a general introduction to design, we turn to the design of curriculum programs. We begin with classic advice described almost seven decades ago. Ralph Tyler's *Basic Principles of Curriculum and Instruction* (1949) is an excellent place to begin. This short book acquired and maintained its classic status due to the clarity and simplicity of the following four fundamental questions:

- What educational purposes should the school seek to obtain?

This is one of the critically important questions to ask in the initial phase of designing a curriculum. Several sources should inform the discussion: national and state standards for science, studies of students' needs and interests, science-related issues in contemporary society, research on learning, and recommendations from specialists outside of the science education community.

- How can we select learning experiences that are likely to be useful in attaining these objectives?

The term *learning experiences* refers to interactions between the learner and situations in the environment—natural phenomena in science. The assumption is that learning takes place through active engagement, both mentally and physically, by the student. Consider, for example, the type of activities required for different objectives: developing the knowledge and abilities of scientific investigations, developing the conceptual foundations of science disciplines, acquiring scientific vocabulary, and maintaining and expanding scientific attitudes and interests.

- How can these education experiences be organized effectively?

Learning takes time. While some content may be recalled immediately, learning science concepts, developing habits of mind, and cultivating interests and attitudes take more time and different educational experiences. Criteria for the organization of experience in a curriculum include coherence, sequence, and a clear progression in horizontal and vertical dimensions of the curriculum. We return to this principle and instruction in the next chapter.

- How can the effectiveness of learning experiences be evaluated?

Evaluation is essential for determining the extent to which education aims and objectives have been attained. Both formative and summative assessments should be included in the design of cur-

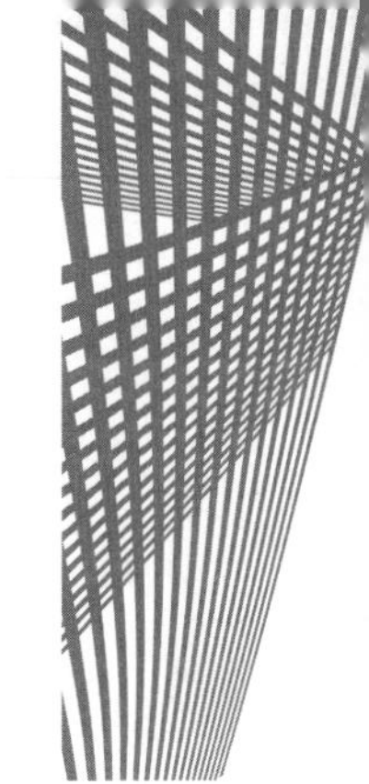

riculum and instruction. While some evaluations may inform teachers about students' progress, it also is important to consider evaluation results as feedback for program improvement.

In the nearly seven decades since Tyler published his principles, we have learned more about how change occurs in school and the essential role of professional development in curriculum reform (Hall and Hord 2001; Fullan 2001). With all due respect for Ralph Tyler's original questions, we would add the following to his original list of basic principles:

- How can the learning experiences of a new curriculum be effectively implemented in schools?

First, change in school programs takes at least three to five years. Second, effective implementation of a new program requires people—teachers and administrators—to change. Third, implementation requires leadership. Fourth, change in school programs is best achieved by teams of school personnel. Finally, if teachers have adequate and appropriate professional development and the support of colleagues, then change in curriculum programs is highly probable.

In 1988, F. James Rutherford and Andrew Ahlgren addressed the challenge of rethinking the science curriculum (Rutherford and Ahlgren 1988). Their chapter in the ASCD Yearbook extended the work of the American Association for the Advancement of Science (AAAS) Project 2061. The rethinking centered on specific questions of strategy, scope, emphasis, and criteria for revising a science curriculum. In the chapter, they stated a number of principles to consider and address. We think the specificity of the principles is especially important, so, with some adaptations to the 1988 suggestions, we present the principles in Figure 12.1 (p. 168). We also recommend another Project 2061 publication applicable to this chapter; *Designs for Science Literacy* (AAAS 2000) is a detailed presentation of many themes important for designing and developing curriculum programs.

Previous discussions identified social forces, science subject matter, and student learning as three factors influencing decisions about curriculum reforms. In this era, national and state standards for science education usually address social forces such as the needs for a 21st-century workforce as conveyed by business and industry and the requirements of science subject matter as included in *A Framework for K–12 Science Education* (NRC 2012) and the *Next Generation Science Standards* (*NGSS*; NGSS Lead States 2013), both published by the National Research Council. So, we address the third factor, student learning, in the next section.

How Students Learn Science

Enhancing student achievement relies on well-designed curricula based on research that has advanced our understanding of how students learn science. The National Research Council (NRC) reports *How People Learn: Brain, Mind, Experience, and School* (Bransford, Brown, and Cocking 2000); *How People Learn: Bridging Research and Practice* (Bransford, Pellegrino, and Donovan 1999); and the more recent *How Students Learn: Science in the Classroom* (NRC 2005) present a major synthesis of research on human learning. We also note *Learning Science and the Science of Learning*, a book edited for NSTA that presented these findings for science teachers (Bybee 2002).

Figure 12.1. Redesigning the Science Curriculum

STRATEGIES

Principle 1: To build an effective science curriculum, it is first necessary to identify what students should learn.

Principle 2: Learning goals for different student populations (college preparatory, 21st-century careers) and for students of varied interests and abilities should build on the foundation designed for all students.

Principle 3: To reduce or eliminate the tendency to redistribute the existing curriculum program into new categories, learning outcomes should be expressed conceptually and procedurally rather than as a list of topics.

Principle 4: Statements of learning outcomes for the curriculum program should be accompanied by suggested teaching practices that contribute to the realization of those outcomes.

SCOPE

Principle 5: The content of the school science curriculum should represent the basic disciplines of physical, life, and Earth science.

Principle 6: School science learning outcomes should incorporate mathematics when that content is conceptually or historically linked to and complements the content of science and technology.

Principle 7: The science curriculum should include an appropriate integration of content relating to technology and engineering.

EMPHASIS

Principle 8: The science curriculum should deal with science in contexts that include science as a way of knowing, a world view, a way of conducting inquiry and solving problems, and in relation to social and cultural issues.

Principle 9: The science curriculum should include content that deals with the role of science and technology in human affairs.

Principle 10: The selection of content for a school science program should be based on explicitly stated educational criteria.

Source: Adapted from Rutherford and Ahlgren 1988.

Three findings from the NRC reports have both a solid research base and clear implications for a science curriculum. The following findings are from *How People Learn: Bridging Research and Practice:*

1. *Students come to the classroom with preconceptions about how the world works. If their initial understanding is not engaged, they may fail to grasp the new concepts and information that are taught, or they may learn them for the purposes of a test but revert to their preconceptions outside the classroom. (Bransford, Pellegrino, and Donovan 1999, p. 10)*

Based on this research, the design of curricula should first include activities that engage students' current conceptions (i.e., preconceptions) as they attempt to explain phenomena. The activities

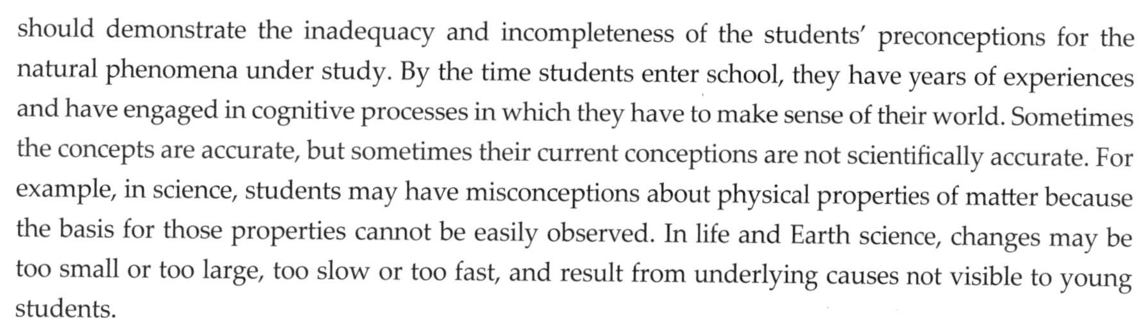

should demonstrate the inadequacy and incompleteness of the students' preconceptions for the natural phenomena under study. By the time students enter school, they have years of experiences and have engaged in cognitive processes in which they have to make sense of their world. Sometimes the concepts are accurate, but sometimes their current conceptions are not scientifically accurate. For example, in science, students may have misconceptions about physical properties of matter because the basis for those properties cannot be easily observed. In life and Earth science, changes may be too small or too large, too slow or too fast, and result from underlying causes not visible to young students.

One of the now-classic examples of students' misconceptions is their persistent belief, even at the college and university levels, that the cause for seasons is the Earth's distance from the Sun rather than the angle of incoming solar radiation due to the tilt of the Earth's axis. This misconception likely is based on students' numerous actual experiences with fires, stoves, and heating as they grow and develop. In time, these experiences contribute to the reasonable explanation that the closer one is to heat, the warmer it is; hence the misconception that when the Earth is closer to the Sun, it is warmer—summer. So, the challenges for curriculum design are to draw out students' current understanding, help them realize the inadequacy of their current ideas, and provide activities and time to help them reconstruct their ideas so they are consistent with basic scientific explanations.

Here is the second finding about how students learn:

> 2. *To develop competence in an area of inquiry, students must (a) have a deep foundation of factual knowledge, (b) understand facts and ideas in the context of a conceptual framework, and (c) organize knowledge in ways that facilitate retrieval and application. (Bransford, Pellegrino, and Donovan 1999, p. 12)*

A science curriculum should incorporate fundamental knowledge and be based on, and contribute to, the students' development of a strong conceptual framework. Research comparing performance of novices and experts, as well as research on learning and transfer, shows that experts draw on a richly structured information base. Although factual information is necessary, knowledge of many disconnected facts is not sufficient. Essential to expertise is the mastery of concepts that allows for deep understanding and a framework that organizes facts into usable information.

In science, most students will begin a unit of study as novices expressing varied informal and perhaps inadequate ideas about science-related experiences. The challenge of designing a curriculum program can be viewed as helping students progress from a novice explanation to higher levels of formal understanding (i.e., expertise). This implies a deepening factual knowledge and development of a conceptual framework for the disciplines of science.

Here is the third major finding on student learning:

> 3. *A "metacognitive" approach to instruction can help students learn to take control of their own learning by defining learning goals and monitoring their progress in achieving them. (Bransford, Pellegrino, and Donovan 1999, p. 13)*

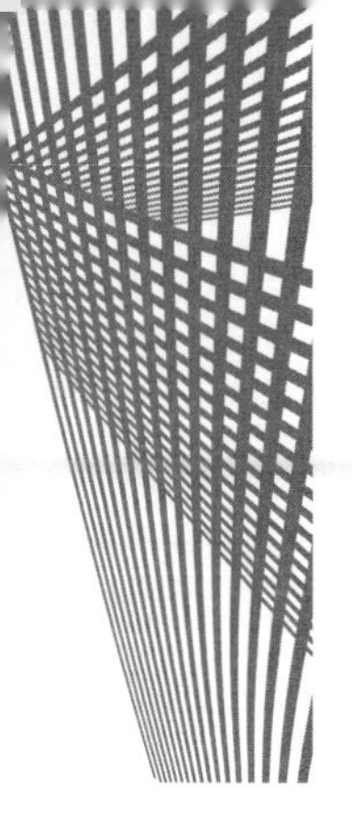

This finding addresses students' ability to consider their thinking and learn strategies that will help them monitor their progress in a problem-solving situation. Research on the performance of experts suggests that as they progress through an investigation, for example, they reflect on and monitor their understanding of the findings and questions involved. For example, they note any requirement for additional information, the alignment of new information with what is known, and the use of analogies that may provide insights and advance their understanding. For experts, there are often internal conversations grounded in the processes of scientific inquiry. This finding has clear implications for the design of curriculum and teaching science practices.

In summary, designing a science curriculum should acknowledge the fact that students already have ideas about objects, organisms, and phenomena, and that many of these ideas do not align with contemporary science concepts and explanations. The challenge for curriculum developers and classroom teachers is not the fact that students have misconceptions, but how to help students reconstruct their current concepts into adequate and accurate explanations. In contrast to many contemporary science programs, research on learning indicates that curriculum and instruction should include a clear conceptual framework complemented by facts and information.

Designing and Developing a Science Curriculum

Whether carefully designed by professional curriculum developers or compiled by teachers, any science curriculum provides an answer to this general question: "What should students know and be able to do?" And echoing Ralph Tyler (1949), answers address these specific questions as well: "What learning experiences can be selected that are likely to be useful in attaining these learning outcomes?" and "How can the learning experiences be organized for effective instruction?"

Based on research, the general answers to these questions seem clear. Students should learn both facts and concepts. And—this is just as important—the curriculum should be structured using a conceptual framework. The goal of learning with understanding depends on factual knowledge placed in a conceptual framework. We underscore the complementary nature of these two ideas, because many contemporary programs and assessments give much greater emphasis to facts without attention to the underlying concepts. Do we recommend emphasizing big ideas or many facts? Our answer to this question is that you should appropriately balance both.

A science curriculum should be structured using a framework that includes ideas central to disciplines, such as the structure and properties of matter, biological evolution, and geochemical cycles. In addition, students should learn about science; that is, ideas fundamental to the practices of science should be part of the curriculum. The content of school science should include scientific inquiry and the nature of science and fundamental ideas such as the empirical nature of science and the role of evidence in scientific explanations.

You likely have noted that this discussion only answers part of the question, "What should students know and be able to do?" Still to be addressed is the issue of the abilities, procedures, and practices of inquiry that should be developed in a school science curriculum based on contemporary,

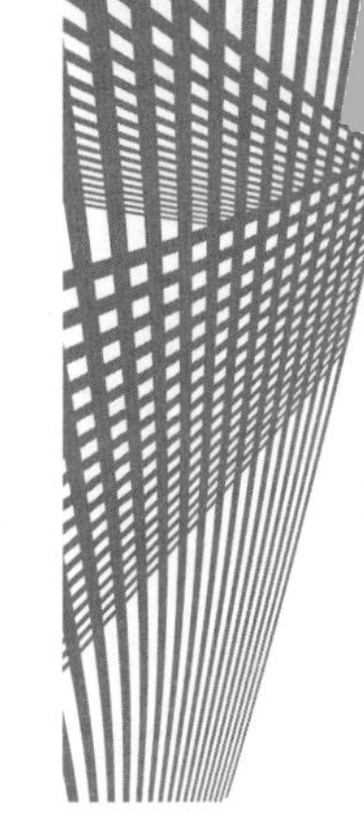

national, and state standards. Following are our nominations that generally address the question of what students should be able to do:

- Identify questions or problems that guide scientific investigations or engineering design.
- Think critically and logically to make the relationships between evidence and explanations.
- Formulate and revise scientific explanations and models using logic and evidence.
- Recognize and analyze alternative explanations and predictions.
- Communicate and defend a scientific argument.

Of course, these abilities will have to be developed in the context of investigations, experiments, and natural phenomena.

Table 12.1 summarizes many of these points and extends them to practical issues for teaching and learning. The latter are addressed as what teachers need to know and do. This table is an adaptation of an earlier work (Powell, Short, and Landes 2002).

Table 12.1. Designing a Science Curriculum

What We Know About Student Learning	What to Consider in Designing a Science Curriculum	What Science Teachers Need to Know	What Science Teachers Need to Do
Students have current concepts about objects, organisms, and phenomena. The current concepts may be misconceptions.	• Sequence should elicit current conceptions. • Curriculum should provide opportunities and time for conceptual change.	• Students' misconceptions • Processes of conceptual change	• Challenge students' misconceptions • Provide time and opportunities for conceptual change.
Learning requires both a conceptual framework and facts.	• Curriculum should be based on concepts and practices fundamental to science. • Curriculum should connect facts to concepts. • Curriculum should organize knowledge in ways that facilitate retrieval and application.	• Conceptual understanding of fundamental science • Practice of scientific inquiry	• Make continual reference to conceptual and procedural content. • Connect facts to concepts.
Learning is facilitated by self-reflection and monitoring.	• Make learning outcomes explicit. • Incorporate time and opportunities for metacognitive skills in curriculum experiences.	• Strategies to enhance metacognition • Content and processes that indicate a learning progression	• Include students' goal-setting in class time. • Teach skills of self-reflection.

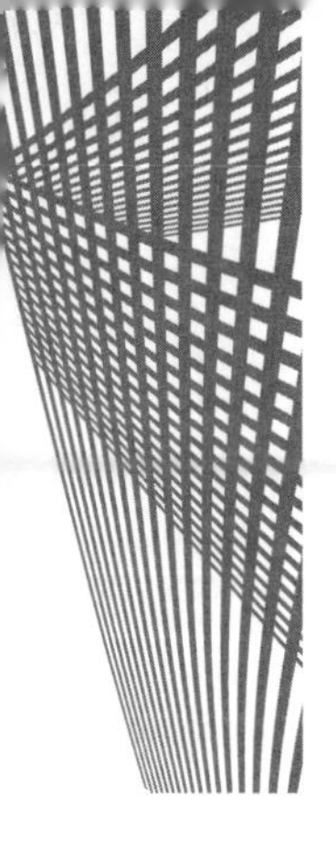

By this point in the chapter, you should understand that designing a science curriculum is a complex process. This discussion only scratches the surface of the complex and interrelated issues that must be addressed. A coherent and rigorous science curriculum consists of a conceptual framework that includes major practices of scientific inquiry as content and a meaningful context.

In addition to the research foundations described in prior sections, we note other important elements of curriculum design such as coherence, alignment of activities with learning outcomes, scaffolding for learning progressions, equitable opportunities for all students to learn, and multiple and varied experiences for students to learn the valued content—science concepts and practices.

Coherence

Near the beginning of an essay on "Coherence in High School Science," F. James Rutherford (2000) defined the term:

> *In general, the notion of coherence itself is simple enough. It has to do with relationships. Things are coherent if their constituent parts connect to one another logically, historically, geographically, physically, mathematically, or in some other way to form a unified whole. Coherence calls for the whole of something to make good sense in the light of its parts, and the parts in light of the whole. (Rutherford 2000, p. 21)*

To further clarify coherence, Roseman, Linn, and Koppal (2008) describe instructional materials as "presenting a complete set of interrelated ideas and making connections among those explicit" (p. 14). What, one might ask, does a curriculum designer use to make connections as described in both of these quotations? The answer is through a storyline (Roth and Garnier 2006/2007) and conceptual flow (DiRanna et al. 2008) tying important concepts and processes together. There is a process for analyzing coherence of science curriculum materials (Gardner et al. 2014).

Although the need for coherence in curriculum programs has wide support in the science education community (see, e.g., Bybee 2003; Schmidt 2003; Schmidt and McKnight 1998; Newman et al. 2001), the evidence indicates most curriculum materials lack coherence and are fragmented (Kesidou and Roseman 2002).

Alignment of Activities and Learning Outcomes

Related to coherence, this design specification requires clear links between stated goals and among learning activities such as investigations, simulations, and readings. Many curricula have interesting videos and simulations, for example, but the connection to learning goals is marginal (Seidel, Rimmele, and Prenzel 2005). Selection of content and sequencing activities should be done with the aim of developing student understanding and abilities that may be applied to new situations (Edelson 2001).

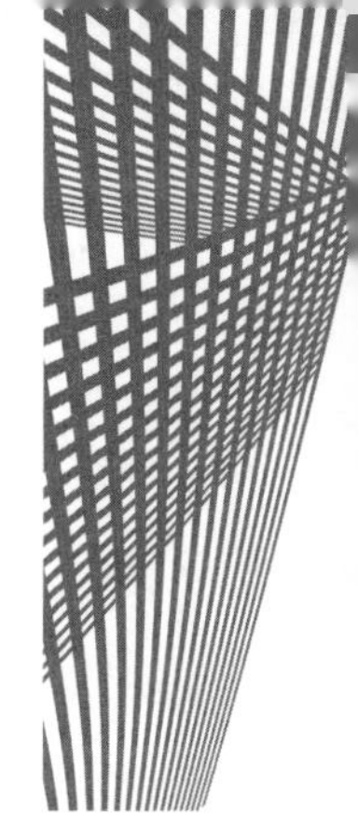

Scaffolding for Learning Progressions

The sequence of activities within a course curriculum and across grades should scaffold learning progressions for both science concepts and processes. Various goals such as cultivating evidence-based scientific explanations (McNeill and Krajcik 2012), developing formative assessments (Black and Wiliam 1998, 2009), and applying engineering practices (NGSS Lead States 2013) all are important elements of curriculum design.

Multiple and Varied Opportunities to Learn

Students need different activities and varied contexts to develop their knowledge and skills. This recommendation for instructional materials is closely tied to several of the other features (NRC 2006), so this becomes a design specification for curriculum programs.

Many of the aforementioned design specifications are related to both curriculum materials and instructional practices. We discuss the latter in the next section in Chapters 13 and 14.

Equitable Opportunities for Learning in *NGSS*-Based Curriculum Programs

The *NGSS* offer a vision of science teaching and learning that presents both opportunities and demands for *all* students. The *NGSS* highlight issues related to equity and diversity and offer specific guidance for fostering science learning for diverse groups. Opportunities for learning only occur with adequate resources, including instructional time, equipment, materials, and well-prepared teachers. In addition, instructional resources include the degree to which instruction is designed to meet the needs of diverse students and to identify, draw on, and connect with the advantages their diverse experiences give them for learning science (Pellegrino et al. 2014). The convergence of core ideas, practices, and crosscutting concepts offers multiple entry points to build and deepen understanding for all students. The science and engineering practices offer rich opportunities for language learning while they support science learning for all students. Issues related to equity and diversity become even more important when standards are translated into instructional materials and assessments (NRC 2012).

All students bring their own knowledge and understanding about the world when they come to school. Their knowledge and understanding are based on their experiences, culture, and language (NRC 2007). Their science learning will be most successful if curriculum, instruction, and assessments draw on and connect with these experiences and are accessible to students linguistically and culturally (Rosebery et al. 2010; Rosebery and Warren 2008; Warren et al. 2001). Researchers who study English language learners also stress the importance of a number of strategies for engaging those students, and they note that these strategies can be beneficial for all students. For example, techniques used in literacy instruction can be used in the context of science learning. These strategies promote comprehension and help students build vocabulary so they can learn content at high levels while their language skills are developing (Lee et al. 2013).

Here are some key points regarding instructional materials that support equitable opportunities for learning under the *NGSS*:

- The text recognizes the needs of English language learners and helps them both access challenging science and develop grade-level language. For example, materials might include annotations to help with comprehension of words, sentences, and paragraphs and give examples of the use of words in other situations. Modifications to language should neither sacrifice the science content nor avoid necessary language development.
- The language used to present scientific information and assessments is carefully considered and should change with the grade level and across science content.
- The materials should provide the appropriate reading, writing, listening, and/or speaking modifications (e.g., translations, front-loaded vocabulary word lists, picture support, graphic organizers) for students who are English language learners, have special needs, or read below grade level.
- The materials should enable extra support for students who are struggling to meet the performance expectations.

Curriculum Programs and Contemporary Reform

Publication of the *NGSS* signaled a new era of reform in science education. Almost immediately, the new standards exhibited the power to influence changes in assessments, teacher education, curriculum, and instruction. The discussion here, however, is on the science curriculum. As related to the *NGSS*, what are the changes to the fundamental aims of scientific knowledge, scientific methods, applications of science in personal and social perspectives, and career awareness?

In the two decades of work on first- and second-generation standards, the science education community has come to recognize the long-term positive influence of national standards. First, national standards can influence *all* of the fundamental components of the education system. Second, standards clarify the most basic goals—the learning outcomes—for *all* students. Third, standards at the national level are necessary for *equality* of educational opportunity. Fourth, standards have the potential to *reduce variations* among national, state, and local science education standards. Finally, standards can bring *coherence* among curriculum, instruction, and assessments.

How should one think about the implementation of the *NGSS*? The fundamental idea underlying standards is to describe clear, consistent, and challenging goals for science education. Then, based on the standards, we can reform school science programs and classroom practices to enhance student learning. An adequate implementation of standards as a basis for reform rests on the effects that the standards have on three channels of influence: curriculum, teacher development, and assessment and accountability. These, in turn, influence teachers, teaching practices, and, ultimately, student learning (NRC 2002).

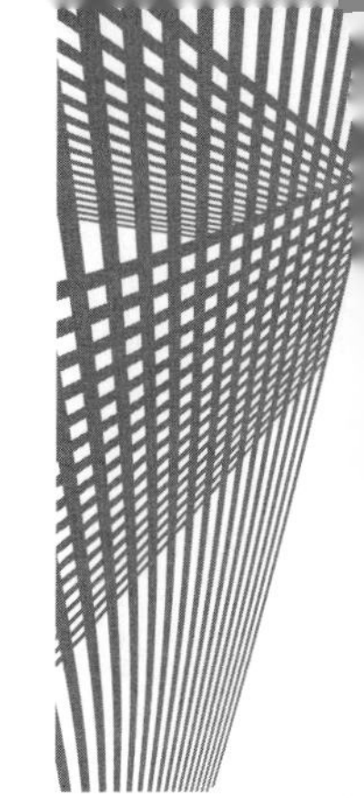

Although the process just described may be the ideal, in the United States, history shows numerous and varied reform efforts that have had little effect on teaching and learning in classrooms. Richard Elmore and Milbrey McLaughlin expressed the fundamental issue quite insightfully:

> *Reforms that deal with the fundamental stuff of education—teaching and learning—seem to have weak, transitory, and ephemeral effects; while those that expand, solidify, and entrench school bureaucracy seem to have strong, enduring, and concrete effects. (Elmore and McLaughlin 1988, p. v)*

In 1988, Elmore and McLaughlin identified a problem that persists to this day. Science educators simply must get to the essential components of reform. The *NGSS* move the conversation to curriculum programs and classroom practices and the potential to improve science education at a significant scale.

Translating *NGSS* to Curriculum Programs

We have the *NGSS*. Now, we must turn to the task of translating the standards from education policies to curriculum programs and classroom practices. *A Framework for K–12 Science Education: Practices, Crosscutting Concepts, and Core Ideas* (*Framework*; NRC 2012) presented a clear, coherent, and challenging vision for science education. The *Framework* defined three dimensions of content: science and engineering practices; crosscutting concepts; and the core ideas in disciplinary areas of physical, life, and Earth and space sciences, as well as engineering, technology, and the application of science. By design, the *Framework* was to be the basis for the *NGSS*. The process required translation of content from the *Framework* to the *NGSS*. The next step, implementation of the *NGSS*, requires translations of the *NGSS* to the aforementioned channels within which national and state standards influence the education system—curriculum, teacher development, and assessment and accountability (NRC 2002).

The task of translating standards to curriculum programs and classroom practices has some characteristics in common with the process of translating a book from one language to another. The challenges of translating include expressing the original language and ideas in another language while retaining the intentions and meanings of that text. One must note that fluency in both languages is required, and there must be an understanding and sensitivity for the ideas, historical period, and cultural subtleties that cannot be directly translated but must be conveyed or expressed in the translation. In the case of the *NGSS*, there is the challenge of translating policies with significant complexity and no small amount of subtlety.

In translating standards, there are gains, losses, and trade-offs—something will be lost and something gained. From the beginning, one must realize that the only thing that exactly aligns with the *NGSS* are the *NGSS*. Still, the *NGSS* should be the guide and the translations must represent the intentions of the standards. With a clear understanding of the *NGSS*, the results should be an accurate, reasonable, and responsible translation. This said, the *NGSS* present an intricate set of concepts

and practices combined in performance expectations that are guides for curriculum, instruction, and assessments.

We conclude this section by returning to themes from the historical examples. First, social forces such as international economic competitiveness, national security, and the need for a 21st-century workforce with deep technical skills have been at the forefront of justifications for new standards and a contemporary reform of education. Closely related to these justifications, one finds science and technology as the basis for many innovations with clear links to the economy, national security, and jobs.

Concerning goals for the curriculum, the *NGSS* have maintained an emphasis on concepts central to the disciplines (i.e., disciplinary core ideas in life, Earth, and physical sciences) and the major conceptual themes that unify the sciences (i.e., crosscutting concepts).

Likewise, there is an emphasis on the goal of methods (i.e., science and engineering practices). The translation of science and engineering practices into curriculum has resulted in the individual practices (e.g., developing models, planning and carrying out investigations, argumentation) being translated into instructional strategies used for the purpose of learning scientific ideas and concepts.

The applications of science have a role, but this aspect continues to be less emphasized compared to the goals of knowledge and methods.

The question "Who will change the instructional materials and curriculum programs?" presents new and significant challenges to those in leadership positions. For more than 50 years, the National Science Foundation (NSF) has provided funding to develop innovative instructional materials. In this era, however, the NSF has new priorities and only very limited support for instructional materials based on the *NGSS* innovations.

It seems the reform of instructional materials will be left to curriculum groups such as Lawrence Hall of Science (LHS), National Science Resources Center (NSRC), Lab Aids, Education Development Center (EDC), Biological Sciences Curriculum Study (BSCS), and others; commercial textbook publishers; science coordinators; and classroom teachers. At best, this leaves the reform of science education in a tenuous situation and presents a significant challenge for leaders.

A PERSONAL PERSPECTIVE

SCIENCE CURRICULUM DEVELOPMENT

Rodger Bybee

I developed the point of view expressed here during my tenure at the Biological Sciences Curriculum Study (BSCS). My work at BSCS covered 18 years and included responsibilities as principal investigator for National Science Foundation (NSF) grants to develop new curriculum programs for elementary school, middle school, high school, and community college. These experiences included facilitating advisory board meetings, designing the various curricula, working with BSCS teams to actually field-test and revise materials (often over two- or three-year periods), and working with publishers on production and BSCS teams on implementation. I wish to be very clear and acknowledge the contributions of others, especially the BSCS developers, editors, art department, and support staff. So, based on these experiences, here is my perspective on science curriculum development.

A SYSTEMS VIEW OF CURRICULUM DESIGN AND DEVELOPMENT

To begin, the term *curriculum* covers multiple components, such as syllabi, textbooks, materials, and school science programs. Mostly, how one uses the term *curriculum* signals that individual's perspective. For example, science teachers use the term to mean what they teach, and administrators may associate the term with textbooks and materials for elementary, middle, and high school science. Examining historical definitions of *curriculum* provides some help—for instance, "Curriculum is all of the experiences children have under the guidance of teachers" (Caswell and Campbell 1935), or "Curriculum encompasses all learning opportunities provided by the school" (Saylor and Alexander 1974) or "Curriculum [is] a plan or program for all experiences which the learner encounters under direction of the school" (Oliva 1982; all cited in Jackson 1992). Taking the time to analyze the difference between these definitions may be an insightful academic activity, but it is not very helpful when one is pressed for time and must develop a particular science curriculum within a budget, within a time limit, and according to a set of criteria and constraints. For instance, materials for introductory physical science must be developed over three years, or a local K–12 school district has one year to develop a philosophy statement and a district syllabus and provide curriculum materials aligned with *A Framework for K–12 Science Education.* This perspective is Joseph Schwab's "practical" view of curriculum making (Schwab 1970), but he also identifies the "impractical" position of each science teacher making his or her own curriculum.

What do I mean by *science curriculum*? Although a science curriculum includes science content, materials, teaching strategies, and learners in complex ways, one can identify an idea that unifies the

intricate interactions of these components. The science curriculum represents a group of interdependent relationships among science concepts, practices of scientific inquiry, and contextual factors. In short, there are science concepts, practices, and topics that are organized in programs representing the science concepts and various emphases that textbook writers, curriculum developers, teachers, and assessment specialists give to the concepts, practices, and contexts. This definition varies from the common view of curriculum as an operational plan that guides learning, courses of study, and the experiences of learners. Such definitions are limited in that they usually are reduced to syllabi, frameworks, or textbooks. I recommend a more dynamic and systemic view of the science curriculum, one that includes science content, the actions and strategies of teachers and learners, and the various technologies used in teaching. Thus, the teacher's strategies introduce dynamic qualities to the organized relationships that result in effective teaching and student learning.

A systems perspective includes other features of a science curriculum: structure, function, and feedback. *Structure* consists of the relationships among science concepts identified in materials, such as syllabi, frameworks, workbooks, laboratory manuals, tests, software, and textbooks. In general, structure is usually thought of as curriculum materials. It sometimes is referred to as the intended curriculum (Glatthorn 1987; Murnane and Raizen 1988; Posner 1992). Structured materials have been rationally thought out and planned by curriculum developers and likely express a particular emphasis (Roberts 1982, 1995), scope, sequence, and organization.

Function consists of the many ways science teachers introduce dynamic qualities through adapting, modifying, adjusting, and changing the structured materials to accommodate classroom situations involving a diversity of individuals and groups of learners. Function includes actions, strategies, and behaviors of teachers and learners—for instance, questions, teacher-directed discussions, inquiry-oriented investigations, project-based laboratories, and use of educational technologies. This discussion elaborates on the idea of enacted or taught dimensions of science curriculum.

Feedback involves assessment of teaching and learning experiences and student attainment of the learning outcomes. Taking the end-of-lesson or course perspective, this is the achieved or learned curriculum. But I use *feedback* in the systemic sense. The feedback should serve to modify the structural and functional aspects of the curriculum. It certainly also should serve to identify the degree to which learners have achieved science knowledge, abilities, and understandings identified in state standards or the *NGSS*, for example.

SOME PRINCIPLES OF SCIENCE CURRICULUM DEVELOPMENT

The following educational principles are based on my experiences with curriculum development. These principles do not represent a theory of curriculum development; rather, they are practical guides for consideration by those with the opportunities and challenges of developing a science curriculum.

1. A science curriculum should be based on our best understanding of how students learn and develop, both individually and in groups.

2. A science curriculum should represent science as expressed in national and/or state standards and associated assessments. Such standards represent the valued outcomes, aims, and goals toward which students' educational experiences should be directed.
3. A science curriculum should incorporate integrated instructional sequences (i.e., coordinated lessons with varied activities all in a coherent organization). The instructional sequence should provide adequate time and varied opportunities for student learning.
4. A science curriculum should provide materials that support teachers' professional development as they implement the program.
5. A science curriculum should be designed and developed using the process of backward design.

LEADERSHIP RESPONSIBILITIES OF THE CURRICULUM DEVELOPER

Regardless of their perspective—local, state, or national—science curriculum developers have responsibilities to several groups, including students, teachers, scientists and engineers, parents, and citizens.

Curriculum developers have a responsibility to provide all students with the best possible opportunities to learn. On the surface, this idea may seem simple enough; it involves developmentally appropriate activities aligned with learning goals, often standards. On further analysis, however, one may recognize more elusive ideas, such as incorporating engaging and interesting experiences that have meaningful connections for learners and activities that encourage those who often do not study and achieve in science, such as students representing diverse populations.

Curriculum developers should design materials that optimize teachers' knowledge and abilities. Curriculum materials should be understandable, manageable, and usable in the classroom. Additional teacher resources can assist continuing professional development through background readings in science and pedagogy and through suggestions for additional experiences. Curriculum developers should also provide suggestions and resources for teachers who wish to improve the science curriculum through adaptation based on their unique qualities and their understanding of the students they teach. The availability and thoroughness of teacher resource materials is a critical test of locally developed programs.

Curriculum developers have an obligation to represent science accurately and thoroughly. For example, the science curriculum should adequately represent the domains of science and—where appropriate—technology, engineering, and mathematics. Those who are developing curriculum have some responsibility to the major conceptual ideas that form the science disciplines and an understanding of the use of scientific inquiry. This responsibility is not, however, usually a problem. Rather, there is a problem of incorporating *too much* science content into a curriculum.

Science curriculum developers also have a responsibility to accurately represent science as a way of knowing that is based, for example, on empirical evidence, logical argument, and skeptical review.

As appropriate, developers should defend against those who would introduce into the science curriculum topics that are non-science or positions that deny legitimate and accepted scientific findings. In recent years, various groups have confronted curriculum developers, publishers, adoption committees, and science teachers with curriculum materials or topics that they think should be included in the science curriculum. Such groups claim the materials or topics are science, but on analysis they are not. A specific example includes a situation that I continually encountered at BSCS: the proposal to include creationism or intelligent design in a biology curriculum in addition to, or instead of, evolution. A more contemporary example is the omission of content that recognizes the scientific evidence supporting climate change. In my view, those who develop science curriculum demonstrate their leadership when they maintain the integrity of science and support those who understand that non-science positions have no place in the science curriculum.

Finally, a note on the leadership and responsibilities of professional curriculum developers: In an ideal situation, all science teachers would have the time, budget, expertise, and personnel to develop science curriculum for their unique teaching style, students, and community. To point out the obvious, science teachers do not work in such ideal situations. They work in an education system with constraints on time and budget and variations in students' knowledge, abilities, and motivations. In this era, I fear that the burdens for standards-based curricula will fall on science teachers and that they may succumb to the weight, reducing the potential of this contemporary reform, if not bringing it to a halt.

Those who design and develop science curriculum provide a valuable resource for teachers, but like most resources, science curriculum materials must be adapted to meet unique needs of teachers and students. The important point here is the adaptation of curriculum materials by science teachers. Teachers often adopt and use science curriculum materials without personal adaptation. The process of adaptation by science teachers—whether the materials were developed at the national, state, or local levels—is one critical aspect of improving the science curriculum.

I defend the work of groups such as the BSCS, LHS, American Chemical Society (ACS), EDC, NSRC, and Technical Education Resource Center (TERC). I do emphasize the importance of specialized groups for curriculum development. I am not the first to speak in favor of groups that specialize in the development of science curriculum and provide high-quality materials to meet various requirements for local and state curricula (Karplus 1971). Most science teachers, local school districts, and state agencies do not have the technical capabilities, personnel, time, and money to develop science curricula and meet criteria of field testing in schools, which can include coordinating content, incorporating sound pedagogy, and aligning formative and summative assessments with standards. The most important benefit of locally developed science curricula is ownership of the program and subsequent use by school personnel. Local ownership can be achieved through professional development that focuses on the adaptation of materials originally developed outside the local district.

This perspective on science curriculum development is mine. It was developed during my tenure in a national organization with responsibilities for designing and producing instructional materials.

OUR COMMON PERSPECTIVE AND LEADERSHIP OPPORTUNITIES

The leadership opportunities for designing curriculum materials are plentiful and vitally important. Certainly, the need for curriculum programs presents one of the grand challenges in this era of education reform.

The history of curriculum reform in science education is impacted by the social forces for changes in goals and different priorities for the goals. However, the actual changes and programs are not as clear. It seems much was left to classroom teachers when it came to development of new curriculum programs for science education.

The era of standards-based reform presents the challenge of translating standards to curriculum programs. While this may well be a grand challenge, it is possible for school personnel to design and develop instructional materials. This requires leadership and insights relative to decisions about the adoption of new programs or adaptation of current materials.

Although published in the late 1990s, Ralph Tyler's classic on curriculum and instruction is an excellent introduction for those confronting the challenge of reforming a science curriculum.

ISSUES AND QUESTIONS FOR DISCUSSION

1. After reviewing historical changes in science curriculum, how would you describe the processes of curriculum reform?
2. What is the primary difference between historical and contemporary examples of social forces, theories of learning, and science subject matter?
3. We used "planning a dinner" to exemplify the process of designing and developing a science curriculum. Can you suggest another better example of the process? If so, what is it, and what are the details?
4. What are your reactions to the strategies, scope, and emphasis for redesigning the science curriculum (see Figure 12.1)?
5. What do you perceive as the curriculum implications of contemporary learning theory?
6. To clarify the need to translate standards as policies to curriculum programs, we used the example of translating a book from one language to another. What do you consider the appropriate strengths and obvious limitations of this analogy?
7. How would you approach curriculum reform in your school, district, or state?

REFERENCES

American Association for the Advancement of Science (AAAS), Project 2061. 2000. *Designs for science literacy.* New York: Oxford University Press.

Black, P., and D. Wiliam. 1998. Inside the black box: Raising standards through classroom assessment. *Phi Delta Kappan* 80 (2): 139–148.

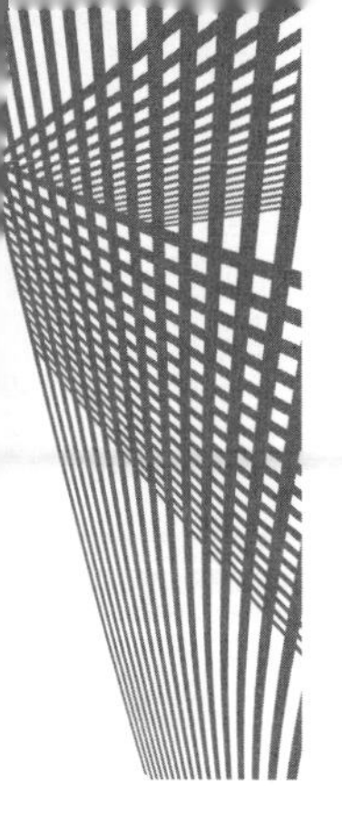

Black, P., and D. Wiliam. 2009. Developing the theory of formative assessment. *Educational Assessment, Evaluation and Accountability* 21 (1): 5–31.

Bransford, J. D., A. L. Brown, and R. R. Cocking, eds. 2000. *How people learn: Brain, mind, experience, and school.* Washington, DC: National Academies Press.

Bransford, J. D., M. Pellegrino, and W. Donovan, eds. 1999. *How people learn: Bridging research and practice.* Washington, DC: National Academies Press

Bybee, R. W., ed. 2002. *Learning science and the science of learning.* Arlington, VA: NSTA Press.

Bybee, R. W. 2003. The teaching of science: content, coherence, and congruence. *Journal of Science Education and Technology* 12 (4): 343–358.

Caswell, H., and D. Campbell. 1935. *Curriculum development.* New York: American Book Company.

DiRanna, K., E. Osmundson, J. Topps, L. Barakos, M. Gearhart, K. Cerwin, D. Carnahan, and C. Strang. 2008. *Assessment-centered teaching: A reflective practice.* Thousand Oaks, CA: Corwin Press.

Edelson, D. 2001. Learning-for-use: A framework for the design of technology-supported inquiry activities. *Journal of Research in Science and Teaching* 38 (3): 355–385.

Elmore, R., and M. McLaughlin. 1988. *Steady work: Policy, practice, and the reform of American education.* Report No. 3574-NIE/RC. Washington, DC: National Institute of Education.

Fullan, M. 2001. *The new meaning of educational change.* New York: Teachers College Press, Columbia University.

Gardner, A., R. Bybee, L. Enshan, and J. Taylor. 2014. Analyzing the coherence of science curriculum materials. *Curriculum and Teaching Dialogue* 16 (1 and 2): 65–85.

Glatthorn, A. 1987. *Curriculum renewal.* Alexandria, VA: Association for Supervision and Curriculum Development (ASCD).

Hall, G. E., and S. M. Hord. 2001. *Implementing change: Patterns, principles, and potholes.* Boston: Allyn and Bacon.

Jackson, P. 1992. Conceptions of curriculum and curriculum specialists. In *Handbook of research on curriculum,* ed. P. Jackson, 4–5. New York: MacMillan Publishing Company.

Karplus, R. 1971. Some thoughts on science curriculum development. In *Confronting curriculum reform,* ed. E. W. Eisner, 56–61. Boston: Little, Brown, and Company.

Kesidou, S., and J. E. Roseman. 2002. How well do middle school science programs measure up? Findings from Project 2061's curriculum review. *Journal of Research in Science Teaching* 39 (6): 522–549.

Lee, O., H. Quinn, and G. Valdes. 2013. Science and language for English language learners in relation to *Next Generation Science Standards* and with implications for *Common Core State Standards* for English language arts and mathematics. *Educational Researcher* 42 (4): 223–233.

McNeill, K. L., and J. Krajcik. 2012. *Book study facilitator's guide: Supporting grade 5–8 students in constructing explanations in science: The claim, evidence and reasoning framework for talk and writing.* New York: Pearson.

Murnane, R., and S. Raizen, eds. 1988. *Improving indicators of the quality of science and mathematics education in grades K–12.* Washington, DC: National Academies Press.

National Research Council (NRC). 2002. *Investigating the influence of standards: A framework for research in mathematics, science and technology education.* Washington, DC: National Academies Press.

National Research Council (NRC). 2005. *How students learn: Science in the classroom.* Washington, DC: National Academies Press.

National Research Council (NRC). 2006. *America's lab report: Investigations in high school science.* Washington, DC: National Academies Press.

National Research Council (NRC). 2007. *Taking science to school.* Washington, DC: National Academies Press.

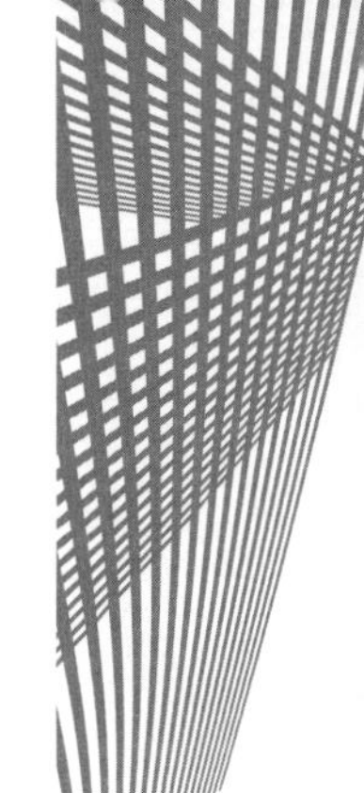

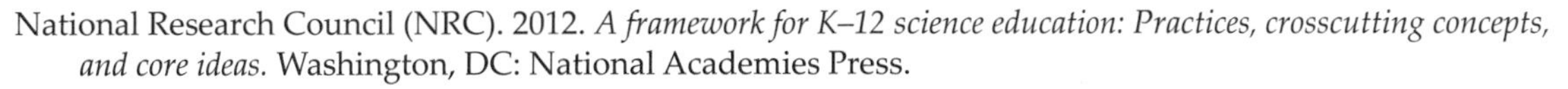

National Research Council (NRC). 2012. *A framework for K–12 science education: Practices, crosscutting concepts, and core ideas.* Washington, DC: National Academies Press.

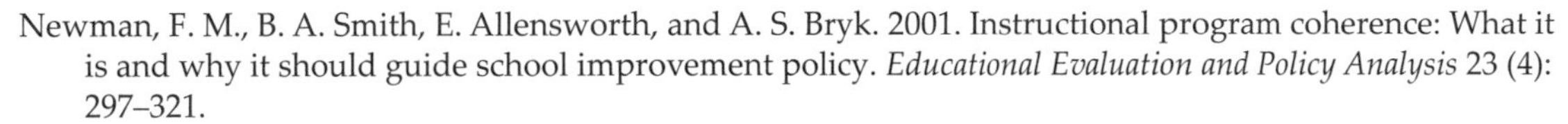

Newman, F. M., B. A. Smith, E. Allensworth, and A. S. Bryk. 2001. Instructional program coherence: What it is and why it should guide school improvement policy. *Educational Evaluation and Policy Analysis* 23 (4): 297–321.

NGSS Lead States. 2013. *Next Generation Science Standards: For states, by states.* Washington, DC: National Academies Press.

Pellegrino, J., M. Wilson, J. Koenig, and A. Beatty, eds. 2014. *Developing assessments for the* Next Generation Science Standards. Washington, DC: National Academies Press.

Posner, G. J. 1992. *Analyzing the curriculum.* New York: McGraw Hill.

Powell, J. C., J. B. Short, and N. M. Landes. 2002. Curriculum reform, professional development, and powerful learning. In *Learning science and the science of learning*, ed. R. W. Bybee, 121–136. Arlington, VA: NSTA Press.

Roberts, D. 1982. Developing the concept of "curriculum emphasis" in science education. *Science Education* 66 (2): 243–260.

Roberts, D. 1995. Scientific literacy: The importance of multiple curriculum emphases. In *Redesigning the science curriculum: A report on the implications of standards and benchmarks for science education,* ed. R. W. Bybee and J. D. McInerney, 75–80. Colorado Springs, CO: Biological Sciences Curriculum Study (BSCS).

Rosebery, A., M. Ogonowski, M. DiSchino, and B. Warren. 2010. "The coat traps all your body heat": Heterogeneity as fundamental for learning. *Journal of the Learning Sciences* 19 (3): 322–357.

Rosebery, A., and B. Warren, eds. 2008. *Teaching science to English language learners: Building on students' strengths.* Arlington, VA: NSTA Press.

Roseman, J. E., M. C. Linn, and M. Koppal. 2008. Characterizing curriculum coherence. In *Designing coherent science education: Implications for curriculum instruction, and policy*, ed. Y. Kali, M. Linn, and J. E. Roseman, 13–36. New York: Teachers College Press, Columbia University.

Roth, K. J., and H. E. Garnier. 2006/2007. What science teaching looks like: An international perspective. *Educational Leadership* 64 (4): 16–23.

Rutherford, F. J., and A. Ahlgren. 1988. Rethinking the science curriculum. In *Content of the curriculum: 1988 ASCD yearbook*, ed. R. Brandt, 75–90. Alexandria, VA: Association for Supervision and Curriculum Development (ASCD).

Rutherford, J. J., and A. Ahlgren. 1989. *Science for all Americans.* New York: American Association for the Advancement of Science (AAAS).

Rutherford, F. J. 2000. Coherence in high school science. In *Making sense of integrated science: A guide for high schools.* Colorado Springs, CO: Biological Sciences Curriculum Study (BSCS).

Saylor, G., and W. Alexander. 1974. *Planning curriculum for schools.* New York: Holt, Rinehart, and Winston.

Schmidt, W. 2003. The quest for a coherent school science curriculum: The need for an organizing principle. *Review of Policy Research* 20 (4): 569–584.

Schmidt, W., and C. McKnight. 1998. What can we really learn from TIMSS? *Science* 282 (5395): 1830–1831.

Schwab, J. 1970. *The practical: A language for curriculum.* Washington, DC: National Education Association.

Seidel, T., R. Rimmele, and M. Prenzel. 2005. Clarity and coherence of lesson goals as a scaffold for student learning. *Learning and Instruction* 15 (6): 539–556.

Tyler, R. W. 1949. *Basic principles of curriculum and instruction.* Chicago: The University of Chicago Press.

Warren, B., C. Ballenger, M. Ogonowski, A. Rosebery, and J. Hudicourt-Barnes. 2001. Rethinking diversity in learning science: The logic of everyday language. *Journal of Research in Science Teaching* 38 (5): 529–552.

SECTION VII

CLASSROOM PRACTICES FOR SCIENCE TEACHING

SECTION VII

The title "Classroom Practices for Science Teaching" refers to the specific actions, methods, and strategies educators use as they implement a science program. This is the most unique and fundamental level of applying purposes to the policies, programs, and, ultimately, practices of teaching.

Teaching science, or any subject, is an incredibly complex process. Teaching requires multiple decisions made from moment to moment, all based on teachers' understanding of the subject and skills of teaching and perceptions of students' current knowledge, interests, attitudes, and motivations. Yes, there are even other considerations—the social dynamics of 25 or more students, the management of materials and equipment, adequate and appropriate interpersonal relations, and the selection of available strategies and methods, to mention only a few.

The responsibility for establishing a positive environment for learning ultimately lies with the teacher. Teachers also have remarkable freedom to select topics, materials, activities, and strategies they think will contribute to student learning.

So, when, where, and how do teachers develop the knowledge and skills required to make sense of and act effectively within the complexities of classrooms? What do teachers need to know and be able to do? Are some individuals just naturally effective? The first chapter of this section explores these questions. The second chapter centers on research relative to effective classroom practices.

Interestingly, the *Science* issue on grand challenges in science education (AAAS 2013) included very little on the challenges associated with classroom practices. Figure VII.1 presents the challenges from that issue. The article by Suzanne Donovan addressed the challenge of shaping research around practice (see the first topic in the following figure). The challenges identified in the section on professional development also relate to classroom practices. We point out the fact that reforms ultimately require changes in the classroom practices of science teachers.

Figure VII.1. Challenges of Improvement Through Research and Development in the Education System

Shift incentives to encourage education research on the real problems of practice as they exist in school settings. The current incentive structures in universities and in agencies that fund education support theoretical research and its "translation" for practice. Those incentives must shift to encourage research on the problems of practice in school settings.

Create a set of school districts where long-standing, multidisciplinary teams work together to identify effective improvements. Because the problems of practice are multidimensional, addressing them effectively will require the creation of settings in which teams of researchers, practitioners, and education designers are supported to follow the contours of a problem to identify effective improvements.

Create a culture within school systems that allows for meaningful experimentation. Because contextual factors are critical to the effective scaling of improvements, experimentation in real school contexts is critical. This will require a major shift from the belief that experimentation is inappropriate in school systems to the belief that experimentation provides new opportunities for the success of education professionals and the students they serve.

Source: Adapted from AAAS 2013.

REFERENCE

American Association for the Advancement of Science (AAAS). 2013. Special issue, *Science* 340 (130).

SUGGESTED READINGS

Bruner, J. 1968. *Toward a theory of instruction.* New York: W. W. Norton. The collection of essays offers the insights and wisdom of this great educator. He discusses, for example, education as a social invention, the will to learn, and notes on a theory of instruction.

Bybee, R. 2015. *The BSCS 5E Instructional Model: Creating teachable moments.* Arlington, VA: NSTA Press. An introduction to this widely used instructional model, the book includes applications of the model to the *Next Generation Science Standards,* STEM, and 21st-century skills.

Crawford, B. A. 2014. From inquiry to scientific practices in the science classroom. In *Handbook of research in science education: Volume II,* ed. N. Lederman and S. Abell, 515–542. New York: Routledge. A review of seminal statements and peer-reviewed articles clarifying both the history of and research on inquiry.

Osborne, J. 2014. Scientific practices and inquiry in the science classroom. In *Handbook of research in science education: Volume II,* ed. N. Lederman and S. Abell, 579–599. New York: Routledge. A clear and deep rationale for the scientific practices central to the *Next Generation Science Standards.*

Schwab, J., and P. Brandwein. 1966. *The teaching of science.* Cambridge, MA: Harvard University Press. This volume contains one essay from each of these two leaders. Given as the Inglis Lecture (Schwab) and Burton Lecture (Brandwein) at Harvard in 1961, these essays present similar ideas on classroom practices and the teaching of science.

Stigler, J., and J. Hiebert. 1999. *The teaching gap.* New York: The Free Press. Based on a video study of teaching mathematics in Germany, Japan, and the United States, the authors present cultural differences in teaching and systemic views for reform. There are challenging insights that apply to science teaching.

CHAPTER 13

SCIENCE TEACHING: ART OR SCIENCE?

The question expressed in the title of this chapter has a long history. To begin this exploration, we note that the question, stated in *either/or* form, betrays a false dichotomy. Are some individuals just born teachers? We can be clear about this claim: The knowledge and abilities for successful teaching are not innate. There are no sets of genes for effective teaching.

We can consider another fallacy often expressed in statements such as, "Those who can, do; those who can't, teach." So, it seems, anyone can teach. Except most of us know this is just not true. There are less-than-productive teachers. Unfortunately, some think teaching means having adequate knowledge of a subject. Here, the problem is equating teaching with talking—in particular, telling about a subject. This fallacy can be expressed ironically: If teaching were telling, it would be easy and we would have accomplished our goals long ago.

This discussion leads us to the reasonable conclusion that effective science teaching is both art and science. So, the issue is, how much of each? What do you think? Is science teaching more art or science? How would you justify your answer? We will not suggest a percentage for each. However, we do propose that the science education community does have a knowledge base that contributes to the science of teaching.

Those interested in a philosophical, historical, and psychological discussion similar to the introductory paragraphs will find *The Practice of Teaching* by Philip Jackson (1986) quite interesting. Jackson addresses the underlying complexity beneath a deceptive simplicity of teaching. He discusses many of the challenges today's educators face, even though the volume was published in the 1980s.

We continue the exploration of science teaching with a review of research from the cognitive sciences and students' learning. Although this repeats some prior discussion, here we place the discussion in the context of science teaching.

CHAPTER 13

HISTORICAL PERSPECTIVE

Learning Science and Cognitive Development

Research on student learning has long been an important factor in any consideration of effective teaching. Let's review some of that research.

In the late 1950s, the 1960s, and the 1970s, science teachers looked to Jean Piaget's theory of cognitive development (Piaget 1952, 1977; Bybee and Sund 1982). Piaget's original work pursued answers to questions about the origin and nature of knowledge; the questions were epistemological in nature. The model of learning focused on two major conclusions of Piagetian investigations. First, Piaget proposed that learning occurs through an individual's interaction with the environment. This interaction was described as a student assimilating new information and ideas from various educational experiences and then accommodating the new information with previously held concepts. This model of learning establishes a consistency between the individual's cognitive structure and everyday experience. Second, Piaget proposed that each individual passes through different stages of development that are characterized by the ability to perform various cognitive tasks.

Piaget's notion of learning as an interaction with the environment has been generally supported (Lawson, Abraham, and Renner 1989; Renner, Abraham, and Birnie 1986) and, in fact, was a foundation for constructivist explanations of students' conceptual understanding and change. The concept of stages—for example, concrete and formal reasoning—however, has been criticized and revised. Studies (see, e.g., Chiapetta 1976; Renner, Grant, and Sutherland 1978; Wavering, Perry, and Birdd 1986) have demonstrated that the majority of secondary students are at the concrete stage of reasoning, as measured by their performance on cognitive tasks. There also is evidence that performance on such tasks is strongly influenced by context, mode language of task presentation, and subject matter (Golbeck 1986; Brandwein 1979). Other studies have demonstrated that even young children are capable of abstract thought in certain situations (Chi and Koeske 1983).

Constructivism Emerges

Science educators' research progressed to a model—*constructivism*—to help understand students' learning. The theoretical basis for constructivist research comes from several sources, including David Ausubel (1968). The essence of Ausubelian theory (Ausubel, Novak, and Hanesian 1978) holds that a learner's prior knowledge is an important factor in determining what is learned in a given situation. L. S. Vygotsky (1968) is a second important source for constructivism. He wrote of student and teacher conceptions and how students and teachers might use similar words to describe concepts, yet have different personal interpretations of those concepts. Vygotsky's work implies that classroom practices should take into account the differences between teachers' and students' conceptions and provide student-student interaction so that learners can develop understanding of concepts from those whose understandings and interpretations are closer to their own.

Early work in constructivist research focused on identifying students' conceptions about scientific phenomena and how such conceptions differ from accepted scientific conceptions. Student con-

ceptions have been referred to by various labels, such as misconceptions, alternative conceptions, alternative frameworks, and naïve theories. We think such labels demean students and believe some conceptions are reasonable because they are based on everyday experiences. It is much more productive, and even constructive, for the teacher to recognize students' *current* conceptions and emphasize that teaching is aimed at reconstructing those conceptions so they are more aligned with those recognized as scientific. Several early reviews of research on students' conceptions (Driver, Guesne, and Tiberghien 1985; Osborne and Freyberg 1983; Duit 1987) provide support for the role of students' current conceptions and the challenges teachers face in changing those conceptions.

In the constructivist model, students construct knowledge by interpreting new experiences in the context of their current conceptions, experiences, and images. Students' construction of knowledge begins at an early age, so that by the time they encounter formalized study of science, they have developed stable and highly personal conceptions for many natural phenomena. Given this view, one goal of classroom practice must be to facilitate change in students' conceptions of the natural world. Researchers (Posner et al. 1982; Smith, Carey, and Wiser 1985) have likened this process of conceptual change to the process by which scientific theories undergo change and restructuring. In fact, some studies have demonstrated that students' beliefs and conceptions often parallel early scientific theories, dating back to Aristotle and Jean-Baptiste Lamarck (Caramazza, McCloskey, and Green 1981; Champagne, Klopfer, and Gunstone 1982; Wandersee 1986). However, we caution against drawing too strong a parallel between students' conceptions and the history of science, largely because student conceptions are not nearly as comprehensive as, say, Aristotelian theories.

Posner and colleagues (1982) proposed four conditions for conceptual change. First, for students to change their conceptions of a given phenomenon, they must be *dissatisfied* with their current conception. This dissatisfaction presumably comes about through repeated exposure to experiences they cannot explain using current conceptions. Second, the new conception must be *intelligible* in terms of prior experiences and knowledge. Third, the new conception must be *plausible*, in that it can explain a number of prior experiences and observations. Finally, the new conception must be *fruitful*, opening up new areas of inquiry, primarily through predictions about future events.

Constructivist research also suggests other strategies to promote conceptual change. You should be aware that students have conceptions of the world and that they often do not differentiate concepts (Trowbridge and McDermott 1981). Students need time to make their ideas explicit, and they should have opportunities to apply their conceptions of the world in different contexts (Minstrell 1989).

Research associated with cognitive sciences and constructivism has had an important influence on science teaching. Although not exhaustive, there are many examples of this early research (Hewson and Hewson 1984; Perkins and Simmons 1988; Eylon and Linn 1988; Burbules and Linn 1988; Linn et al. 1989; Guzzetti et al. 1993).

CHAPTER 13

CONTEMPORARY PERSPECTIVES

How Students Learn Science

The National Research Council (NRC) reports *How People Learn: Brain, Mind, Experience and School* (1999a); *How People Learn: Bridging Research and Practice* (1999b); and *How Students Learn: Science in the Classroom* (2005) synthesize a body of research on how students learn. These reports present an essential response to a significant challenge in science education.

We realize the following research-based summaries repeat those discussed in Chapter 12. These messages and our discussion, however, have clear and direct implications for classroom practices. The previous chapter described implications for curriculum design. Let's again consider the three research-based findings. This time, we will discuss their implications for classroom practices.

- Students come to the classroom with preconceptions about how the world works. If their initial understanding is not engaged, they may fail to grasp new concepts and information presented in the classroom, or they may learn them for purposes of a test but revert to their preconceptions outside the classroom.

Educators, especially classroom teachers, realize that students have current conceptions about many objects, organisms, and phenomena in the world. Because learning is a natural process, one that has continued since birth, school-age students have developed what are, for them, meaningful and reasonable representations of their world. However, many of these current conceptions, as students express them in science classrooms, do not align with concepts fundamental to science disciplines. Here are several implications of this finding:

- Effective instruction requires science teachers to engage and draw out students' current conceptions of the phenomena under study. Furthermore, science teachers should understand that students' current conceptions are remarkably resilient and largely unexamined.
- Effective instruction encourages students to confront the inadequacy of their current conceptions and make meaningful links between their current explanations and new science-based knowledge. Students should have opportunities and time to reconstruct their explanations so they accurately reflect the concepts, knowledge, and procedures of science.

One significant challenge for science teachers resides in the observation that not all 25 to 30 students in a class have the same current conception (i.e., misconception) at the same time! Providing opportunities for students to visualize and express their ideas through interactions with peers is one way to accommodate this problem.

Here is the second finding on student learning:

- To develop competence in an area of learning, students must have both a deep foundation of factual knowledge and a strong conceptual framework.

This finding gives insight into the complexity of teaching. Simultaneously, the teacher must have the knowledge, skill, and wisdom to introduce major scientific concepts and associated facts at the right time and in the appropriate context. No easy task, this. Research on experts in a discipline such as science shows that experts have a deep understanding of concepts and an equally deep base of information. Most of teaching centers on facts and information, with little or no mastery of concepts. Factual information is necessary but not sufficient. Conceptual understanding contributes to the transfer of information from one context to another and the reorganization of information into new and meaningful patterns that, in time, show that situations are applicable to other unique situations.

Implications of this finding for classroom practice include the following:

- Effective instruction attends to both the fundamental concepts of science and appropriate facts and information.
- Effective science instruction also should include generalized and discipline-specific procedures (i.e., the practices of scientific inquiry) for acquiring new explanations.
- Effective classroom practices include sufficient time and opportunities to practice and apply new knowledge and skills to reasonably challenging situations.

The third research-based finding has to do with metacognition:

- Strategies can be taught that allow students to monitor their understanding and progress in problem solving.

A less complex and acceptable way to define metacognition is "thinking about one's thinking." The late Michael Martinez provides a more precise definition: "Metacognition is the monitoring and control of thought" (Martinez 2006, p. 696).

The research for this finding is based on the performance of experts in disciplines such as science. Experts monitor their progress, noting the need for additional information, the alignment of new information with what is known, and the appropriateness of metaphors in the clarification and solution of problems (NRC 1999b). Experts monitor their understanding, to paraphrase the Martinez quotation.

The implications of research on metacognition include the following:

- Effective teaching must include problem solving, debates, and group discussions—opportunities and time for reflection and application of higher-order thinking.
- Effective teaching requires teachers to model metacognitive strategies such as "thinking out loud," "personal reflection," and, to use a term suggested by Michael Martinez, "making thinking an audible."
- Effective teaching involves establishing a "culture of scientific inquiry" as an integral feature of classroom practice.

Another NRC publication, *America's Lab Report: Investigations in High School Science* (2006), analyzed the effectiveness of laboratory experiences. The committee proposed using the phase *integrated instructional units* to describe sequences of instruction that connect laboratory investigations with other types of teaching strategies, such as readings, discussions, lectures, and web searches. Using a framework based on the practices of scientific inquiry, students might be engaged by framing research questions, making observations, designing an investigation, gathering data, and using those data to construct an explanation. Sequences of laboratory experiences combined with other forms of instruction show this approach is effective for achieving three goals: improving mastery of subject matter, developing scientific reasoning, and cultivating interest in science. Furthermore, integrated instructional units appear to be effective in helping diverse groups of students progress toward these goals. Table 13.1 provides a summary of research-based conclusions. What follows is a major conclusion from this NRC report:

> *Four principles of instructional design can help laboratory experiences achieve their intended learning goals if: (1) they are designed with clear learning outcomes in mind, (2) they are thoughtfully sequenced into the flow of classroom science instruction, (3) they are designed to integrate learning of science content with learning about the processes of science, and (4) they incorporate ongoing student reflection and discussion. (NRC 2006, p. 6)*

Table 13.1. Attainment of Educational Goals in Typical Laboratory Experiences and Integrated Instructional Units

Goal	Typical Laboratory Experiences	Integrated Instructional Units
Mastery of subject matter	No better or worse than other modes of instruction	Increased mastery compared with other modes of instruction
Scientific reasoning	Aids development of some aspects	Aids development of more sophisticated aspects
Understanding of the nature of science	Little improvement	Some improvement when explicitly targeted as the goal
Interest in science	Some evidence of increased interest	Greater evidence of increased interest
Understanding the complexity and ambiguity of empirical work	Inadequate evidence	Inadequate evidence
Development of practical skills	Inadequate evidence	Inadequate evidence

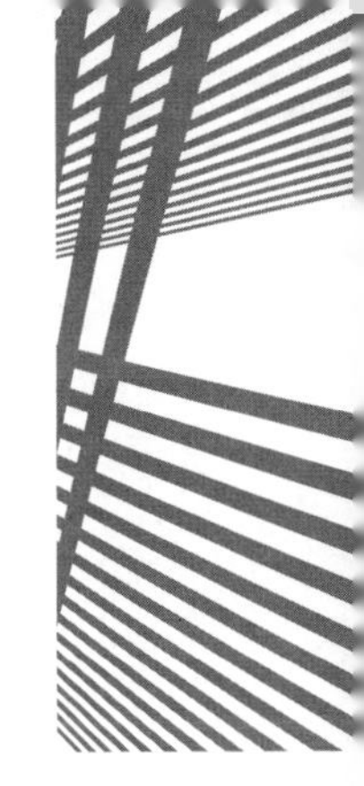

CONCLUSION

When considering classroom practices, it is helpful to recognize that students may already have explanations for many objects, events, and phenomena. The challenge science teachers face is helping students realize the inadequacy of their current conceptions and providing the time and opportunity for them to construct more scientifically accurate concepts.

A PERSONAL PERSPECTIVE

NGSS AND OPPORTUNITIES FOR INSTRUCTIONAL INNOVATIONS

Stephen Pruitt

The *Next Generation Science Standards* (*NGSS*) provide an incredible opportunity for our classroom teachers to broaden their instructional practices. However, as with most opportunities, teachers need to reach out and take them. Our classroom practice has, for the most part, been heavily influenced by three factors: textbooks, our previous experience, and how "large" we perceive our educational world. I know I will need to explain that third one; I will save that for last. Suffice it to say that we have some incredible teachers doing incredible things, but as I have stated previously, the challenge still boils down to the fact that not every student has had those same opportunities to learn science.

I first discuss textbooks and instructional materials. While quality materials are necessary and research has shown that they significantly affect instruction, they cannot continue to rule the day. Instructional materials and textbooks have often been used to the exclusion of student needs. What I mean by this is that we have made a practice of using the book, its sequence, and its materials as our rule for organizing curriculum, as opposed to using our students and best research practices as our rule. A good example of this is the scientific method. At least since the 1990s, science education experts have been trying to explain that the scientific method is actually a variable process more than a linear progression. Yet the scientific method has remained a part of classroom practice as a proxy for teaching students how to think like a scientist. What we know is that actual practices applied by scientists do not happen in a particular order and that the scientific method is a vastly simplified view of scientific processes. So, our classroom practice is going to need to change to meet the *NGSS*. Of course that is true, but it needs to change not simply because of the *NGSS*, but because the practice of the classroom needs to match the practice of science. A simple five-step method cannot begin to replicate the need for students' being able to model, construct scientific arguments and explanations, or even ask questions about natural phenomena.

The other key part not covered by the scientific method that is critical for scientific literacy in the 21st century is understanding the engineering process. Many parts overlap, but the idea of persevering, revising, and using failure to build a better solution is not captured at all.

Another issue with letting the textbook dictate instruction is the focus on content. Of course, the *NGSS* require a deep knowledge of concepts, but not to the exclusion of how students learn best. As has been discussed earlier, the purpose of the *NGSS* is to enable students to build knowledge and apply it. Most traditional textbooks spend page after page discussing content and asking students to discuss content in the questions they are asked. There are typically traditional labs that go along with them, but they are what I would consider validation labs. Redesigning our classroom practice means that students are given opportunities to explain phenomena, not focus on mundane topics. Our best and most innovative teachers understand this. I have had the incredible opportunity to see some of these teachers in action. They take everyday events and build the course around them in a way that builds knowledge. One particular teacher, Tricia Shelton from Boone County, Kentucky, took students' understanding of energy in an ecosystem to a whole other level. In recognizing energy as both content and a crosscutting concept, she had students build a Rube Goldberg device. Now, you may question why this is innovative; lots of teachers have students do this. What they do not do is have students build a device in a biology class so they can develop explanations of how energy transfers through various levels of ecosystems. Students had to explain the need to keep energy flowing, dissipation of energy, and how there is far less energy in the final stages of their designs. Then, they had to use this experience to explain how energy flows through the various trophic levels of an ecosystem. This is instruction at its best. It is here we see that what has actually constrained us as teachers has not been standards; it has been a commitment to textbooks or curriculum materials as the basis for instruction. Our classroom practice under the *NGSS* allows teachers to bring the wonder back to the science classroom.

Next, I turn to previous experience. When I first started teaching, I remember spending time preparing for my first classes thinking about how I learned the material. My first assignment was to teach ninth graders a course called Physical Science. This was basically a survey of chemistry and physics but was supposed to be fairly watered down. In my later years, I found teaching this class was a great way to stay grounded in the needs of all students. I wish I had taken more time to understand this earlier in my career. So, as I prepared, because I was a new teacher, my first thought was, "How did I learn this material? What did my labs, quizzes, and tests look like? What did instruction look like?" And, I am ashamed to say, I wondered, "How can I show my students how much I know?" All of these questions were not good for my students. I started lecturing with a smattering of labs. I remember even having the bizarre idea that ninth graders learning to draw isomers was somehow useful (I loved organic chemistry). The point I am trying to make is that my classroom instruction was all wrong. I had not fully embraced what I had learned in my science education courses, but then I also did not have a lot of opportunity to practice it, either. I had an incredible student teaching supervisor, Jean Prezel, who helped me learn to do a lot more than just lecture. The problem was, when I went out on my own, I saw it was easier to lecture and set up the occasional lab. It took several years to really overcome this, and I am ashamed to say it took way too long. While many teachers are excited about the possibilities and promise of the *NGSS*, it does not mean that these changes in classroom practice will be easy or happen overnight. The practice needed by teachers will be foreign to some. Thinking about constructing instruction in a coherent manner as opposed to thinking of flow is a challenge. The idea of teaching toward phenomena as opposed to teaching a standard is not natural to much of our teaching force. This is not intended as a slight to our teachers; this is the reality of how we were all trained. Our classroom practice needs to change across the board to embrace the idea of helping students understand the world around them.

Finally, our classroom practice really cannot change until we broaden our educational worldview. I believe we choose the size of our world. I can limit my world by staying just within my classroom, or I can enlarge it by being part of the school, district, state, or national discussions. I can choose to just teach my subject, or I can choose to advocate for policies that allow me to be a better teacher. I can listen to what others tell me is the truth, or I can seek it myself by being knowledgeable about the education system and how it works. Because of the first two factors I discussed, teachers have to get outside their comfort zones to change their instruction. There are many ways to do this. For me personally, I think I am a far better teacher today, not because I know more, but because I've experienced so much. I have had the opportunity to spend time with the leading researchers in the field and observe some of the best teachers in the country. All teachers can have many of the same experiences I have had. Quality professional development opportunities such as the National Science Teachers Association (NSTA) programs at the district level, experiences such as assessment review at the state level, and participation in programs such as the Presidential Award for Excellence in Mathematics and Science Teaching can all provide a wider view of science teaching excellence. The reason for this is simple: Teachers need to experience what best practices are beyond their own classroom. I do not think it is an accident that our best teachers seek out these types of opportunities. Like doctors and lawyers, teachers need opportunities to observe other teachers, as they show not only exemplary practice but also the struggles with new practices. Like Vygotsky showed with regard to students and how they learned, teachers also need to feel the cognitive disequilibrium that convinces them to change. We know that best practice for classroom science is hard. As I have said many times before, we cannot hope to achieve best practices in an easy way.

There is a little more I want to comment on as I close out this perspective on the *NGSS* and opportunities for instructional innovations. What are some specifics regarding science classroom practice that we should change? First, we cannot think of teaching a single *NGSS* or state standard as a quality practice. In fact, we should think about the fact that all of the *NGSS* were designed to be met by the end of the year. For this to happen, students need experience with all of the science and engineering practices and must be allowed to both see and make their own connections using the crosscutting concepts. Well-designed classroom practice will allow students to learn rich content, but in a way that is meaningful for them. We can talk about matter being made of particles, or we can have students engage in learning that allows them to use this as an explanation, for instance, of how flags wave. The days of the practice of writing the standard of the day on the board and teaching that standard that day must come to an end. Looking for—and helping students find—connections to prior learning and yet seeing how current learning will lead to future understanding have to be our aims. That does not mean that students have a syllabus that shows the next topic; rather, it means that students are learning in an environment that supports their engagement in an ongoing revision of their own cognitive models. To do this, we need to embrace the concept of coherence trumping flow. It is easy to see in a chemistry class, for instance, the instructional flow of moving from atoms to particles and molecules to substances to equations and reactions and, finally, to stoichiometry. However, this is not how chemists learned these concepts; they actually went from macroscopic observations of effects to microscopic explanations of cause. So, coherence comes not from flow of instruction but from opportunities for students to see those connections that make sense. For instance, while we may talk about weather, climate, and maybe even energy as it relates to weather fronts, how often do we also help students understand the role that forces play in weather fronts? As we consider classroom instruction, we need to understand the deliberateness of practice that results in addressing the metacognitive aspects that must be respected if students are really going to be scientifically literate. That is to say that while many scientists and science teachers were able to make those connections on their own, students need the additional support to learn to think this way.

Another key change in practice must be that classroom instruction should also be coherent in content and in our approach to ensuring students have opportunities to learn to think scientifically. This is part of the reason the *NGSS* scientific and engineering practices and crosscutting concepts are so critical. They provide students with opportunities to see the overall coherence in science, often by asking new questions that lead to new pieces of a puzzle. Actually, that is a pretty good analogy. Throughout school, our classroom teaching must adhere to the idea that each day adds another piece of the puzzle for students. However, it should be the student who puts the puzzle together. Sometimes they work around the edges for a foundation, sometimes they work on sections of the puzzle, and sometimes they need guidance on how to fit the pieces, but the classroom and our teachers need to provide the table, the big picture, and certainly a safe environment in which to try pieces that may not fit together yet but will still fit in the big picture.

Our classrooms must turn to the incredible innovation that all of our great teachers have. The classroom is as complex and changing as any medical practice. So, if we are to change, we must start with that in mind. Innovation means thinking differently than how we have in the past and having the courage and patience to learn new instructional strategies while also helping students learn new science practices, disciplinary ideas, and crosscutting concepts.

REFERENCES

Ausubel, D. P. 1968. *Educational psychology: A cognitive view.* New York: Academic Press.

Ausubel, D. P., J. D. Novak, and H. Hanesian. 1978. *Educational psychology: A cognitive view.* 2nd ed. New York: Holt, Rinehart, and Winston.

Brandwein, P. 1979. A general theory of instruction: With references to science. *Science Education* 63 (3): 285–297.

Burbules, M., and M. Linn. 1988. Response to contradiction: Scientific reasoning during adolescence. *Journal of Educational Psychology* 80 (1): 67–75.

Bybee, R. W., and R. Sund. 1982. *Piaget for educators.* Columbus, OH: Merrill.

Caramazza, A., M. McCloskey, and B. Green. 1981. Naïve beliefs in sophisticated subjects: Misconceptions about trajectories of objects. *Cognition* 9: 117–123.

Champagne, A. B., L. E. Klopfer, and R. F. Gunstone. 1982. Cognitive research and the design of science instruction. *Educational Psychologist* 17 (1): 31–53.

Chi, M. T. H., and R. D. Koeske. 1983. Network representation of a child's dinosaur knowledge. *Development Psychology* 19: 29–39.

Chiapetta, E. L. 1976. A review of Piagetian studies relevant to science instruction at the secondary and college level. *Science Education* 60 (2): 253–261.

Driver, R., E. Guesne, and A. Tiberghien, eds. 1985. *Children's ideas in science.* Philadelphia, PA: Open University Press.

Duit, R. 1987. Research on students' alternative frameworks in science: Topics, theoretical frameworks, consequences for science teaching. Proceedings of the Second International Seminar on Misconceptions and Educational Strategies in Science and Mathematics. Ithaca, NY.

Eylon, B-S., and M. Linn. 1988. Learning and instruction: An examination of form research perspectives in science education. *Review of Education Research* 58 (3): 251–301.

Golbeck, S. L. 1986. The role of physical content in Piagetian spatial tasks: Sex differences in spatial knowledge. *Journal of Research in Science Teaching* 23: 323–333.

Guzzetti, B., T. Snyder, G. Glass, and W. Gamas. 1993. Promoting conceptual change in science: A comparative meta-analysis of instructional interventions from reading education and science education. *Reading Research Quarterly* 28 (2): 117–158.

Hewson, P., and M. Hewson. 1984. The role of conceptual conflict in conceptual change and the design of science instruction. *Instruction Science* 13: 1–13.

Jackson, P. 1986. *The practice of teaching.* New York: Teachers College Press.

Lawson, A., M. Abraham, and J. Renner. 1989. *A theory of instruction: Using the learning cycle to teach science concepts and thinking skills [Monograph, Number One].* Kansas State University, Manhattan, KS: National Association for Research in Science Teaching.

Linn, M. C., C. Clement, S. Pulos, and P. Sullivan. 1989. Scientific reasoning during adolescence: The influence of instruction in science knowledge and reasoning strategies. *Journal of Education Research* 58 (3): 251–301.

Martinez, M. 2006. What is metacognition? *Phi Delta Kappa* 87 (9): 696–699.

Minstrell, J. A. 1989. Teaching science for understanding. In *Toward the thinking curriculum: Current cognitive research, 1989 yearbook of the ASCD*, ed. L. Resnick and L. Klopfer, 129–149. Alexandria, VA: Association for Supervision and Curriculum Development.

National Research Council (NRC). 1999a. *How people learn: Brain, mind, experience and school.* Washington, DC: National Academies Press.

National Research Council (NRC). 1999b. *How people learn: Bridging research and practice.* Washington, DC: National Academies Press.

National Research Council (NRC). 2005. *How students learn: Science in the classroom.* Washington, DC: National Academies Press.

National Research Council (NRC). 2006. *America's lab report: Investigations in high school science.* Washington, DC: National Academies Press.

Osborne, J., and P. Freyberg, eds.1983. Concepts, misconceptions, and alternative conceptions: Changing perspectives in science education. *Studies in Science Education* 10: 61–98.

Piaget, J. 1952. *The origins of intelligence in children.* New York: International Universities Press.

Piaget, J. 1977. *The development of thought: Equilibrium of cognitive structures.* New York: Viking Press.

Perkins, D., and R. Simmons. 1988. Patterns of misunderstanding: An integrative model for science, math, and programming. *Review of Educational Research* 58 (3): 303–326.

Posner, G. J., K. A. Strike, P. W. Hewson, and W. A. Gertzog. 1982. Accommodation of a scientific conception: Toward a theory of conceptual change. *Science Education* 66 (2): 211–237.

Renner, J. W., M. R. Abraham, and H. H. Birnie. 1986. The importance of the form of student acquisition of data in physics learning cycles. *Journal of Research in Science Teaching* 23 (2): 121–143.

Renner, J. W., R. M. Grant, and P. Sutherland. 1978. Content and concrete thought. *Science Education* 62 (2): 215–221.

Smith, C., S. Carey, and M. Wiser. 1985. On differentiation: A case study of the development of the concepts of size, weight, and density. *Cognition* 9: 177–237.

Trowbridge, D. E., and L. C. McDermott. 1981. Investigation of student understanding of the concept of acceleration in one dimension. *American Journal of Physics* 49: 242–253.

Vygotsky, L. S. 1968. *Thought and language.* Trans. and ed. A. Kozulin. Cambridge, MA: MIT Press.

Wandersee, J. H. 1986. Can the history of science help science educators anticipate students' misconceptions? *Journal of Research in Science Teaching* 23: 581–597.

Wavering, M. J., B. Perry, and D. Birdd. 1986. Performance of students in grades 6, 9, and 12 on five logical, spatial, and formal tasks. *Journal of Research in Science Teaching* 23: 321–333.

CHAPTER 14

TEACHING SCIENCE AND THE SCIENCE OF TEACHING

DOES SCIENCE EDUCATION HAVE A THEORY OF INSTRUCTION?

In this chapter, we shift from learning to teaching and ask the question, "What do we know about effective classroom practices?" As it turns out, we know quite a bit about the knowledge requirements for teachers, classroom decisions, and good teaching.

Before turning to summaries of research on science teaching, we explore the idea expressed in the term *science of teaching*. Or, stated as a provocative question, "Does science education have a theory of instruction?"

In science, a theory is a comprehensive explanation of some aspect of the natural world. The explanation has empirical support and is widely accepted by the scientific community. This definition is very different from the everyday use of the term *theory* when it refers to a guess, an opinion, or even a thoughtful, non-empirical description of some phenomena.

So, at this point, what do you think? Does science education have a theory of instruction? Yes? No? Maybe?

In prior discussions, we have differentiated between learning and teaching. And, to be clear, instruction complements curriculum, but it is different. We continue exploring this question by going back in history.

TWO HISTORICAL PERSPECTIVES

How would you describe a theory of instruction for science education? What research topics would you use as the empirical basis for your theory of science teaching? The following discussion will provide a perspective for your ideas.

CHAPTER 14

Criteria for a Theory of Instruction

In 1968, Jerome Bruner outlined the characteristics of an instructional theory in a book titled *Toward a Theory of Instruction.* According to Bruner, a theory of instruction is *prescriptive.* It gives direction and provides guidelines for effective instruction and enables the teacher to evaluate teaching techniques and procedures. A theory of instruction also is *normative*; it is general rather than specific. For example, a theory of instruction would give some criteria for a chemistry lesson on acids and bases but would not give specific guidelines for the lesson.

What help does a theory of instruction provide? What questions will it answer? According to Bruner, an instructional theory has the following major features:

- First, a theory of instruction should help identify experiences that would establish a predisposition toward learning. This feature focuses on factors (e.g., motivational and personal) that establish a desire to learn. The experiences may leave the learner with a curiosity about phenomena, a need to explore possible explanations, or some uncertainty about current knowledge or disequilibrium, to use a Piagetian term.
- Second, a theory of instruction should specify some structure for a body of knowledge. In this case, the structure of knowledge should be such that it will be understood by the learner. The theory should help teachers make decisions about ways to simplify major concepts. It should indicate an optimal structure that aligns with the age and stage of the learners.
- The third feature has to do with the effective sequencing for presentation of the subject. We note here the closeness of the second and third characteristics with curricular features. How should learners progress through a series of activities? This is the question a theory should help answer.
- Finally, the theory should indicate the appropriate time and type of assessments. Bruner used the expression "pacing of rewards and punishments in the process of learning and teaching" (1968, p. 41).

A Theory of Instruction for Science?

Another leader in science education, Paul Brandwein, published an article titled "A General Theory of Instruction: With References to Science" in 1979. While recognizing Bruner's brilliant and insightful proposals in his earlier works *The Process of Education* (1960) and *On Knowing: Essays for the Left Hand* (1969), Brandwein argues—we think appropriately—that Bruner did not after all develop and make a systematic assertion that amounted to a theory of instruction (Brandwein 1979, p. 285).

Brandwein makes the point that a theory of instruction should explain changes in learned behavior and what is identified on measures of achievement. Just as a germ theory of disease explains the diseases that result from various bacteria and viruses, a theory of instruction should explain the results of various classroom practices. Based on the four points stated above, we make the observation that Bruner identifies how to structure a body of knowledge, which is a curricular perspective, and how to sequence the presentation of a subject, which may be curricular or instructional. Brandwein argued for a separation of instruction and curriculum.

After qualifications and an appropriate introduction, Brandwein presents, in simplified form, the theory of instruction:

> *In any specified act of instructed learning, a new environment is created; in responding to the altered environment by initiating activity involving the manipulation and transformation of concepts, values, or skills an individual learner gains demonstrably in capacities not achieved through prior experience but specified in the given act of instructed learning. (Brandwein 1979, p. 291)*

The statement requires some discussion. First, the term *instructed learning* was used by Brandwein to make clear that the process of learning in schools implies instruction or that student learning in the context of classrooms was initiated by instruction. The phrase "a new environment is created" suggests an event or experience, directed by a teacher, that brings about mental activity that requires thinking, valuing, or acting differently by the student. Furthermore, this change is new and recognizable and attributed to the instructional strategy or practice. What do you think? Upon close reading, does Brandwein's statement constitute a theoretical statement? Can you identify research supporting the statement?

Brandwein continued his article with a tentative theory of curriculum. It may help to consider his statement.

> *In the design of a curriculum, reduction of complexity is imposed on a domain of experience by reconciling compatible orders of concepts, values, and skills; in turn, nonrandom ordering is imposed on these compatible orders so as to increase the effectiveness of instructed learning described in objectives defining specified changes of behavior affecting individual learners. (Brandwein 1979, pp. 291–292)*

This statement complements the theory of instruction. Does it qualify as education theory? The remainder of Brandwein's article included analogies (e.g., instruction is like the Newtonian concept; for every action there is a reaction); historical references (e.g., Dewey, Comenius); psychological justifications (e.g., Piaget, Skinner); and practical application (e.g., the nature of daily lesson plans and longer activities).

To conclude this section, we wonder what you think. Does science education have a theory of instruction (and curriculum)?

This chapter continues with reviews of research on instruction that provide evidence supporting claims that we have a science of teaching—or not.

CHAPTER 14

CONTEMPORARY PERSPECTIVE: RESEARCH PERSPECTIVES ON CLASSROOM PRACTICES

Instructional Strategies

In "Research on Instructional Strategies for Teaching Science" (Tobin, Tippins, and Gallard 1994), the authors present a thorough, research-based discussion of various topics associated with science teaching, including cognitive demands, classroom management, textbooks, laboratories, classroom environments, and assessments. This team also reviewed foundational issues such as teacher beliefs, pedagogical content knowledge, the learning cycle, and cooperative learning. The review is prefaced with a contemporary introduction to constructivism as a way of knowing.

David Treagust (2007) published a research summary called "General Instructional Methods and Strategies." Treagust and Chi-Yan Tsui updated this summary in 2014. The 2014 chapter addresses demonstrations, explanations, questioning, scientific reasoning, and representational learning (e.g., analogies, metaphors, models). In a summary, the authors suggest that the various instructional methods and strategies have some empirical support of their effectiveness in enhancing student learning and opportunities for student learning from K–12 schools to universities (Treagust and Tsui 2014, p. 315).

Matthew Kloser (2014) published a timely article called "Identifying a Core Set of Science Practices: A Delphi Expert Panel Approach." While *A Framework for K–12 Science Education* (NRC 2012) and the *Next Generation Science Standards* (*NGSS*; NGSS Lead States 2013) describe ambitious aims and content for science education, the documents provide little practical guidance on questions teachers have, such as "How and when should I employ classroom practices that are core to helping students achieve the *NGSS* goals?"

Kloser used the Delphi methodology and an expert panel in three anonymous and iterative rounds to identify a set of core science practices. He provides an answer to science teachers' questions about which instructional practices best align with the *NGSS*. Figure 14.1 lists the core teaching practices for science that the research identified.

We must note that this core set of practices aligns with a set proposed by Windschitl and colleagues (2012). Their set included constructing the big ideas, eliciting ideas, activity and sense making, and pressing for explanations.

With these contemporary presentations on the theme of teaching science and the science of teaching, let's continue with research on specific factors that underlie teaching, such as teachers' knowledge, instructional decisions, and good teaching.

Content Knowledge and Science Teaching

Lee Shulman, then at Stanford University, reported on the role and development of teachers' knowledge in relation to teaching. Shulman (1986) identified three categories of content knowledge: subject matter, pedagogical, and curricular.

Figure 14.1. Core Teaching Practices for Science

1. Engaging students in investigations
2. Facilitating classroom discourse
3. Eliciting, assessing, and using student thinking about science
4. Providing feedback to students
5. Constructing and interpreting models
6. Connecting science to its applications
7. Linking science concepts to phenomena
8. Focusing on core science ideas, crosscutting concepts, and practices
9. Building classroom community

N/A Adapting Instruction*

*Adapting instruction was not on the final round for panelist rating. It received a mode of 5 (out of 5) in round 2 and thus is included.

Source: Adapted from Kloser 2014.

Science Subject Matter

For science teachers, subject matter is more than concepts (i.e., disciplinary core ideas) and facts about a discipline. Content knowledge of a subject includes what Joseph Schwab (Schwab and Brandwein 1966) called the "substantive and syntactic structures" of a discipline. That is, substantial knowledge for science teachers includes an understanding of the different ways the basic concepts and principles of a discipline are organized. What are the major conceptual schemes in a science discipline? If you had to organize the information and facts of physics, chemistry, biology, or the Earth sciences, what major ideas (not topics) would you identify as basic structures of these disciplines? In biology, for example, one can use the levels of organization approach, studying biology from the smallest particles to larger domains and explaining living processes in terms of molecular activities. One also can use an ecological approach in which the ecosystem is the basic level of study and individual activities are studied in terms of the systems in which they live and interact. Using either of these structures, you can develop basic conceptual schemes of biology, such as energetics, genetics, diversity, and evolution.

The syntax of a discipline is the set of ways scientists establish the truth or falsehood, validity or invalidity, of new or extant knowledge claims. This relates to the nature of science. Science teachers' use of inquiry introduces students to the practices of obtaining new knowledge, such as developing and using models, planning and conducting investigations, constructing explanations, and engaging in argument from evidence. Students also gain the understanding that knowledge must be evaluated.

Pedagogical Content Knowledge (PCK)

Pedagogical content knowledge describes the depth and breadth of knowledge a teacher has about teaching a particular subject. Pedagogical content knowledge is the capacity to formulate and represent science in ways that make it comprehensible to learners. Examples of pedagogical content knowledge include, for example, forms of representing concepts through use of analogies, examples, illustrations, and demonstrations. Effective science teachers have a variety of ways and means of representing ideas such as ionic bonding, density, recombination of DNA, and stellar evolution.

Another dimension of pedagogical content knowledge is the understanding of what makes a concept easy or difficult for a learner to grasp. What current conceptions might students have about phenomena such as heat and temperature, position and velocity, or living and nonliving things? What conceptions do students have for objects and events in the natural world? The science teacher's understanding of pedagogical content knowledge helps establish links between new concepts and the students' current understanding.

Curricular Knowledge

Next, we consider curricular knowledge. The science curriculum includes a full range of materials with which science teachers should be familiar. Materials are designed for a particular subject, at a particular level, to be used with particular students. Each discipline has its own textbooks, kinds of laboratory equipment, and educational software.

In addition to knowledge of curricular materials and how best to use them, curricular knowledge extends to teachers' abilities to relate topics of study to the curricula their students may be studying in other disciplines.

The discussion of teacher knowledge as it relates to content, pedagogy, and curriculum is obviously important, even essential, but you will need more than an adequate knowledge base. Effective science teaching requires many instructional decisions. We discuss instructional decisions next.

Instructional Decision Making

David Berliner (1984) reviewed research on teaching and provided valuable insights about decisions using the categories of pre-instructional decisions, instructional decisions, and post-instructional decisions.

Pre-instructional Decisions

Before beginning a science lesson, teachers should be aware of the effect of certain decisions on student achievement, attitudes, and behaviors. For example, teachers must make *content decisions*. What is the content of the science lesson? Teachers must not only consider national standards and state and district standards and assessments but must also make judgments about issues such as the effort required to teach a subject and the problems they perceive the students will have with the subject. Finally, teachers must take into account the subjects they enjoy teaching. What are the areas within a discipline that they really like? Are teachers excited about introducing students to the nature and history of science? Do they include crosscutting concepts? How about science and engineering practices? Do they consider it most important to have students recognize science-related social issues?

Science teaching involves groups of students. *Grouping decisions* are part of your preparation for a sequence of lessons. What is the best size of a group? How much laboratory equipment is available? Who should (or should not) work together? Should one use cooperative groups? Whether lecturing or working in the laboratory, effective instruction requires making decisions about the students as a group. Even when assigning individual work on projects, experiments, and tests, a group in some sense still exists.

Finally, teachers have to make *decisions about activities*. Laboratory work, for example, has specific functions—that is, it is used to achieve certain goals. In a laboratory, students may learn to design an investigation, manipulate equipment, and use computers. Operations—the rules or norms of conduct for the activity—are also important. Is it okay to be out of one's seat? What type of conversation is acceptable? What rules *must* be followed for safety reasons?

The importance of these kinds of decisions cannot be overstated. Just reading this section should make you aware of the many and varied decisions teachers must make *before they begin teaching even the simplest lesson.*

Instructional Decisions

From the beginning of a lesson, numerous factors determine what students learn. The amount of time students spend on a task—*engaged time*—is directly related to how much students will learn. Recognize that *engaged* can mean physically (hands-on), mentally (minds-on), or both. Although this seems obvious, the amount of engaged time varies from student to student and class to class. Engaged time is especially important for students who are underachievers and of low ability. These students will benefit most from time *on task*, but they also are the students who are most likely to be *off task.*

The *success rate* of students is related to their achievement. Success in the early stages of learning new concepts or skills is especially important for students of low ability and those who have traditionally been unsuccessful. If students do not experience some success in the early stages of lessons, their frustration and lack of understanding can contribute to low achievement.

There are decisions about *questioning* that also influence teaching effectiveness. Science teachers in particular should ask many questions; questions about natural phenomena are the foundation of science. One should first consider the cognitive level of the question. Many teachers ask low-level questions such as, "What do the letters DNA stand for?" "What is the second law of thermodynamics?" or "What is a silicon oxygen tetrahedron?" Although such questions have some benefit, remember that higher-level questions facilitate thinking and learning. Questions that require students to analyze and synthesize will produce higher levels of student achievement. Science teachers should, for example, provide data in graph form and ask the students to analyze and interpret the graph and engage in arguments based on the evidence (NGSS Lead States 2013). Or you could provide information from two separate but related experiments and ask students for their predictions of possible outcomes.

Another point about questioning concerns the importance of *waiting* after asking a question. Research by Mary Budd Rowe (1974) confirms the importance of wait time. Longer wait times (many teachers wait less than one second after asking a question) result in increases in the appropriateness, confidence, variety, and cognitive level of responses. A two- to three-second adjustment in teaching style can result in a much higher return for students.

Post-instructional Decision

Now that the lesson, unit, or semester is over, how much did the students learn? Science teachers usually arrive at an answer through the assessments, grades, and feedback given to students.

Assessment is not the central issue. In fact, we propose that planning the assessment should correlate with your aims and should have been designed *before* developing the instructional sequence and providing students opportunities to learn. Although this may seem backwards, this process allows for designing school science instruction so it is coherent (Wiggins and McTighe 2005).

Grades motivate students to achieve; however, the overuse of grades or their use as coercion can have detrimental effects. Corrective feedback, if properly given, results in positive achievement and attitudes on the part of students. Teachers' decisions to give praise for correct work, recognition for proper behavior, and personally neutral criticism (as opposed to sarcasm) for incorrect responses all can influence student learning.

Research on Good Teaching

Next, we present research on good teaching. After reviewing the research on good teaching, Andrew Porter and Jere Brophy (1988) concluded that the concept of good teaching is changing. In the past, teachers sometimes were viewed as technicians who had to apply how-to lessons, or as weak links in the education system that had to be circumvented with a teacher-proof curriculum. Such approaches did not work. The current concept of effective teaching deals with empowering teachers. How are teachers empowered? A brief answer is through the application of research on effective teaching. Our discussion assumes that student learning within science classes requires good teaching, and good teaching requires science teachers who make appropriate decisions about how to educate students.

A contemporary image of the good teacher is that of a thoughtful professional who works purposefully toward educational goals. Figure 14.2 summarizes many of the points in this section.

Discussions in prior sections revealed insights about students' learning and teachers' knowledge of the subject, pedagogy, curriculum, and various decisions and characteristics of good teaching. Clearly lacking, but essential, was any discussion of teachers' personal perceptions as they relate to interactions with students and classroom practices. We think this aspect of teaching deserves recognition because it is more important than many recognize.

Effective Relationships and Teachers' Perceptions

For many years, Arthur Combs and colleagues (Combs, Avila, and Parkey 1978) investigated the characteristics of effective helping relationships, including teaching. Perceptual psychology was the

Figure 14.2. Highlights of Research on Good Teaching

Good teaching is fundamental to effective schooling. From the studies of the Institute for Research on Teaching and other studies conducted over the past ten years, there is a picture of effective teachers as semiautonomous professionals who

- are clear about their instructional goals,
- are knowledgeable about lesson content and teaching strategies,
- communicate what is expected of students—and why,
- make expert use of instructional materials in order to devote more time to practices that enrich and clarify lesson content,
- teach students metacognitive strategies and give them opportunities to master them,
- address higher- as well as lower-level cognitive objectives,
- monitor students' understanding by offering regular and appropriate feedback,
- integrate their instruction with that of other subject areas,
- accept responsibility for student outcomes, and
- are reflective about their practice.

Source: Adapted from Porter and Brophy 1988.

orientation of their research. The basic idea of perceptual psychology is that all behavior of a person is based on that person's field of perceptions at the moment of his or her behavior. Specifically, behavior is a result of (1) how individuals see themselves, (2) how they see the situation in which they are involved, and (3) the interrelations of these two (Combs 1965). When a person sees herself as a science teacher, she behaves like a science teacher. In another situation, for example, she may behave like a parent. So, it is not external facts that are important to understanding behavior, but the meaning of the situation to the person. To be clear, the orientation is a humanistic approach to psychology (see, e.g., Bybee and Welch 1972).

The research indicates that teachers' perceptions of self, students, and the teaching task are critical to effective instruction. Of course, the effective teacher is well informed about science in general, but there is also a commitment to the scientific enterprise and a belief that science is an important social institution.

Effective teachers perceive other people, particularly their students, as able, friendly, worthy, intrinsically motivated, dependable, and helpful. In the same manner, effective teachers see themselves as identified with other people, an adequate teacher, trustworthy, wanted, cared for, and worthy.

Teachers' purposes also play an important role in effectiveness. Better teachers see their purpose as assisting and facilitating rather than coercing and controlling. They identify with larger issues; are personally involved with issues, problems, and other people; and see the process of education as more important than the achievement of specific goals. In addition, they tend to be altruistic and self-revealing. Effective teachers see their task as helping people rather than dealing with objects. And, generally, they try to understand the perceptions and experiential backgrounds of their students.

Contemporary discussion of the meaning of a learning experience and its association with learning certainly is an application of a basic principle of perceptual psychology. That is, learning is a function of content (e.g., disciplinary core ideas) and the personal meaning that content has for the individual.

Students' Perceptions of the Ideal Science Teacher

Combs's studies delineated the perceptions of effective teachers. What about the students' perceptions of the science teacher? If the concepts of perceptual psychology hold, these perceptions will influence student behavior. Bybee conducted research on the perceptions of the ideal science teacher (Bybee 1973, 1975, 1978). This research used a 50-item Q-sort as the means of collecting data. The 50 items were divided into five categories identified in the left column of Table 14.1.The results of these studies are summarized in Table 14.1.

Table 14.1 shows that adequate personal relations with students and enthusiasm in working with them consistently rank as the most important characteristics for science teachers. With only one exception, these two categories were ranked first or second by all the groups studied.

We think the research on personal dimensions has important implications for science teaching. We wish to underscore the point that knowledge, planning, and use of different teaching methods also are essential. Effective classroom practices require all dimensions. Because the personal dimensions are often overlooked, we have highlighted them in this section.

Table 14.1. Science Educators' Grand Mean Ranking the Ideal Teacher Compared With Other Populations' Data

	Science Educators (*N* = 172)	In-service Teachers (*N* = 76)	Preservice Elementary Majors (*N* = 58)	High School Students–Average (*N* = 44)	High School Students–Disadvantaged (*N* = 106)	High School Students–Advantaged (*N* = 31)	Elementary School Students–Grade 6 (*N* = 25)	Elementary School Students–Grades 4–6 (*N* = 18)
Knowledge of subject matter	4	4	4	3	4	3	3	3
Adequate personal relations with students	1	1	1	1	1	2	1	1
Adequate planning and organization	5	5	5	4	3	4	5	4
Enthusiasm in working with students	2	2	2	2	2	1	2	2
Adequate teaching methods and class procedures	3	3	3	5	5	5	4	5

CHAPTER 14

A PERSONAL PERSPECTIVE

OF COURSE, THE BSCS 5E INSTRUCTIONAL MODEL

Rodger Bybee

More than 25 years ago, a team of colleagues and I created the BSCS 5E Instructional Model. Portions of this essay are from an article I wrote in *Science and Children* in 2014 called "The BSCS 5E Model: Personal Reflections and Contemporary Implications" (Bybee 2014).

In developing the instructional model, we took several things into consideration. First, we began with an instructional model that already had a research base. We began with the Science Curriculum Improvement Study (SCIS) Learning Cycle because it had substantial evidence supporting the phases and sequence (Atkin and Karplus 1962; Renner, Abraham, and Bernie1988; Lawson, Abraham, and Renner 1989). The BSCS additions and modifications to the Learning Cycle also had a research base. For example, we integrated cooperative learning into the model (Johnson and Johnson 1987).

Second, we realized that the constructivist view of learning required experiences to challenge students' current conceptions (i.e., misconceptions) and ample time and activities for the reconstruction of their ideas and abilities.

Third, we wanted to provide a perspective for teachers that was grounded in research and had an orientation for individual lessons. We asked, "What perspective should teachers have for a particular lesson or activity?" In addition, we wanted to express the value of coherence for lessons within an instructional sequence. We asked, "How does one lesson contribute to the next, and what was the point of the sequence of lessons?"

Finally, we tried to describe the model in a manner that would be understandable, usable, manageable, and memorable. This was the origin of the 5Es for the different phases of the model.

I continue with a brief review of the five phases and some contemporary issues, questions, and reflections.

ENGAGING LEARNERS

This phase captures the students' attention and interest. Get the students focused on phenomena or a problem with the content and abilities that are the aims of instruction. From a teaching point of view, asking a question, posing a problem, showing a brief video, or presenting a discrepant event are all examples of strategies to engage learners. If students look puzzled, asking, "How did that happen?" or saying, "I have wondered about that" and "I need to know more to solve that problem," they likely are engaged in a learning situation. Students have some ideas, but the expression of concepts and use of their abilities may not be scientifically accurate and productive. Getting them to express their understanding is a first step toward their construction of a more elaborate concept.

Over the decades, I have come to realize two things about this phase. The engagement need not be a full lesson, but it usually is because students need to express their current conceptions and teachers need to assess students' conceptions and abilities. Teachers might, for example, provide a brief description of a natural phenomenon and ask students how they would explain the situation. The main point is to get the students puzzled and thinking about content related to the learning outcomes of the instructional sequence. The second point about this phase and the next is that they present opportunities for teachers to informally determine misconceptions expressed by the students.

EXPLORING PHENOMENA

In the exploration phase, students have activities with time and opportunities to resolve the disequilibrium of the engagement experience. Exploration lessons provide concrete, hands-on experiences where students express their current conceptions and demonstrate their abilities as they try to clarify puzzling elements of the engage phase.

Exploration experiences should be designed as introductions to the concepts, practices, and skills of the learning outcomes. Students should have experiences and the occasion to formulate explanations, investigate phenomena, observe patterns, and develop their cognitive understanding, physical abilities, and technical skills.

The teacher's role in the exploration phase is to initiate the activity, describe appropriate background, provide adequate materials and equipment, and counter misconceptions. After this, the teacher steps back and becomes a coach with the tasks of listening, observing, and guiding students as they clarify their understanding and begin reconstructing scientific concepts and developing their abilities.

EXPLAINING PHENOMENA

The scientific explanation for phenomena is prominent in this phase. The concepts, practices, and abilities with which students were originally engaged and subsequently explored now are made clear and comprehensible. The teacher directs students' attention to key aspects of the prior phases and first asks students for their explanations.

Using students' explanations and experiences, the teacher introduces scientific or technological concepts briefly and explicitly. Yes, this can be direct instruction. Here, using an *NGSS* example, the disciplinary core ideas—including vocabulary, science and engineering practices, and crosscutting concepts—are presented clearly, briefly, and simply. Prior experiences should be used to provide context for the explanation. Verbal explanations are common, but use of video, the web, or other digital media also may provide explanations.

ELABORATING SCIENTIFIC CONCEPTS AND ABILITIES

The students are involved in learning experiences that extend, expand, and apply the concepts and abilities developed in the prior phases. The intention is to facilitate the transfer of concepts and abilities to related but new situations. A key point for this phase: Use activities that are challenging but achievable for the students.

In the elaboration phase, the teacher challenges students with a new situation and facilitates interactions among students and with other sources such as written material, databases, simulations, and web-based searches.

EVALUATING LEARNERS

At some point, students should receive feedback on the adequacy of their explanations and abilities. Clearly, informal, formative evaluations will occur throughout the instructional sequence, but as a practical matter, teachers must assess and report on educational outcomes.

In the evaluate phase, the teacher should involve students in performance-based experiences that are understandable and consistent with those activities of prior phases and congruent with the scientific explanations. The teacher should determine the evidence for student learning and means of obtaining that evidence as part of the evaluate phase.

ISSUES, QUESTIONS, AND REFLECTIONS

In the years since the model's development, I have received many questions about the BSCS 5Es. This section addresses some of the issues raised by curriculum developers and classroom teachers applying the 5E model to materials, instruction, and, most recently, the *NGSS*.

Does a Recent Synthesis of Research on Learning Support the 5E Model?

When *How People Learn* (Bransford, Brown, and Cocking 1999) was published, the authors recommended an instructional model very close to the BSCS 5Es. The sequence even used several of the same words. Here is a quotation from *How People Learn:*

> *An alternative to simply progressing through a series of exercises that derive from a scope and sequence chart is to expose students to the major features of a subject domain as they arise naturally in problem situations. Activities can be structured so that students are able to explore, explain, extend, and evaluate their progress. Ideas are best introduced when students see a need or a reason for their use—this helps them see relevant uses of knowledge to make sense of what they are learning.* (p. 127)

This quotation summarizes a substantial body of research and generally confirms an instructional sequence very similar to the BSCS 5Es. You should note that the initial *problem situation* followed by *explore, explain, extend,* and *evaluate* clearly parallels the 5E sequences.

The 5E Model Seems to Have Implications for the Design of Curriculum Materials. Is This Observation Correct?

This observation is correct. The designs of curriculum materials and instructional strategies are closely connected. Although we discussed curriculum design in the prior chapter, I review the design implication of the 5E model in Table 14.2.

Table 14.2. The 5E Model and Specifications for Curriculum Materials

5E Model Stages	Curriculum Materials: Specification
ENGAGE	This phase focuses students' interest and thinking on the learning task. Curriculum materials should (1) activate current knowledge and make connections between past and present experiences; (2) anticipate activities of future lessons; and (3) physically and mentally engage students in the concepts, practices, and applications of the unit.
EXPLORE	The lessons should establish a common base of experiences that students use to begin developing scientific and engineering practices, crosscutting concepts, and disciplinary core ideas. Curriculum materials include activities that help students use current knowledge to generate ideas, explore questions and problems, consider possibilities, design investigations, obtain information, conduct web searches, and engage in discourse about their ideas.
EXPLAIN	Curriculum materials provide clear and succinct explanations for science concepts and engineering designs. Materials provide opportunities for group work where students explain their ideas to peers, review current explanations, read, listen to videos, search the web, and listen to the teacher.
ELABORATE	Lessons in this phase have students apply the concepts and practices to new situations. Curriculum materials require the transfer of prior learning within a reasonable range for students.
EVALUATE	Lessons in this phase use the performance expectations as the basis for assessments of student learning.

What Is the Appropriate Use of the Instructional Model?

More specifically, should the instructional model be the basis for one lesson? A unit of study? An entire program? My experience suggests that the optimal use of the BSCS 5E Instructional Model is a unit of two or three weeks, where each phase is used as the basis for one or more lessons (with the exception of the engage phase, which may be less than a lesson). In this recommendation, I assume some cycling of lessons within a phase; for example, there might be two lessons in the explore phase and three lessons in the elaborate phase.

Using the model as the basis for a single lesson decreases the effectiveness of the individual phases due to shortening the time and reducing the opportunities for challenging and restructuring of concepts and abilities. On the other hand, using the phases for an entire program increases the time and experience of the individual phases so much that the perspective for the phase loses its effectiveness because teachers have too much exploration and multiple explanations that may be concentrated.

Can a Phase Be Omitted?

My recommendation: Do not omit a phase. Earlier research on the SCIS Learning Cycle found a decreased effectiveness when phases were omitted or their position shifted (Lawson, Abraham, and Renner 1989). From a contemporary understanding of how students learn, there is integrity to each phase and the sum of the phases as originally designed (Taylor, Van Scotter, and Coulson 2007). This question is often based on prior ideas about teaching that would omit the the engage or exploration phases, for example, and go immediately to the explain phase. Alternatively, some suggest omitting the elaborate phase. The important point about not omitting the elaborate phase centers on the transfer of learning combined with the applications of knowledge.

Can the Order of the Phases Be Shifted?

My response to this question is similar to the prior one on omitting a phase: What would you shift? Move the explain phase prior to explore? The original sequence was designed to enhance students' learning and was subsequently supported by research (see, e.g., Bransford, Brown, and Cocking 1999; NRC 1999; Bybee et al. 2006; Wilson et al. 2010). There also is earlier research on the learning cycle that specifically investigated the question about changing the sequence (Renner, Abraham, and Birnie 1988; Marek and Cavallo 1997). That research indicated reduced effectiveness when the sequence was changed, so I do not recommend shifting phases.

Can a Phase or Phases Be Added?

My colleague Arthur Eisenkraft added two phases by splitting the engage phase to create elicit and engage phases and adding an extend phase after evaluate to underscore the importance of knowledge transfer (Eisenkraft 2003). In principle, I do not have a problem with adding a phase (or two) if the justification is grounded in research on learning, which was the case for Eisenkraft's modification.

Although there is no research support, I believe there is the practical issue of recalling titles, establishing criteria, and differentiating strategies for more phases. I have found the optimum number of phases to be between three and five.

Can Phases Be Repeated?

Yes, it is sometimes necessary for teachers to repeat a phase. This change should be based on the curriculum developer's or teacher's judgment relative to students' need for time and experiences to learn a concept or develop ability. To be clear, an example of repeating a phase would be *engage, explore, explore*—not necessarily changing the order by, for example, placing a second explore phase before evaluate.

Shouldn't Evaluation Be Continuous?

Effective teachers continuously evaluate their students' understanding. In the instructional model, the evaluate phase is intended as a summative assessment conducted at the end of the instructional sequence or unit. Certainly, some evaluation ought to be informal and continuous, but there also is need for a formal evaluation at the end of the unit.

What If I Need to Explain an Idea Before (or After) the Explanation Phase?

This may be necessary, as some ideas are prerequisites to students' understanding the primary concepts of a unit. Teachers will have to make a judgment about the priority and prerequisite nature of the concepts. One should maintain an emphasis on the primary or major concepts and abilities of the unit and not digress with too many less-than-essential explanations.

Can the 5Es Be Used for *NGSS* and the Integration of Three Dimensions?

Yes, I have actually found the 5E model helps solve the challenge of incorporating the multiple dimensions of *NGSS* in the classroom. The phases of instruction certainly can include activities that afford opportunities for students to experience the science and engineering practices, disciplinary core ideas, and crosscutting concepts. In *Translating the* NGSS *for Classroom Instruction* (Bybee 2013), I used the 5Es for examples of the integration of dimensions of *NGSS*.

Do Certain Phases of the 5E Instructional Model Easily Align With and Contribute to Use of the *NGSS* Science and Engineering Practices?

Yes, they do. I have illustrated some potential alignments in Table 14.3 (p. 218). Of course, different content and goals may align with other *NGSS* practices.

Table 14.3. Potential Alignment of Science and Engineering Practices Within the 5E Framework

5E Model Stages	Instructional Strategies: Essentials	Potential Alignment of Science and Engineering Practices
ENGAGE	The teacher assesses learners' current knowledge and facilitates their interest in and attention to new concepts and practices by posing questions, presenting discrepant events, showing a video, or giving a demonstration.	• Asking questions • Defining problems
EXPLORE	The teacher encourages the examination of current concepts and exploration of practices as students encounter scientific questions and engineering problems. Instruction centers on using practices to challenge current ideas and abilities and begin formulating new concepts, abilities, and behaviors.	• Developing and using models • Analyzing and interpreting data • Constructing explanations • Engaging in argument
EXPLAIN	Teachers directly explain concepts and practices and guide learners toward in-depth understanding. Instruction includes asking for clarification, providing definitions, and using students' current explanations as the basis for more accurate scientific explanations and definitions.	• Constructing explanations • Designing solutions
ELABORATE	Teachers encourage the use of formal labels, definitions, and explanations and provide these if they are not expressed. Instruction has students use evidence for explanations and requires use of logic in formulation of arguments.	• Planning and carrying out investigations • Obtaining, evaluating, and communicating information • Using mathematics and computations thinking
EVALUATE	Teachers may observe students and assess their understanding of concepts and practices and determine the degree to which they met performance expectations.	• Evidence-based arguments • Constructing explanations

The BSCS 5E Instructional Model was originally designed as an instructional sequence that would help teachers approach instruction in a meaningful and effective way, one that enhanced student learning. I still hold this goal. Many within the science education community have recognized the model's practical value and incorporated it into school programs, state frameworks, and national guidelines. This widespread use has been very rewarding. I certainly encourage continued use of the model and appropriate adaptations by school personnel.

OUR COMMON PERSPECTIVE AND LEADERSHIP OPPORTUNITIES

The themes of these chapters explore ideas central to the science education community—improving the practice of science teaching.

Effective science instruction recognizes students' current conceptions and facilitates the reconstruction of those concepts to align with scientific concepts. This statement expresses the essential point of historical and contemporary research on student learning. The challenge is implementing teaching practices that accommodate what we understand about how students learn.

Effective science instruction explicitly integrates the practices of scientific inquiry with the conceptual ideas and facts that are fundamental to a science discipline. The key word here is *integrates.* Past programs and teaching have often separated science concepts and processes. Contemporary standards recommend the integration of concepts and practices (i.e., the three dimensions of the *Framework* and *NGSS*) in curriculum and instruction.

Effective science instruction includes work in structured groups where students reflect on, discuss, and argue their ideas and procedures. Contemporary science teaching requires that students present scientific explanations, incorporate argumentation, and communicate their evidence-based claims reasonably and clearly. Working in groups is a part of these processes.

Science instruction should be based on an integrated instructional sequence. Research on student learning and effective instruction supports the use of sequences of coherent lessons that center on the development of concepts and practices.

ISSUES AND QUESTIONS FOR DISCUSSION

1. In one sentence (or maybe two), how would you describe the essential qualities of effective science teaching?
2. How could you use technology to identify the initial conceptions (i.e., misconceptions) of a class of 25 students?
3. Some claim that science teaching is an art, not a science. Others claim teaching is a science. How would you respond to these claims?
4. Many—in fact, most—discussions of good, effective, and core practices for science teaching do not address personal perspectives (e.g., adequate interpersonal relations with students, enthusiasm for teaching). How would you explain this observation? What could be done to remedy the situation?
5. Does science education have a theory of instruction? How would you justify your answer?

6. In his essay, Rodger Bybee describes the widely used BCS 5E Instructional Model. Research on the model is presented in his 2015 book on that model. Does the BSCS 5E model represent a practical application of a theory of instruction? If yes, how so? If no, why not?
7. If you were to design a research study on classroom practices for science teaching, what question would be central to your investigation? What methodology would you use?

REFERENCES

Atkin, J., and R. Karplus. 1962. Discovery of invention? *The Science Teacher* 29 (5): 45–51.

Berliner, D. 1984. The half-full glass: A review of research on teaching. In *Using What We Know About Teaching*, ed. Philip Hosford, 51–84 Alexandria, VA: Association for Supervision and Curriculum Development.

Brandwein, P. 1979. A general theory of instruction: With references to science. *Science Education* 63 (3): 285–297.

Bransford, J., A. Brown, and R. Cocking, eds. 1999. *How people learn: Brain, mind, experience, and school.* Washington, DC: National Academies Press.

Bruner, J. S. 1960. *The process of education.* Cambridge, MA: Harvard University Press.

Bruner, J. S. 1968. *Toward a theory of instruction.* New York: W. W. Norton & Company.

Bruner, J. S. 1969. *On knowing: Essays for the left hand.* Cambridge, MA: Harvard University Press.

Bybee, R. 1973. The teacher I like best: Perceptions of advantaged, average and disadvantaged science students. *School Science and Mathematics* 73 (5): 384–390.

Bybee, R. 1975. The ideal elementary science teacher: Perceptions of children, pre-service and in-service elementary science teachers. *School Science and Mathematics* 75 (3): 229–235.

Bybee, R. 1978. Science educators' perceptions of the ideal science teacher. *School Science and Mathematics* 78 (1): 13–22.

Bybee, R. W. 2013. *Translating the* NGSS *for classroom instruction.* Arlington, VA: NSTA Press.

Bybee, R. 2014. The BSCS 5E Instructional Model: Personal reflections and contemporary implications. *Science and Children* 51 (8): 10–13.

Bybee, R. 2015. *The BSCS 5E Instructional Model: Creating teachable moments.* Arlington, VA: NSTA Press.

Bybee, R., J. Taylor, A. Gardner, P. Van Scotter, J. Carlson, A. Westbrook, and N. Landes. 2006. *BSCS 5E Instructional Model: Origins and effectiveness.* A report prepared for the Office of Science Education, National Institutes of Health. Colorado Springs, CO: Biological Sciences Curriculum Study.

Bybee, R., and D. Welch. 1972. The third force: Humanistic psychology and science education. *The Science Teacher* 39 (8): 18–22.

Combs, A. 1965. *The professional education of teachers: A perceptual view of teacher education.* Boston: Allyn and Bacon.

Combs, A., D. Avila, and W. Parkey. 1978. *Helping relationship: Basic concepts for the helping profession.* Boston: Allyn and Bacon.

Eisenkraft, A. 2003. Expanding the 5E model. *The Science Teacher* 70 (6): 56–59.

Johnson, D. W., and R. T. Johnson. 1987. *Learning together and alone.* Englewood Cliffs, NJ: Prentice Hall.

Kloser, M. 2014. Identifying a core set of science teaching practices: A Delphi expert panel approach. *Journal of Research in Science Teaching* 51 (9): 1185–1217.

Lawson, A., M. Abraham, and J. Renner. 1989. *A theory of instruction: Using the learning cycle to teach science concepts and thinking skills.* Manhattan, KS: National Association for Research in Science Teaching.

Marek, E., and A. Cavallo. 1997. *The learning cycle: Elementary school science and beyond.* Portsmouth, NH: Heinemann.

National Research Council (NRC). 1999. *How people learn: Bridging research and practice.* Washington, DC: National Academies Press.

National Research Council (NRC). 2012. *A framework for K–12 science education: Practices, crosscutting concepts, and core ideas.* Washington, DC: National Academies Press.

NGSS Lead States. 2013. *Next Generation Science Standards: For states, by states.* Washington, DC: National Academies Press. *www.nextgenscience.org/next-generation-science-standards.*

Porter, A., and J. Brophy. 1988. Synthesis of research on good teaching: Insights from the work of the Institute for Research on Teaching. *Educational Leadership* 45 (8): 74–85.

Renner, J. W., M. R. Abraham, and H. H. Birnie. 1988. The necessity of each phase of the learning cycle in teaching high school physics. *Journal of Research in Science Teaching* 25 (1): 39–58.

Rowe, M. B. 1974. Wait time and rewards as instructional variables: Their influence on language, logic, and fate control. Part one, wait time. *Journal of Research in Science Teaching* 11: 81–94.

Schwab, J., and P. Brandwein. 1966. *The teaching of science.* Cambridge, MA: Harvard University Press.

Shulman, L. S. 1986. Those who understand knowledge growth in teaching. *Educational Researcher* 15 (2): 4–14.

Taylor, J., P. Van Scotter, and D. Coulson. 2007. Bridging research on learning and student achievement: The role of instructional materials. *Science Educator* 16 (2): 44–50.

Tobin, K, D. Tippins, and A. Gallard, 1994. Research on instructional strategies for teaching science. In *Handbook of research on science teaching and learning*, ed. D. Gabel, 45–93. New York: Macmillan.

Treagust, D. 2007. General instructional methods and strategies. In *Handbook for research and science education, Volume I*, ed. S. Abell and N. Lederman, 373–391. Mahwah, NJ: Laurence Erlbaum Associates.

Treagust, D., and C.-Y. Tsui. 2014. General instructional methods and strategies. In *Handbook of research on science education, Volume II*, ed. N. Lederman and S. Abell, 303–320. New York: Routledge.

Wiggins, G., and J. McTighe. 2005. *Understanding by design.* Alexandria, VA: Association for Supervision and Curriculum Development.

Wilson, C., J. Taylor, S. Kowalski, and J. Carlson. 2010. The relative effects and equity of inquiry-based and commonplace science teaching on students' knowledge, reasoning, and argumentation. *Journal of Research in Science Teaching* 47 (3): 276–301.

Windschitl, M., J. Thompson, M. Braaten, and D. Stroupe. 2012. Proposing a core set of instructional practices and tools for teachers of science. *Science Education* 96 (5): 878–903.

SECTION VIII

PROFESSIONAL DEVELOPMENT FOR TEACHERS OF SCIENCE

SECTION VIII

When states, school districts, and schools establish new standards for science education, it is essential to provide teachers with the knowledge, skills, and materials to implement those standards. Ultimately, the influence of standards rests with teachers and teaching practices in classrooms.

In some cases, teachers may express resistance or concerns. Positively, teachers' responses to the challenges of change may be expressions of the need for professional development. Yes, there is need for instructional materials, assessments, facilities, and equipment, but the classroom teacher is the essential link between the intended aims, policies, and programs and the students. Classroom teachers present what science is, how it is practiced, and why science is important. So, when the science education community develops new state or national standards, the need for teachers to understand and implement changes becomes significant.

We do not hesitate to identify professional development for teachers of science as a challenge for the science education community. In fact, the special issue of *Science* called "Grand Challenges in Science Education" included a review of professional development for science teachers (see Figure VIII.1; Wilson 2013). Furthermore, the introduction of new standards for science education indicates major changes in key components such as curriculum, instruction, and assessments. The implication here is that professional development should be an organized set of learning experiences—a program. Professional development for teachers of science should not be an event, such as a single introduction to standards. To be clear, there are some educational changes for which a workshop may be appropriate, as in the introduction of a new skill such as "wait time" or a new technology. But for major changes, such as implementing instructional materials based on newly adopted state standards, a professional development program will be required.

Figure VIII.1. Grand Challenges in Professional Development

Identifying the underlying mechanisms that make some teacher professional development (PD) programs more effective than others. Rigorous research on effective PD for science teachers is gradually accumulating, but we need a stronger theoretical base that reflects the complex ecology in which teachers work and learn. We also need better measures and interventions that are more highly specified to meet particular teacher needs.

Identify the kind of PD that will best prepare teachers to meet the challenges of the *Next Generation Science Standards* (*NGSS*). The *NGSS* specify an entirely new way of teaching science in the United States. This new way of teaching will require a considerable investment of resources to develop appropriate instructional materials and the tools needed to support teachers and students in using those materials. We must realign the considerable resources spent on PD with the demands teachers will face in an *NGSS* classroom, and we cannot afford such broad experimentation without funding the research required to learn from it.

Harness new technologies and social media to make high-quality science PD available to all teachers. Online PD has the potential for providing "just in time assistance," and it is potentially more scalable than PD that relies on limited local resources. To date, there exists little research to help us understand the affordances and limits of these venues.

Source: Adapted from AAAS 2013.

Setting the stage for a discussion of professional development is relatively easy. A majority of states now have new standards. Those standards were either direct adoptions of the *NGSS* or adaptations based on *A Framework for K–12 Science Education* (*Framework*) and the *NGSS*. In states not adopting the *NGSS*, increasing numbers of school districts are adopting or adapting these standards, hence our contention that professional development is a challenge.

Teachers recognize the challenge they have and are responding. A 2015 *NSTA Reports* article summarized results of a survey that centered on teachers' summer activities. The results support teachers' recognition of professional development. Virtually all respondents indicated they would participate in professional development activities such as a workshop or seminar—40% indicated they would spend 26–50% of their summer on these activities (*NSTA Reports* 2015, p. 8). When asked what they were planning to do differently next year, 51% of respondents indicated they planned to implement a different teaching strategy. While this survey and the results did not center on implementing standards, the teachers clearly indicated an interest in professional development and improving their instruction. Now it is up to the science education community to respond with programs for professional development.

REFERENCES

American Association for the Advancement of Science (AAAS). 2013. Grand challenges in science education. Special issue, *Science* 340 (6).

National Science Teachers Association (NSTA). 2015. Using summer activities to enhance a new school year. *NSTA Reports* 27 (2): 8–9.

Wilson, 2013. Professional development for science teachers. *Science* 340 (6130): 310–313.

SUGGESTED READINGS

Hall, G., and S. Hord. 2014. *Implementing change: Patterns, principles, and potholes.* New York: Pearson.

Loucks-Horsley, S., K. Stiles, S. Mundry, N. Love, and P. Hewson. 2010. *Designing professional development for teachers of science and mathematics.* 3rd ed. Thousand Oaks, CA: Corwin.

Luft, J., and P. Hewson. 2014. Research on teacher professional development. In *Handbook of research on science education: Volume II*, ed. N. Lederman and S. Abell, 889–909. New York: Routledge.

Penuel, W., B. Fishman, R. Yamaguchi, and L. Gallagher. 2007. What makes professional development effective? Strategies that foster curriculum implementation. *American Educational Research Journal* 44 (4): 921–958.

Whitworth, B. A., and J. L. Chiu. 2015. Professional development and teacher change: The missing leadership link. *Journal of Science Teacher Education* 26: 121–137. A thorough review of the literature and support for school and district leadership.

Wilson, S., H. Schweingruber, and N. Nielsen, eds. 2015. *Science teachers' learning: Enhancing opportunities, creating supportive contexts.* Washington, DC: National Academies Press.

CHAPTER 15

PROFESSIONAL DEVELOPMENT AND SCIENCE EDUCATION

As in any profession, the background, initial training, and motivations of science teachers vary considerably. Think, for example, of the differences between elementary and secondary teachers relative to science content and teaching strategies. We mention this example to emphasize the fact that professional development presents science educators with a complex design problem. The connections between professional development and science education include the historical purposes, contemporary policies, program reforms, and development of new classroom practices.

RESEARCH PERSPECTIVES

In our view, one question is central to research on professional development of science teachers: What are the form and function of professional development that affects classroom practice? One study (Desimone et al. 2002) provides an initial answer. The study was longitudinal, so it included repeated observations of subjects across time with respect to the variables under study. This national study of professional development activities for science and mathematics instruction provided a few insights.

Professional development is more effective in changing teachers' classroom practices when it is organized using the *collective participation of teachers* (e.g., from the same school, department, or grade levels), *focused on active learning activities* (e.g., teachers learn about a curriculum and apply what they are learning), and *coherent* (e.g., aligns activities with teachers' professional knowledge, as well as with state standards and assessments). While the study could be criticized for weaknesses (e.g., data were based on self-reported surveys, learning outcomes were not reviewed, and it was not a random sample of teachers), the results are still worth consideration, as other studies build on this knowledge and propose a preliminary answer to the question posed in the first sentence of the section.

In a more recent summary of key findings related to professional development in science, William Penuel of the University of Colorado presented six research-based characteristics that should be considered by science education leaders. Here is the summary of findings:

- Professional development should be focused on disciplinary core ideas and practices as students encounter them in science classrooms.
- Professional development should be of extended duration.
- The process of designing and adapting curriculum materials—when supported by subject matter and curriculum experts—can be a powerful form of professional development.
- Professional learning communities (PLCs) can extend professional development under some conditions.
- Formative assessment can be an effective focus of professional development when it helps teachers elicit, interpret, and make use of information about student thinking.
- The effectiveness of professional development depends on vertical and horizontal coherence in systems—the degree of alignment among standards, curriculum, assessments, and PD, as well as support from leaders (Penuel 2016).

In the past few decades, science educators have established a body of evidence supporting answers to questions about effective professional development, especially as it pertains to changing classroom practices (see, e.g., Wilson, Schweingruber, and Nielson 2015; Luft and Hewson 2014).

In 2016, a research team from the Biological Sciences Curriculum Study (BSCS) reported the effects of a high school science professional development program on student achievement. The study was unique in its attempt to link professional development and student achievement. The quasi-experimental study tested the effects of a three-year professional development program and found an initial dip in student achievement. However, the dip was followed by steady improvement, with increasing effect sizes of the intervention in years 6, 7, and 8 (Bintz, Taylor, and Stuhlsatz 2016).

CONTEMPORARY PERSPECTIVES

In the following sections, we introduce general ideas that underscore the design of professional development programs. In Chapter 16, we provide a more specific discussion of professional development as it relates to contemporary standards, particularly the *Next Generation Science Standards* (*NGSS*).

Understanding the Purposes of Professional Development for Science Education

The literature on professional development in science education is expansive and deep. Providing a brief introduction presented us with a challenge, to say the least, so we appealed to *Designing*

Professional Development for Teachers of Science and Mathematics (Loucks-Horsley et al. 2010). Our discussion draws directly from this book, as it is the most comprehensive synthesis by leaders in the profession. Here, we acknowledge an extensive use of the ideas presented in this discussion and refer those interested in deeper understanding to the aforementioned book.

We begin by paraphrasing five guiding ideas that establish an understanding of professional development. These ideas are based on knowledge from research, theoretical discussions, and the wisdom of practicing professional developers.

1. **Professional development experiences must center on all students and their learning.** Science education reforms share a commitment to the achievement of all students. This is an explicit statement to recognize the different perspective students bring to science and the implication of those views for the science content, teaching strategies, and design of professional development programs.
2. **Professional development experiences should develop teacher pedagogical content knowledge.** As a complement to science concepts and practices of scientific inquiry, professional learning should include knowing how to teach specific concepts to students at varying developmental levels. The latter is pedagogical content knowledge, commonly referred to as PDK (Shulman 1987).
3. **Professional development experiences should mirror the principles recommended for the improvement of student learning.** If we claim that students should engage in active learning and three-dimensional learning in contemporary contexts, then learning opportunities for science teachers also should reflect this approach to teaching and learning.
4. **Professional development experiences should both honor the teachers' current knowledge and skills and represent new knowledge and skills from other sources of expertise.** At any stage in their careers, teachers have varying levels of expertise and abilities. Others individuals—such as scientists, education researchers, policy makers, and experienced program developers—also have made contributions for the continuing improvement of teaching and learning.
5. **Professional development experiences must align with system-based changes for science education.** Too often, professional development has been displaced by other critical components in the education system. Specifically, professional development should support changes in science education standards, curriculum, instruction, and assessment.

Designing Professional Development Programs for Science Education

Designing effective professional development programs requires consideration of numerous different components. In Figure 15.1 (p. 230), we present a framework adapted from *Designing Professional Development for Teachers of Science and Mathematics* (Loucks-Horsley et al. 2010). In addition, we recommend "Research on Teacher Professional Development Programs in Science" (Luft and Hewson 2014). The figure represents a number of different elements that should be considered in the design, development, and implementation of a new program or the analysis of an extant program. The following discussion examines Figure 15.1.

Figure 15.1. A Framework for Designing Professional Development

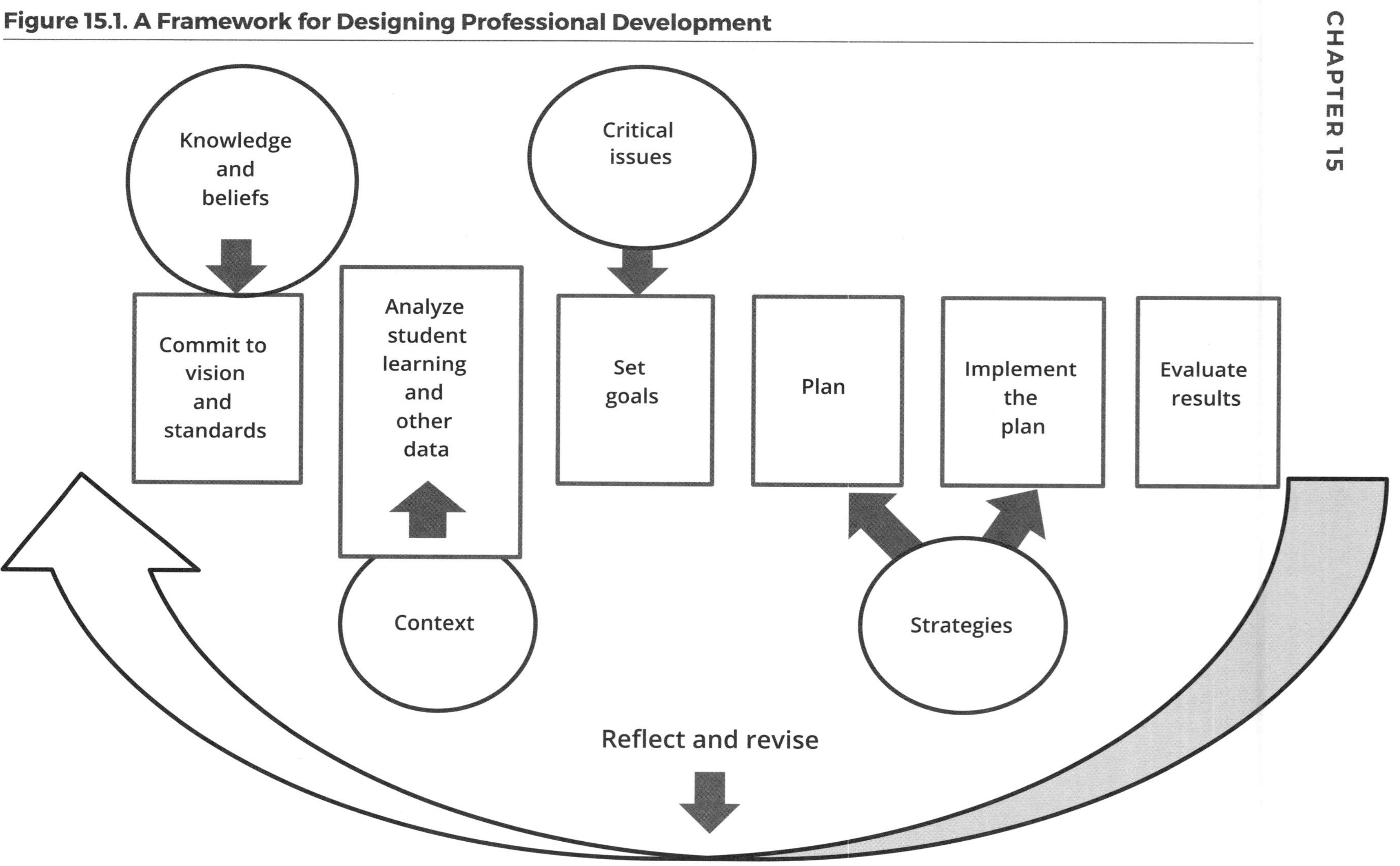

Source: Adapted from Loucks-Horsley et al. 2010.

The six boxes that form the central parts of the framework represent common processes of planning and implementing a professional development program. Anyone who has engaged in the process of designing a program understands the logical sequence as an ideal; in reality, the process is iterative. We note here a variation to the logical sequence. We propose using backward design (Wiggins and McTighe 2005) in this process.

1. **A vision and standards.** There is a need to commit to a vision and standards. A contemporary example of a vision with a major effect on science education is *A Framework for K–12 Science Education: Practices, Crosscutting Concepts, and Core Ideas* (NRC 2012). The commitment may have been made by the state or district to new standards such as the *Next Generation Science Standards* (*NGSS*; NGSS Lead States 2013) or adaptations of those standards.
2. **Analysis of data on student learning.** The initial process should include analysis of student learning and other important data. This analysis may include prior results on state science assessments, local results on literacy and mathematics, and information about student interests and needs.
3. **Goals.** The program goals should be clear, brief, and achievable. The goals should drive all discussions about other elements of the program.
4. **Evaluation design.** Here is the variation on using backward design we propose. At this point, there needs to be consideration of an evaluation of the program. The team should ask, "How will we know if the professional development program was effective? What evidence can we collect to demonstrate the effectiveness?" Clearly, the evaluation is connected to the goals, and it influences discussions about the selection of strategies and implementation of the program.
5. **Plan.** The plan includes careful consideration of the various components and how they are coordinated across the program.
6. **Implementation.** How will the program be implemented? What are the resources, schedule of events, and requirements for the program's execution?
7. **Evaluation and feedback.** Finally, there is an evaluation and the use of the evidence and feedback for the program's revision and improvement.

This framework for design is meant to be iterative, repeating itself as needed over the course of designing, developing, and implementing the program.

In addition to the basic steps in the process, the framework includes four inputs for the design process. Those designing professional development programs for science need to consider these four inputs:

1. **Knowledge and beliefs.** Knowledge should form a foundation under professional development. Research suggests that learners construct their own understandings and that certain teaching strategies—such as building on prior knowledge and active exploration of concepts—facilitate student learning. The same can be said for teachers.

Effective professional development involves active study, over time, of science content and pedagogy in ways that model effective learning and make direct connections with teachers' practice. Research on change indicates the importance of attending to individual teachers' needs and concerns.

The term *beliefs* refers to ideas, knowledge, and current conceptions (which may be misconceptions) about important matters, such as science, teaching, and learning in this case. It should be clear that beliefs should be considered because they are important and, in many cases, do not change easily. Important domains for consideration by those designing professional development include

- learners and learning;
- teachers and teaching;
- the concepts, practices, and nature of science and technology;
- professional development and adult learning; and
- the change process.

2. **Context.** Teachers teach and students learn in a variety of contexts. A thorough examination of factors will help designers identify criteria and constraints for a program. The needs and backgrounds of students and teachers; the resources, facilities, and instructional materials; and the requirements and expectations of school boards, parents, and administrators are all important considerations. Here is a list of contexts that may be reviewed:

    - Students' needs
    - Teachers' needs
    - Current practices—curriculum, instruction, and assessment
    - Policies—national, state, and local
    - Instructional resources
    - Organizational culture and learning community
    - Organizational structures
    - Families and community

3. **Critical issues.** In states, districts, and schools, there are any number of issues, some of which are no doubt critical. Here are common issues that designers face:

    - Building capacity and sustainability
    - Making time for professional development
    - Developing leadership at local to national levels
    - Ensuring equity
    - Building a professional learning community
    - Scaling up the program
    - Including and maintaining public understanding and support

4. **Strategies.** After making commitments to standards, analyzing contexts for teaching and learning, and setting goals, designers turn to planning and implementing the program. This is the appropriate time to consider strategies. Actually, the array of strategies is vast. *Designing Professional Development for Teachers of Science and Mathematics* (Loucks-Horsley et al. 2010) describes 4 categories and 16 strategies for professional development. We list those categories and strategies and refer you to the aforementioned book for in-depth discussions of them.

    - Immersion in content, standards, and research
        - Curriculum topic study
        - Immersion in inquiry and the science practices
        - Content courses
    - Examining teaching and learning
        - Examination of student work and thinking
        - Demonstration lessons
        - Lesson study
        - Action research case discussion
        - Coaching
        - Mentoring
    - Aligning and implementing curriculum
        - Instructional materials selection
        - Curriculum implementation
    - Professional development structures
        - Study groups
        - Workshops, institutes, and seminars
        - Professional networks
        - Online professional development

The challenge here is the selection and implementation of those strategies that are most appropriate for the unique situation.

CHAPTER 15

A PERSONAL PERSPECTIVE

PROFESSIONAL DEVELOPMENT: DOES IT REALLY MATTER?

Stephen Pruitt

Professional development, professional learning, professional learning communities—whatever you call it, we have all gone through it. When I reflect on the professional development I experienced during my classroom days in the 1990s, I wish I could say I looked forward to them, but that would not be honest. I remember having different types of experiences; most were not focused on science but on general education. Some were meant to be inspirational, but I was a bit of a cynic, so those did not go over well. I remember one particular time a person was brought in to talk about the "power of words." She called people to the front of the room and had them hold their arms out from their body. She would tell them how good they are and how much influence they had on the people around them while pushing down on their arms between their shoulder and elbow. The arms would not go down. She tried to show through this how empowering words strengthened a person. She would then have them assume the same pose and yell at them about what miserable failures they were while pushing down near their wrist. Of course the arms went down, and she talked about how she drew the strength right out of them. I pointed out to the people near me that this was a demonstration on the physics of levers, not of empowerment. Of course, the people around me got a kick out of that, but we all concluded it was another wasted day when we could have been in our classrooms working to prepare for students.

You may ask, "Why start a section on the importance of professional development with a terrible story like this?" Well, this story sets up some points I need to make. All too often in those days, professional development (PD) was set up so that it was ineffective at best, or a waste of time at worst. I think this happens when we go for general PD experiences and not PD specific to the needs of teachers. Not unlike students, our teachers must also have their needs met. The best PD I experienced when I was in the classroom was conducted by my science supervisor or other science teachers and was in the context of science classrooms. Even general PD opportunities that provided learning in a science context were more meaningful. This is why I always enjoyed our state science teachers' conference and the National Science Teachers Association (NSTA) meetings. In those experiences, I could usually find something for my classroom. We have come a long way since then. We have better research on what effective PD looks like. We have found that job-embedded PD that extends over time does change classroom practice for the better. The idea of professional learning communities has led to a better dialogue between teachers about effective instruction and a better collaboration with regard to planning instruction; we have known this for a while. In 1999, Deborah Ball and David Cohen stated that professional learning should be long-term, ongoing, active engagements that provide opportunities for teachers to see the connection between what they are learning and implementing and what the students are able to perform as a result (Ball and Cohen 1999). Furthermore, teachers need to have the opportunity to practice their new knowledge in classrooms and have the ability to follow up after they have tried their new practice (Peressini et al. 2004). Huffman, Thomas, and Lawrenz (2003) conducted a study in which five types of professional development strategies were used. These strategies included (1) teachers' immersion by working with a scientist or mathematician; (2) teachers using and refining the use of instructional materials; (3) teachers creating new materials; (4) teachers examining practice through discussing the real-world classroom; and (5) teachers' use of study groups, peer coaching, and mentoring. The researchers found that curriculum development and examining practice were most related to standards-based instructional strategies. These two strategies were the best predictors of the future use of standards-based curriculum in the classroom.

Finally, I will share that since leaving the classroom, I have witnessed firsthand the power of discussion in PD. As much as we talk about students not getting what they need from "sit and get," neither do teachers. In watching PD on standards and EQuIP (Educators Evaluating the Quality of Instructional Products), I have seen where teachers feel more comfortable sharing both their wisdom and their need for learning. The power of the discussion prompts them to want to repeat it in their schools. We all know that once a person can discuss an issue, they have owned the issue and can engage in it in a different way. The power of discussion in PD cannot be overstated. In a time of change, discussion must be front and center. From a leadership perspective, this will take some courage. You have to be willing to receive hard questions and allow the conversation to take on a natural flow without interference. If this is done, PD can be a powerful tool.

So why does that work? Well, I think it has to do with several factors, some of which I am now ashamed to admit I was not on board with when I was in the classroom. I believe there are five factors that affect the quality of PD, especially with regard to the *NGSS*. First, PD must be clearly aligned with standards. Second, PD must be focused on the needs of teachers. Third, PD must be firmly grounded in research. Fourth, PD must ultimately be connected to classroom and student context. Fifth, and perhaps the most important, teachers must realize they need quality PD and will not accept less. I discuss each of these in turn.

PD ALIGNMENT WITH THE STANDARDS

I think this is the first issue that must be considered when either selecting or designing PD. I wish I had more confidence that this happens, but I do not. All of the hours spent on PD will be useless if the PD is not rooted in the standards and does not emphasize a quality understanding of those standards. Some of my points overlap, and it is difficult to disentangle them. For the *NGSS*, it is almost impossible to consider PD without grounding it in the *NGSS* and how that affects classrooms, as the *NGSS* are all about what the student does. I have attended full PD sessions that claimed to be "aligned to," "correlated to," or "based on" the *NGSS* and they simply were not. They were, in my opinion, the same old activities with new color schemes and words, but they missed the intent of the standards. As I have said many times, three-dimensional learning is hard; to pretend it is not is dishonest.

The intent of the *NGSS* is to have students engage in the three dimensions of the *NGSS* to explain phenomena or design solutions, so I am not sure how doing a 10-minute activity without addressing how it builds or fits into an instructional plan is helpful. I also found that the activities often were very traditional in the sense that the participants were doing activities but not being required to bring the three dimensions together. So, when teachers and leadership begin to look for PD in the *NGSS*, there needs to be a close look at the alignment. I would actually suggest that a team of teachers and district leaders do an evaluation of the proposed PD to ensure the material meets the intent of the *NGSS*. This task must involve taking a deeper look rather than just seeing if the right words are used or the dimensions are listed. In fact, it will be critical to have teachers who have worked to make their own connections with the *NGSS* and the *Framework* make these determinations. For instance, if activities are done in isolation or if the activities focus solely on a single dimension, the activities probably do not support the *NGSS*. If the PD claims to be two dimensions and participants are only doing investigations, the actvitity probably cannot be considered to have an *NGSS* focus. If a crosscutting concept is mentioned but not used to make connections to other concepts, the activity probably does not fit with the *NGSS*. Finally, if the idea of students engaging with phenomena is not there, the activity definitely does not support the *NGSS*. So, in short, do not trust—verify.

PD MUST BE FOCUSED ON THE NEEDS OF TEACHERS

Teachers need PD that focuses on their needs. When beginning the implementation of new standards, there is a need to build common understanding of the new standards and their benefits and challenges. Teachers must own their PD in many ways, and one is to force themselves to identify their own needs. This is not as simple as saying, "Well, I am not as clear on ecosystems, so I want PD on ecosystems." I would say the challenge is far more involved, especially with the *NGSS*. I have seen teachers immediately dive into content, which is a natural reaction, as few of us learned the way the *NGSS* require us to teach. So, it will take a while. I was involved in every stage of development, yet I still learn and need to continue learning. The PD has to challenge the participants beyond the simple content. I say simple, but I really should not. To many, the content is about topics. The reality is that even the way we approach content in the *NGSS* is different. The *NGSS* look at the disciplinary core ideas and crosscutting concepts. For example, the topic of bonding is actually explained by the concepts of energy, fields, and forces, so the PD needs to be focused on how these concepts help participants know how to leverage them to give students that deeper understanding. However, as I said, we cannot stop there. The idea of creating an environment that facilitates students using three-dimensional learning is key.

As plans for PD around science are made, teachers must be engaged. They should be included in the planning but they should also be outspoken and take ownership in their own PD (see my fifth point). As district and school leaders plan, they should provide opportunities for teachers to give feedback through needs assessments and other evaluations. At the same time, teachers should be willing to give that feedback and not fear expressing their needs. Teachers cannot be apathetic or not give feedback, then complain when the PD does not address their needs. Unfortunately, there was a time in my early career when I was that way. I just went with the flow and loved to join in the critique after. It took me a few years to realize that the problem was not always the PD; sometimes it was actually me. How could leadership know what I needed if I did not engage or share my needs, or frankly even admit I had needs?

PD MUST BE FIRMLY GROUNDED IN RESEARCH

This may be obvious, but it often is overlooked. As I have already said, we know that job-embedded and focused PD is critical. So what additional factors has research found to be important for sustaining quality science instruction? These factors include (a) duration of professional development activity and continuance of follow-up support (Jeanpierre, Oberhauser, and Freeman 2005; Johnson 2007, 2009; Kimble, Yager, and Yager 2006; Penuel et al. 2009; Suppovitz and Turner 2000); (b) increasing the depth of science content and knowledge of practice (Jeanpierre et al. 2005; Kimble et al. 2006; Suppovitz and Turner 2000); (c) administrative support (Johnson 2009, 2010; Suppovitz and Turner 2000); (d) the level of involvement of the teachers in designing the curriculum (Huffman, Thomas, and Lawrenz 2003; Penuel et al. 2009); (e) attitudes of the teachers toward reform and congruence with the reform philosophy (Penuel et al. 2009; Suppovitz and Turner 2000); (f) situating professional development activities in the teacher's own classroom context (Johnson 2007); and (g) the establishment of a collaborative professional development community with more than one teacher participating in a building (Butler et al. 2004; Jeanpierre, Oberhauser, and Freeman 2005; Johnson 2007; Pruitt and Wallace 2012). In a large, nationwide study at sites across the United States, Suppovitz and Turner (2000) found that the amount of time teachers spent in professional development activities correlated significantly with their engagement of students in scientific interpretations and explanations. In general, teachers needed more than 80 hours of professional development–based practice to sustain long-term changes. However, the greatest factor in school-level science achievement scores continued to be the proportion of students in poverty, despite teacher improvement. Finally, there is tremendous evidence in documents from the National Research Council such as *Taking Science to School* (NRC 2007), *America's Lab Report* (NRC 2006), and *A Framework for K–12 Science Education* (*Framework*; NRC 2012) that bring to bear the need for three-dimensional learning. These documents, in particular the *Framework*, show that to really make a difference in scientific literacy, teacher PD must be engaged in three-dimensional learning as well.

Given that most of us were not taught in this manner, the learning that goes on in PD should mirror what we are asking our students to do. However, it cannot solely stay there. The point of the PD, I believe, was well stated by Huffman, Thomas, and Lawrenz (2003). They wrote, "The emphasis is on a continuous cycle of exploring new issues and problems, creating cognitive dissonance, engaging in collaborative discussions, constructing new understanding, and improving professional practice" (p. 379). Quality PD should leverage the research on adult learning and student learning, which actually overlap significantly. The study mentioned above does both. Simply doing activities will not change practice. Simply teaching content will not change practice. However, the research shows that bringing those areas together with a focus on professional practice can make a huge difference. Additionally, PD that does not challenge teachers' current understanding will not result in changed practice either.

So, when evaluating PD for its connection to the *NGSS*, consider having people on the review committee who are familiar with the research and can inform decisions that lead to *NGSS*-based PD. It is easy to think we know what we need; I would argue it's more important to base that knowledge on proven, researched techniques.

PD MUST BE CONNECTED TO CLASSROOM AND STUDENT CONTEXTS

Unfortunately, PD has gotten a pretty bad rap over the years. Equally unfortunate, it has deserved this bad rap. As I described earlier, PD can be not only completely content free but often is context free as well. I have had two experiences that gave me a more firsthand view of the importance of connections to classrooms. First, when I was working on my doctorate, I had a wonderful opportunity to teach a course called Classroom Management. The members of the class were preservice teachers who were in their final semester and working as student-teachers. We met once a week. The name of the course was a bit of a misnomer because we did not focus on classroom management per se; rather, we focused on effective teaching strategies and issues these students would face as full-time teachers. What struck me about this class was that many of the ideas we discussed they had learned in other courses, but they were simply not meaningful to them at the time. These students had no real context on which to base this knowledge. As a result, I would get comments such as, "Wow, this really works," or "I did not see any way for this to be effective, but now I cannot teach without it." I remember thinking back to my own undergraduate experience and listening to my professors and thinking there was no way that "theory stuff" would work. I could probably even name a few things that I intentionally refused to try because I just knew they would not work. Over time, however, I began to realize those theories were good, but I needed to see them in connection to my classroom. I could talk about inquiry with the best of them, but I could not always pull it off in a classroom beyond simply doing labs. I think if I had more PD while I was in a classroom that allowed me to practice applying these theories, report back, and maybe have professional discourse about my practice, I would have been a far better teacher much earlier in my career. But the PD I received, at least the general PD, had no connection to my classroom, so I was bored and found I was not a very willing participant. I am sure some of my PD providers would chuckle at some of my first attempts at providing PD because some of my participants probably paid me back for the grief that I gave or my lack of interest.

My second example of making connections to my classroom, or at least connections to my job, was critical was during the time I was working on my administrator certification. We had a mixed class of individuals who were either practicing administrators or planned to become administrators. I found differences in the conversations striking between those of us who were practicing some sort of administrative duties and those who were just learning about these responsibilities. I was already at the Georgia Department of Education and was very much acting in an administrative capacity. I found my conversations with practicing principals around issues such as budget, instructional evaluation, compliance, and special education needs to be far different and more in-depth than conversations with those who were still in classrooms. This was not their fault; that is not at all what I am saying. I am pointing out that they had no connection or context on which to hang the new knowledge. I think that is why education often feels so much like on-the-job training. We all went to school, so it could be we think we know more than we actually do. In my case, I was a third-generation teacher and had been exposed to some incredible instruction, or so I thought. As such, I thought I knew more than I did. What I did not realize was the incredible instruction I received worked for me, but it probably did not for everyone else. For this, I needed more training, but I could not see that until I had my own classroom to connect with my learning needs. Just as students need to build on prior learning, I needed that connection to my classroom to see how it would work in my context. I needed to see the context for my students.

Even though the study was on the *National Science Education Standards*, I believe Steve Olson and Susan Loucks-Horsley got it right in *Inquiry and the National Science Education Standards* (NRC 2000). They stated,

> *To teach their students science through inquiry, teachers need to understand the important content ideas in science—as outlined, for example, in the Standards. They need to know how the facts, principles, laws, and formulas that they have learned in their own science courses are subsumed by and linked to those important ideas. They also need to know the evidence for the content they teach—how we know what we know. In addition, they need to learn the "process" of science: what scientific inquiry is and how to do it. (NRC 2000, p. 92)*

I think there are two ideas I would add that we have learned since this study. First, we know that crosscutting concepts are the key feature that allows for knowledge to be transferred. Second, we understand how critical it is to have three-dimensional learning.

For true scientific understanding, students need to be involved and discuss observations; teachers need to be engaged in the same type of activity to best serve their students. In short, PD needs to be constructed to allow teachers to experience the student context. We all know the old adage that we teach as we were taught. Why is that? I believe it is that way because we also learned it better that way. Many of us, I think especially in the math and sciences, forget that not all students have had the preparation to learn as we did. I also think that even the way we learned might have been fine for the facts and figures we learned, but not the innovative thinking skills we should have learned. With *NGSS*, this is particularly true. Three-dimensional learning is not easy. It is not easy for so many reasons, but perhaps the first is the fact that most of us did not learn that way. At least, we may not have realized it. Crosscutting concepts in particular tend to be the dimension we learn almost subliminally. PD needs to reflect the obvious and overt nature of the crosscutting concepts and how their use allows students to create context so they can connect concepts and retain them throughout their lifetime. Scientific and engineering practices should be used to deepen understanding of scientific concepts. As such, the PD cannot be a simple activity that overlooks the connections among the three dimensions of the *NGSS*. If the connections are made, it is my contention that student context will be addressed. The teacher must experience the use of explanations, models, and arguments just as the students need to if they are to address the student-learning context. Even the disciplinary core ideas are tough to understand in student context. I think this is why we have resorted to memorizing facts so often. No one loves bonding and gas laws more than me, but I am not sure that our PD around those laws gets much past the words and formulas. The *NGSS* require students to be able to explain phenomena like these based on forces, interaction, and energy. Oddly enough, that means they must use the other two dimensions to do this. And, in doing so, students are provided a context to make a connection between two topics that may seem completely separate.

Finally, PD must also facilitate the teacher putting themselves into the shoes of students who did not learn as they did or who have not had the same opportunities. Only through putting teachers into this situation can we hope to engage all students in all standards. I will also mention here my earlier assertion that discussion between teachers must be a centerpiece. I have seen some great PD, and what made it great was I when I heard "Aha!" after teacher discussions. I have seen the EQuIP training up close. What made the training work was not that we told people this is good or bad in terms of instructional materials, but that the participants had an opportunity to discuss and come to grips with their own misunderstandings. A lot of times, I would hear, "Well, what I would do is …" or "Of course a teacher would do…" It was not until the discussion that people began to see how quality instructional materials actually exemplify three-dimensional learning. I learned through this experience that quality PD on standards really cannot be about the standards; it must be about the discussions around things affected by standards for teachers to really understand how to implement them. I know some people would say, "If the standards were written better, they would not need all this PD." I would say if they were written to the degree that no discussion is needed, they are not standards—they are curriculum, they are scripted, and they take away from the teachers' natural ability to be creative.

QUALITY PD MUST BE DEMANDED BY TEACHERS

I will end my discussion of PD with this observation: PD will not change if our teachers do not demand it. I know it is tough to implement new standards, especially those that actually challenge our own teaching. But we must also embrace that to really make a difference in our students' lives, we need to change as our students have. No one, especially a teacher, wants to admit they are anything less than spectacular. To be clear, I believe our teachers are spectacular, which does not mean we should be intimidated about discussing and implementing change. We have to know that becoming an expert in anything takes time and repetition. As such, we need our PD opportunities to force this. Districts, schools, and teachers are inundated with promises of great PD fully aligned with the *NGSS*. I would guess there are many out there who have not actually changed more than the colors used in their materials. Teachers must demand the PD to be aligned and well researched. Using every tool in their toolbox, they should demand that any provider show more than a crosswalk or checkbox of their PD. Poor PD can only change if teachers demand it. I have been proud to see the amount of support and commitment teachers have had to the *NGSS*. I know that if given the opportunities, teachers can use good judgment on PD offerings. I encourage you all to be skeptics, demand evidence of *NGSS* alignment and PD effectiveness, and engage deeply in those quality opportunities.

REFERENCES

Ball, D., and D. Cohen. 1999. Developing practice, developing practitioners: Toward a practice-based theory of professional development. In *Teaching as the learning profession: Handbook of policy and practice*, ed. G. Sykes and L. Darling Hammond, 3–32. San Francisco: Jossey-Bass.

Bintz, J., J. Taylor, and M. Stuhlsatz. 2016. The effect of a high school science professional development program on student achievement. Paper presented at the annual meeting of the American Educational Research Association, Washington, DC.

Butler, D. L., H. Novak Lauscher, S. Jarvis-Selinger, and B. Beckingham. 2004 Collaboration and self regulation in teachers' professional development. *Teaching and Teacher* Education 20: 435–455.

Desimone, L. M., A. C. Porter, M. S. Garet, K. S. Yoon, and B. F. Birman. 2002. Effects of professional development on teachers' instruction: Results from a three-year longitudinal study. *Educational Evaluation and Policy Analysis* 24 (2): 81–112.

Huffman, D., K. Thomas, and F. Lawrenz. 2003. Relationship between professional development, teachers' instructional practices, and the achievement of students in science and mathematics. *School Science and Mathematics* 103 (8): 378–387.

Jeanpierre, B., K. Oberhauser, and C. Freeman. 2005. Characteristics of professional development that effect change in secondary science teachers' classroom practices. *Journal of Research in Science Teaching* 42 (6): 668–690.

Johnson, C. C. 2007. Whole-school collaborative sustained professional development and science teacher change: Signs of progress. *Journal of Science Teacher Education* 18: 629–661.

Johnson, C. C. 2009. An examination of effective practice: Moving toward elimination of achievement gaps in science. *Journal of Science Teacher Education* 20: 287–306.

Kimble, L. L., R. E. Yager, and S. O. Yager. 2006. Success of a professional-development model in assisting teachers to change their teaching to match the more emphasis conditions urged in the national science education standards. *Journal of Science Teacher Education* 17: 309–322.

Loucks-Horsley, S., K. Stiles, S. Mundry, N. Love, and P. Hewson. 2010. *Designing professional development for teachers of science and mathematics.* 3rd ed. Thousand Oaks, CA: Corwin.

Luft, J., and P. Hewson. 2014. Research on teacher professional development programs in science. In *Handbook of research on science education: Volume II*, ed. N. Lederman and S. Abell, 889–909. New York: Routledge.

National Research Council (NRC). 2000. *Inquiry and the* National Science Education Standards. Washington, DC: National Academies Press.

National Research Council (NRC). 2006. *America's lab report: Teaching high school science.* Washington, DC: National Academies Press.

National Research Council (NRC). 2007. *Taking science to school: Learning and teaching science in grades K–8*, ed. R. Duschl, H. Schweingruber, and A. Shouse. Washington, DC: National Academies Press.

National Research Council (NRC). 2012. *A framework for K–12 science education: Practices, crosscutting concepts, and core ideas.* Washington, DC: National Academies Press.

NGSS Lead States. 2013. *Next Generation Science Standards: For states, by states.* Washington, DC: National Academies Press. *www.nextgenscience.org/next-generation-science-standards.*

Penuel, W. 2016. Some key findings related to effective professional development in science: 1996–2014. Unpublished paper prepared for the Committee on Professional Learning, Council of State Science Supervisors.

Penuel, W. R., H. McWilliams, C. McAuliffe, A. E. Benbow, C. Mably, and M. M. Hayden. 2009. Teaching for understanding in Earth science: Comparing impacts on planning and instruction in three professional development designs for middle school science teachers. *Journal of Science Teacher Education* 20: 415–436.

Peressini, D., H. Borko, L. Romangnano, E. Knuth, and C. Willis. 2004. A conceptual framework for learning to teach secondary mathematics: A situative perspective. *Educational Studies in Mathematics* 56 (1): 67–96.

Pruitt, S., and C. Wallace. 2012. The effect of a state department of education teacher mentor initiative on science achievement. *Journal of Science Teacher Education* 23 (4): 367–385.

Shulman, L. 1987. Knowledge and teaching: Foundations of the new reform. *Harvard Educational Review* 57: 1–22.

Suppovitz, J. A., and M. Turner. 2000. The effects of professional development on science teaching practices and classroom culture. *Journal of Research in Science Teaching* 37 (9): 963–980.

Wiggins, G., and J. McTighe. 2005. *Understanding by design.* 2nd ed. Alexandria, VA: Association for Supervision and Curriculum Development (ASCD).

Wilson, S., H. Schweingruber, and N. Nielson, eds. 2015. *Science teachers' learning: Enhancing opportunities, creating supportive contexts.* Washington, DC: National Academies Press.

CHAPTER 16

CONTEMPORARY STANDARDS AND PROFESSIONAL DEVELOPMENT

One challenge that has been a theme of our discussions is implementing the *Next Generation Science Standards* (*NGSS*; NGSS Lead States 2013) for school programs and associated professional development. We understand that not all states have adopted these standards and that some states never will. But the *NGSS* and *A Framework for K–12 Science Education* (*Framework*; NRC 2012)—on which the *NGSS* are based—have had, and will continue to have, a significant effect on district and state standards. The innovations described in the *Framework* and *NGSS* imply fundamental changes in the way science is taught, and changing instructional practice presents a significant implementation challenge. The subsequent implications for professional development are, to say the least, considerable.

CHALLENGES

What Is New and Different in Contemporary Standards?

In a prior chapter, we described the *NGSS* innovations. We briefly restate them here, with an emphasis on professional development.

1. **Teaching that includes three dimensions—science and engineering practices, crosscutting concepts, and disciplinary core ideas.** This innovation has direct implications for science teachers and classroom instruction. How do science teachers provide opportunities for students to learn all three dimensions? No easy task, this. The implications for professional development are significant. Changing instruction requires intensive and coherent learning experiences for teachers. Those experiences should include curriculum materials, assessments, and opportunities to interact with other science teachers. These observations are not new (see, e.g., Smith and O'Day 1991).

2. **Teaching students with engaging natural phenomena and design problems.** This innovation is a variation on "hands-on" investigations and activities. However, it implies more than a lesson and is complicated by the fact that some phenomena are not immediately available for hands-on activities because the objects of study may be too small, too large, too distant, too fast, and so on.
3. **Teaching science practices and crosscutting concepts in ways that include engineering and the nature of science.** Both the practices and crosscutting concepts present opportunities to include the nature of science and engineering design.
4. **Teaching should build a coherent learning progression.** Teaching a unit or yearlong program should involve students in increasingly more abstract concepts and complex practices. There is a larger challenge for state and district leaders, as there are implications for K–12 school science programs.
5. **Teaching should make connections to math and literacy standards.** While the new science standards are not directly related to the *Common Core State Standards*, they do provide opportunities to enhance math and literacy skills.

Several science educators and organizations have responded to the need for standards-based professional development. Todd Hutner and Victor Sampson (2015) have suggested five indicators of good standards-based science teaching. These, we think, have direct implications for professional development. We highlight the teacher's role.

- Teachers create a need to learn by giving students a challenge, question, problem, or engaging phenomena.
- Teachers make student thinking visible by having them make a prediction, then present their ideas or explanations orally.
- Teachers have students engage in an activity before delving into content by conducting an investigation, collecting data, or making a prediction.
- Teachers have students participate in the practices of science by planning investigations, analyzing data, and justifying claims using evidence and reasoning.
- Teachers have students negotiate meaning through questions, discourse, and scaffolding.

In a paper prepared for an Invitational Research Symposium on Science Assessment, Brian Reiser (2013) answered this question: "What professional development strategies are needed for successful implementation of the *Next Generation Science Standards*?" Again, we direct your attention to the demands for teacher learning.

- The focus on disciplinary core ideas implies more emphasis on explaining phenomena and less emphasis on facts.
- The focus on science and engineering practices implies more emphasis on science learning that involves engaging students in practices to build and use knowledge and less emphasis on inquiry as a separate activity.

- The focus on learning progressions involves more emphasis on building coherent story lines across time and less emphasis on too many topics in disconnected lessons.

Reiser concludes with three research-based recommendations for professional development. The discussion combines the demands for teacher learning, elaborated above, with research-based characteristics that hold great promise for education leaders.

1. **Structure teacher sense-making around rich images of classroom enactment.** Teachers need to analyze examples of teaching with the goal of determining what can be applied to their own classrooms. They should examine rich cases that reflect the complexity of teacher-student interactions and contexts that reveal the justification for certain strategies. Reiser recommends analysis of video cases as a fruitful strategy.
2. **Structure teachers' work to be collaborative efforts to apply the *NGSS* to their own classrooms.** This recommendation relates to designing professional development for active learning and the importance of teachers working together to understand, apply, and reflect on the various aspects of standards-based recommendations.
3. **Capitalize on cyber-enabled environments.** The expanse of professional development efforts in both space and time raises the importance of using technology. Use of digital video of classrooms, video presentations, and online PD (Bates, Phalen, and Moran 2016) provides potential responses to the problems of scale.

Suzanne Wilson, Heidi Schweingruber, and Natalie Nielson completed a report called *Science Teachers' Learning: Enhancing Opportunities, Creating Supportive Contexts* for the National Research Council (2015). The report suggests that few K–12 science teachers have the expertise needed to teach the science required in the *NGSS*. This finding is especially true for elementary teachers and those individuals working in schools with students from low-income families.

On a more positive note, the report declares that the *NGSS* is the type of response called for in *A Nation at Risk* and that the "*NGSS* represents a fundamental change in the way science is taught and, if implemented well, will ensure that all students gain mastery over core concepts of science that are foundational to improving their scientific capacity" (Wilson, Schweingruber, and Nielson 2015, p. 1). Following this uplifting vision, the report concludes that teachers will need new scientific knowledge and skills, new instructional approaches, and materials to implement those strategies in their classrooms. Then comes a point basic to this chapter: "To enable teachers to acquire this kind of learning will in turn require profound changes to current systems for supporting teachers' learning across their careers, including induction and professional development" (Wilson, Schweingruber, and Nielson 2015, p. 2).

The report's recommendations highlighted districts and schools as the means to improve learning opportunities for teachers. The recommendations aligned with the report's title—*Science Teachers' Learning: Enhancing Opportunities, Creating Supportive Contexts.* Here are the recommendations:

1. **Take stock of the current status of learning opportunities for science teachers.** Administrators should identify professional development offerings. Particular attention should

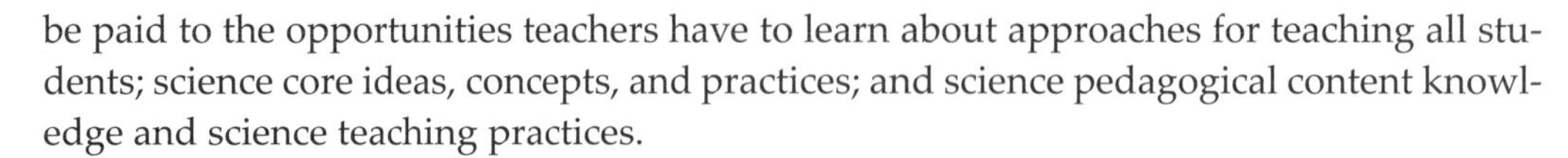

be paid to the opportunities teachers have to learn about approaches for teaching all students; science core ideas, concepts, and practices; and science pedagogical content knowledge and science teaching practices.

2. **Design a portfolio of coherent learning experiences for science teachers that attend to teachers' individual and context-specific needs in partnership with professional networks, institutions, and the broader scientific community as appropriate.** This recommendation centers on responses to four questions: In the context of the schools' or districts' goals, what learning opportunities will teachers need? What expertise will be needed to support the learning opportunities? Where is that expertise? What arrangements are required to enable the work?
3. **Consider both specialized professional learning programs outside of school and opportunities for science teachers' learning embedded in the work day.** Accommodating this recommendation may require some restructuring of teachers' work and providing resources such as time, facilities, materials, incentives, and, very important, budget.
4. **Design and select learning opportunities for science teachers that are informed by the best available research.** Learning opportunities for a school or district should include specific science concepts and practices, a specific focus on student needs, links to classroom instruction and an analysis of instruction, opportunities to practice new teaching strategies and interact with peers, and collecting and analyzing data on student learning.
5. **Develop internal capacity in science while seeking external partners with science expertise.** The need to build capacity for science teachers' learning seems clear. Consideration might be given to having science specialists in elementary schools, providing release time for high school department heads, and improving the relevant expertise of those providing professional development.
6. **Create, evaluate, and revise policies and practices that encourage teachers to engage in professional learning related to science.** This recommendation implies opportunities for professional development during the salaried work week and year. It also, for example, suggests parity among science, math, and language arts for elementary teachers.
7. **The potential of new formats and media should be explored to support science teachers' learning.** The use of technology and online resources should be reviewed and used as appropriate. Such resources may be useful for long-term support, opportunities in rural areas, and flexible scheduling, among other challenges and opportunities.

We will conclude this discussion on professional development aligned with contemporary standards by directing attention to science teachers and recognizing the fact that they often select professional learning experiences. Julie Luft, Eunjin Bang, and Peter Hewson (2016) present a research-based set of questions for science teachers in the process of selecting a professional development program (PDP). We present their list in Figure 16.1.

Figure 16.1. Questions for Science Teachers Reviewing Professional Development Programs

DESIGN

- Is the professional development program (PDP) professionally appropriate (e.g., beginning teacher or master teacher)?
- Are there opportunities to work with other science teachers?
- Does the PDP align with the school, district, and state standards and curriculum?
- Should the PDP include opportunities to examine student work? If so, does this PDP provide opportunities to do this?
- Are the duration and delivery appropriate for the intended outcome of the PDP?

KNOWLEDGE DEVELOPMENT

- Will the knowledge that is important in the PDP be explicitly taught after an active learning experience?
- Do the program instructors know how to support science teacher learning?
- Will the PDP involve practice with students or in a classroom?

INSTRUCTION

- Do the organizers of the PDP recognize that changing a teacher's practice will take time and that teachers implement the advocated instructional practices differently?
- Does the instruction advocated in the PDP align with the local curriculum and school culture?
- What type of support is offered to implement the advocated instruction with students?

Source: Adapted from Luft, Bang, and Hewson 2016.

CHAPTER 16

REALITIES OF REFORM AND PROFESSIONAL DEVELOPMENT FOR SCIENCE TEACHERS

Rodger Bybee

Standards-based education reform has an ultimate reality: Classroom teachers have to change their instructional practices. By September 2016, 17 states and the District of Columbia had adopted the *Next Generation Science Standards* (*NGSS*; NGSS Lead States 2013). In addition, many other states are developing standards based on *A Framework for K–12 Science Education* (*Framework*; NRC 2012) or adapting the *NGSS*. There are 13 other states that either are developing or will adopt the *NGSS*. These numbers certainly pass a tipping point that indicates a reality for contemporary standards-based reform in science education. Understanding this reality opens the door to recognizing and acting on the essential role of profession development for science teachers.

Charles Thompson and John Zeuli succinctly summarize my point:

> *It is now widely accepted that, in order to realize recently proposed reforms in what is taught and how it is taught (as described in national standards document) . . . , teachers will have to unlearn much of what they believe, know, and know how to do (Ball 1988) while also forming new beliefs, developing new knowledge, and mastering new skills. The proposed reforms constitute, if you will, a new curriculum for teacher learning. If they do not specify precisely what teachers should know and be able to do, they do outline it rather clearly and exemplify aspects of it with a nearly literary vividness. (Thompson and Zeuli 1999, p. 1)*

Standards cannot directly change behavior or beliefs, but they can point the way by defining desirable goals, stimulating movement toward the goals, and reducing conflicts among the policies faced by educators at each level. They can begin to create images of what different components of the education system would look like were they to be aligned with the desired outcomes for students. Some have argued that policies (including standards) might be viewed as an effort to teach those who must carry them out, rather than solely as an effort to induce or constrain them to behave differently (Cohen and Barnes 1993); however, teaching both students and adults requires far more than attending to policy statements. There must be substantial effort to translate policies into school programs and classroom practices (Bybee 2013; Bybee et al. 2003). Without professional development, this cannot be accomplished. I believe the position that the ultimate reality of reform is defensible and supported by research (see, e.g., Wilson, Schweingruber, and Nielson 2015). However, I advise those providing professional development to consider several proximate realities of reform.

THE REALITIES OF TEACHERS AND TEACHING

One could write a book on this topic. I hope several examples may make my point. In any school, one may observe a range of teachers, from those experiencing their first year to those considering retirement. The instructional skills and abilities likely will vary from those demonstrating a variety of strategies with enthusiasm and deep understanding of the science content and pedagogy to those with limited methods, interest, and science knowledge. Common knowledge and experience confirm the difference between elementary and secondary science teachers and the additional variations of middle school teachers. Teachers of different grade levels differ in their levels of scientific knowledge and application of instructional strategies. The lesson here has implications for professional development. In general, professional development for elementary teachers should include experiences that enhance their understanding of science content and practices, while PD for secondary teachers should include a greater use of instructional strategies.

THE REALITIES OF STUDENTS AND LEARNING

Yes, one could also write a book on this topic. Here is my point: Each teacher confronts his or her own students and the variations here include those ready to learn, those with learning difficulties, those learning a new language, those unmotivated, and . . . the list could easily be extended. Teachers clearly recognize their singular situations when they report the various difficulties of teaching English language learners, those with limited math skills, or those perceived to be destined for an Ivy League university (or small, selective liberal arts colleges).

THE REALITIES OF CONTEMPORARY STATE STANDARDS

Implementing the *NGSS* or state standards adapted from the *Framework* requires educators to address five innovations. I suggest using the term *innovations* because, compared to the 1996 *National Science Education Standards*, there are new and different characteristics in the *NGSS*. To be blunt, educators are cautioned to not review the *NGSS* and say, "We are already doing that" because they probably aren't. We can capture the innovations in a few words: The *NGSS* integrate three *dimensions*; emphasize *experiences* with phenomena; present *extensions* of the nature of science and engineering from traditional content; describe learning *progressions* for kindergarten though grade 12; and make *connections* among science, math, and English language learning. Those innovations have been introduced in prior discussions.

The complexities of implementing a reform based on the *NGSS* make this a challenge. In fact, at least one critic thinks it is a reform that has gone past the line where educators must admit the expectations for teachers are unreasonable (Coppola 2016). I will return to this critical view.

THE REALITIES OF PROFESSIONAL DEVELOPMENT OPPORTUNITIES

This reality may be unusual but, considering the many providers of professional development, I don't think it is. When asked to provide professional development, providers often confront a situation with set parameters, some of which do not align with recommendations made by those quite knowledgeable about professional development. Here are some examples: Present a keynote on the *NGSS* but do not mention *NGSS*; conduct a workshop on one dimension of the *NGSS*, such as disciplinary core ideas; discuss the nature of science and why it is not more prominent in the *NGSS*; provide a workshop on the *NGSS* and how a district's current instructional materials can be adapted; or respond to questions about state assessments, accountability, and college and career readiness. There are other issues, to be sure. My point is that the PD provider does not always get to set the agenda. The reality is one that requires flexibility.

I have described different realities to make two points. First, the implied changes are not simple; indeed they are quite complex. Anybody who suggests otherwise either is not telling the truth or has a severely limited understanding of what the professional development of science teachers really requires. Second, professional development for science teachers must recognize the unique circumstances of both teachers and those providing professional development.

Developing and releasing the *NGSS* fulfills the first step of a much longer process of improving science teaching and student learning. Between the science standards and student learning stand teachers of science and the need for professional development. Here is a major paradox, as I see the situation. The science standards are designed for all of science education, yet the changes implied by the standards must occur with individual teachers at K–12 grade levels and recognize the many realities teachers face. Recommending personal professional development for solitary teachers would be unrealistic, given that there would be more than three million separate realities. Yet, presentations for groups tend to leave some with the feeling that their unique concerns have not been addressed.

So, you ask, what do I suggest? How do I think about and approach these issues associated with professional development in this era of standards-based reform? Here are my recommendations, which are divided into two sets: five on preparation and five on professional development activities.

1. **Stay focused on the instructional core.** I have said this in prior chapters, and it is worth repeating. The professional world of teachers centers on instruction and their work in classrooms. Yes, this is obvious, but it is not always where professional development begins.

As a reminder, the instructional core has three fundamental features: First, the core includes the content of school science programs. The content is expressed in state and local standards and assessments. Second, curriculum materials that complement the standards engage students and provide opportunities to learn. The third element includes the teachers' knowledge and skills.

As leaders plan professional development, there is a complex set of concepts to address. Clearly, everything cannot be addressed at the same time. So, what should you consider?

2. **Begin with an awareness of the standards, state or *NGSS*.** Science education standards and associated documents such as curriculum frameworks are often long and complex. Developing a deep and broad understanding takes time. This is why I use the term *awareness*. In the initial reviews, leaders should identify the key innovations that will require instructional changes. What is new and different in these standards? As one identifies the innovations, consider the implications for instructional materials and teachers' knowledge and skills. Using the *NGSS*, for example, one identifies the integration of science and engineering practices, crosscutting concepts, and disciplinary core ideas, referred to as three-dimensional teaching. Next, there is the role of phenomena as a central part of instruction. Third, the new standards incorporate the nature of science and engineering design as part of science and engineering practices and crosscutting concepts. Finally, the standards cover how to make connections to English language arts and math as part of science instruction.
3. **Identify the critical changes for assessments.** Here, I am referring to both formative and summative assessments in the classroom. In the *NGSS*, for example, the standards are stated as performance expectations that integrate the three dimensions mentioned in the prior recommendations.

As a leader, you also should be aware of requirements in national, state, and local assessments. For example, how will your district implement the new requirements of the Every Student Succeeds Act (ESSA) for English language arts, math, and science?

4. **Clarify the knowledge and skills required to implement the standards-based reform.** As you examine state standards and assessments, consider what teachers must know and be able to do when they translate the standards and establish new approaches to classroom instruction. This process becomes important when you introduce new science content and new instructional approaches. You also should identify the changes implied for instructional materials.
5. **Review curriculum materials in light of the standards.** The innovations identified in the new standards will require changes in curriculum materials. The changes may be minor, but the likelihood of changing instructional materials is high. Based on time, availability, budget, and expertise, one will have to decide on the review of new materials, adaptation of current materials, or development of materials.

The next five recommendations address professional development activities for science teachers. In 2015–2016, I chaired an advisory panel for the Gottesman Center for Science Teaching and Learning at the American Museum of Natural History. James Short was director of the center. The leadership team also included Jody Bintz of Biological Sciences Curriculum Study (BSCS) and Kathy DiRanna of the WestEd K–12 Alliance. The following discussion is my variation on the tools developed by this project.

6. **The initial translation of the *NGSS* for classroom instruction.** The first activity has participants plan an instructional sequence based on a performance expectation from the *NGSS*. Participants select a performance expectation at their grade level and from a discipline with which they are most comfortable. The activity immediately sets a context that responds to teachers' expressed concerns—namely, "Are there curriculum materials I can use to implement the standards?" As participants begin planning their instructional sequence, they refer to selections from the *Framework* and the *NGSS*. The plan for an instructional sequence may be rough and general, but the first tool sets a meaningful stage for developing the knowledge and abilities needed to implement new standards such as the *NGSS*. This activity challenges participants to integrate the three dimensions of practices, crosscutting concepts, and disciplinary core ideas.
7. **Use the performance expectations to design classroom assessments.** The second activity responds to teachers' second major concern: Will assessments change? Using performance expectations from the *NGSS*, participants plan assessments for classroom instruction, specifically the instructional sequence they designed. They read sections from *Developing Assessments for the* Next Generation Science Standards (NRC 2014). Here, participants are challenged to determine what counts as evidence for student learning and how to design methods and materials to collect evidence, in both formative and summative contexts, that students have indeed learned the science and engineering practices, crosscutting concepts, and disciplinary core ideas expressed in a performance expectation.
8. **Improving the instructional sequence.** Now, a personal favorite is introduced. The BSCS 5E Instructional Model (Bybee 2015) is central for this activity. Introducing the model provides a concrete and meaningful response for science teachers. The professional world of teachers centers on instruction, and introducing the BSCS 5E Model builds on the initial sequence of lessons. The activity further extends the initial experiences of planning an instructional sequence.
9. **Completing a unit of instruction.** This brings the prior experiences with the design of instruction together as participants use the original design from the first activity, the assessments from the second, and the instructional model to develop an instructional sequence for a unit. By placing emphasis on what teachers and students do in each phase of the instructional model, participants gain an understanding of what three-dimensional teaching is and how to implement it in classrooms.
10. **Creating a summative assessment.** In the final activity, participants return to the performance expectation used as the basis for their unit and create a performance assessment that would serve as a summative evaluation for this unit. I encourage the use of backward design (Wiggins and McTighe 2005) as one feature of the teachers' professional development. The assessment designed in step 7 can be used as part of an iterative process for developing this assessment.

In conclusion, the realities of reform assert a variety of constraints on the criteria for professional development. With that realization, here is what I tried to do in this discussion. First, I directed this discussion toward the instructional core and concerns of classroom teachers. Second, I tried to reduce the complexity of the *NGSS* (or new state standards) by respecting the innovations and addressing the issues teachers face. Finally, the discussion addressed a sequence individuals might use to prepare and conduct professional development for teachers of science.

OUR COMMON PERSPECTIVE AND LEADERSHIP OPPORTUNITIES

Professional development should be an organized set of experiences, not an event. Professional development is an organized, continuing process. It is not a simple workshop.

Professional development should align with contemporary standards and address the needs of science teachers—change in curriculum, instruction, and assessments. Address the concerns and needs of teachers and you will get the most benefit from the professional development.

Consider the use of technologies as part of continuing professional development. Technologies can provide low-cost, immediate access to needed materials and resources.

Base professional development on what we have learned from research, especially as it addresses classroom practices. Our theme is "the intentional core."

Use a model for professional development—for example, *Designing Professional Development for Teachers of Science and Mathematics* (Loucks-Horsley et al. 2010). You do not need to invent professional development programs. Use research-based approaches to maximize your time, effort, and themes.

ISSUES AND QUESTIONS FOR DISCUSSION

1. What has been your experience with professional development?
2. What is your response to the guiding ideas for professional development? Would you modify the ideas? Add another guiding idea?
3. What was your response to the framework for designing professional development?
4. Would you add anything to the inputs? If yes, what?
5. How would you approach PD in the context of the *NGSS*?

REFERENCES

Ball, D. 1988. Unlearning to teach mathematics. *For the Learning of Mathematics* 8 (1): 40–48.

Bates, M., L. Phalen, and C. Moran. 2016. Online professional development: A primer. *Phi Delta Kappan* 97 (5): 70–73.

Bybee, R. 2013. *Translating the* NGSS *for classroom instruction.* Arlington, VA: NSTA Press.

Bybee, R. 2015. *The BSCS 5E Instructional Model: Creating teachable moments.* Arlington, VA: NSTA Press.

Bybee, R., J. Short, N. Landes, and J. Powell. 2003. Professional development and science curriculum implementation: A perspective for leadership. In *Leadership and professional development in science education: New possibilities for enhancing teacher learning*, ed. J. Wallace and J. Loughran, 155–176. London: Routledge and Falmer.

Cohen, D., and C. Barnes. 1993. Pedagogy and policy. In *Teaching for understanding: Challenges for policy and practice*, ed. D. K. Cohen, M. W. McLaughlin, and J. E. Talbert, 207–239. San Francisco, CA: Jossey-Bass Publishers.

Coppola, B. 2016. Where is the line? In *Reconceptualizing STEM education: The central role of practices*, ed. R. Duschl and A. Bismack, 131–142. New York: Routledge.

Hutner, T.. and V. Sampson. 2015. New ways of teaching and observing science class. *Phi Delta Kappan* 96 (8): 52–56.

Loucks-Horsley, S., K. Stiles, S. Mundry, N. Love, and P. Hewson. 2010. *Designing professional development for teachers of science and mathematics.* 3rd ed. Thousand Oaks, CA: Corwin.

Luft, J., E. Bang, and P. Hewson. 2016. Help yourself, help your students. *The Science Teacher* 83 (1): 49–53.

National Research Council (NRC). 1996. *National Science Education Standards.* Washington, DC: National Academies Press.

National Research Council (NRC). 2012. *A framework for K–12 science education: Practices, crosscutting concepts, and core ideas.* Washington, DC: National Academies Press.

National Research Council (NRC). 2014. *Developing assessments for the* Next Generation Science Standards. Washington, DC: National Academies Press.

NGSS Lead States. 2013. *Next Generation Science Standards: For states, by states.* Washington, DC: National Academies Press. *www.nextgenscience.org/next-generation-science-standards.*

Reiser, B. 2013. What professional development strategies are needed for successful implementation of the *Next Generation Science Standards*? Paper prepared for The Center for K–12 Assessment & Performance Management at ETS and presented at the Invitational Research Symposium on Science Assessment, Washington, DC.

Smith, M., and J. O'Day. 1991. Systemic school reform. In *The politics of curriculum and testing: The 1990 yearbook of the Politics of Education Association*, ed. S. Fuhrman and B. Malen, 233–267. London: Falmer Press.

Thompson, C., and J. Zeuli. 1999. The frame and the tapestry: Standards-based reform and professional development. In *Teaching as the learning profession: Handbook of policy and practice*, ed. L. Darling-Hammond and G. Sykes, 341–375. San Francisco, CA: Jossey-Bass.

Wiggins, G., and J. McTighe. 2005. *Understanding by design.* 2nd ed. Alexandria, VA: Association for Supervision and Curriculum Development (ASCD).

Wilson, S., H. Schweingruber, and N. Nielson, eds. 2015. *Science teachers' learning: Enhancing opportunities, creating supportive contexts.* Washington, DC: National Academies Press.

SECTION IX

ASSESSMENT AND ACCOUNTABILITY IN SCIENCE EDUCATION

SECTION IX

In the classroom, assessments measure what students know and can do. They also provide feedback about science teachers' effectiveness and help teachers adapt their curriculum and instruction to enhance student learning.

Assessments—whether a science teacher's informal judgment about a student's work or a policy maker's response to standardized state and national tests—have long been part of education. For example, in his classic *Basic Principles of Curriculum and Instruction* (1949), Ralph Tyler included a chapter called "How Can the Effectiveness of Learning Experiences Be Evaluated?" Jerome Bruner, in *Toward A Theory of Instruction* (1966), titled his concluding chapter "Evaluation in the Context of Curriculum." In a more contemporary perspective, in April 2013, *Science* included assessment as one of the "grand challenges" in science education (see Figure IX.1).

Figure IX.1. Grand Challenges for Science Education Assessments

Design valid and reliable assessments reflecting the integration of practices, crosscutting concepts, and core ideas in science. The performance expectations of the *Next Generation Science Standards* (*NGSS*) pose significant challenges for assessment design. Considerable research and development will be needed to create and evaluate assessment tasks and situations that can provide adequate evidence of the proficiencies implied in the *NGSS*. This research must be carried out in instructional settings where students have had an adequate opportunity to construct the integrated knowledge envisioned by the National Research Council's *A Framework for K–12 Science Education* (*Framework*) and the *NGSS*.

Use assessment results to establish an empirical evidence base regarding progressions in science proficiency across K–12. Much of what is assumed in the *NGSS* regarding learning progressions needs to be validated through empirical research. This validation requires assessment tasks and situations that can be used across multiple age and grade bands so that we can determine how proficiency changes over time with appropriate instruction. The empirical results can then be used to support the design of more effective curriculum materials and instructional practices.

Build and test tools and information systems that help teachers effectively use assessments to promote learning in the classroom. For teachers to effectively implement assessment as part of their pedagogy, they need tools for presenting tasks and collecting and scoring student performance. They also need smart systems that provide actionable information about the meaning and implications of student performance relative to instruction and student learning. Such systems will need to be designed in collaboration with learning scientists and teachers to ensure their validity, usability, and utility.

Source: Adapted from AAAS 2013, p. 323.

Assessments provide teachers, administrators, and parents with information about students' learning. Results from state, national, and international assessments give some indication of the education system's status. In recent years, however, the character and importance of educational assessment has changed. Changes based on the 2001 No Child Left Behind Act (NCLB) brought statewide accountability that included all schools and all students in upper elementary and middle grades. The law required challenging state standards in reading and mathematics, annual testing of all students in grades 3–8, and annual statewide progress goals ensuring that all groups of students attain proficiency within 12 years of schooling. With the explicit statement of equity and eventual inclusion of science testing, beginning during the 2007–2008 school year, the law also demonstrated the classic unintentional consequences of a well-intended act. One of the unintentional consequences was reduced emphasis on science teaching in elementary grades and increased attention to reading and math. In addition, the NCLB Act reinforced the importance of academic standards, the role of state assessments, and connections between student achievement and accountability. Time will tell whether the reauthorization of the Elementary and Secondary Education Act (ESEA) as the Every Student Success Act (ESSA) of 2015 will remedy any of the adverse consequences for science education.

REFERENCES

American Association for the Advancement of Science (AAAS). 2013. Grand challenges in science education. Special issue, *Science* 340 (6).

Bruner, J. 1966. *Toward a theory of instruction.* New York: W. W. Norton & Company.

Tyler, R. 1949. *Basic principles of curriculum and instruction.* Chicago, IL: The University of Chicago Press.

SUGGESTED READINGS

Atkin, J. M., and J. Coffey. 2003. *Everyday assessment in the science classroom.* Arlington, VA: NSTA Press. A practical guide for science teachers.

Black, P., and J. M. Atkin. 2014. The central role of assessment in pedagogy. In *Handbook of research in science education, volume II*, ed. N. Lederman and S. Abell, 775–790. New York: Routledge. A basic introduction to assessments by two leaders in science education.

Black, P., and D. Wiliam. 1998. Inside the black box: Raising standards through classroom assessment. *Phi Delta Kappan* 80 (2): 139–148.

Britton, E., and S. Schneider. 2014. Large-scale assessments in science education. In *Handbook of research in science education, volume II*, ed. N. Lederman and S. Abell, 791–808. New York: Routledge. This, too, is a basic introduction to large-scale assessments. Both authors have a history of leadership.

Pellegrino, J., N. Chudowasky, and R. Glaser, eds. 2001. *Knowing what students know: The science and design of educational assessment.* Washington, DC: National Academies Press. A foundation of our contemporary understanding and application of assessments.

Pellegrino, J., M. Wilson, J. Koenig, and A. Beatty, eds. 2014. *Developing assessments for the* Next Generation Science Standards. Washington, DC: National Academies Press. The *NGSS* bring challenges to assessment, and this book provides solutions to those challenges.

Schleicher, A. 2011. Is the sky the limit to education improvement? *Phi Delta Kappan* 93 (2): 58–63. Insightful discussion by an individual who has years of experience with international assessments, and PISA in particular.

CHAPTER 17

CLASSROOM, STATE, AND NATIONAL ASSESSMENTS

Teachers of science hear about various types of assessments, their uses, and their results, including formative and summative classroom assessment, state tests, college placement examinations, tests for teacher certification, and national and international assessments. These and other tests have varying purposes and influences within the education system. Space does not permit us to describe all tests and their specific purposes. We leave you with the idea that the use of tests to make education decisions about achievement and ability falls into three categories: initial evaluation (diagrams, placement), formative evaluation (prescription), and summative evaluation (attainment). These categories are similar to those proposed in the early 1970s (Bloom, Hastings, and Madaus 1971) and represent different content and uses of test results.

PERSPECTIVES

Here, we direct attention to assessment related to classroom and curricular domains. Assessment techniques are a great source of data. The better the evaluation instruments, the greater the information available to the teacher for improving teaching. Classroom assessment usually falls into one of three categories: diagnostic, formative, or summative.

Initial Evaluation

Diagnostic evaluation normally precedes instruction but may be used during instruction to uncover student learning problems and current conceptions and abilities. Such evaluations can provide information to teachers about the knowledge, attitudes, and skills of the students entering a course and can be used as a basis for individual remediation or special instruction.

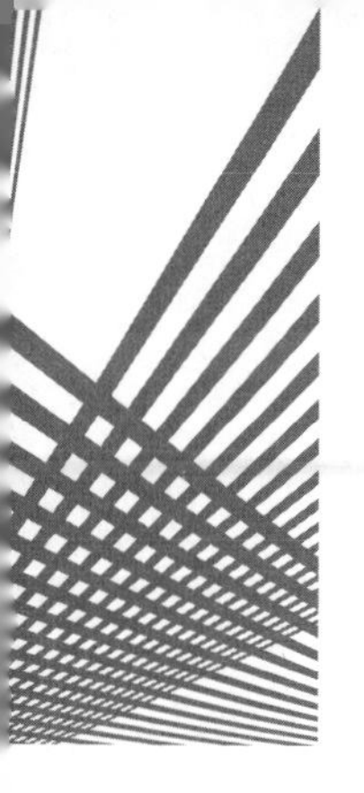

In a less formal way, classroom teachers continuously diagnose and decide on next questions, activities, and experiences for students. In some cases, such as for students with special needs, formal diagnostic evaluations may be used.

Formative Assessments

Formative evaluation may be used during instruction to provide feedback to students and teachers on how effectively content and practices are being taught and learned. Because teaching is a dynamic process, formative evaluation provides useful information that teachers can use to modify instruction and improve teaching effectiveness for greater learning. Black and Wiliam (1998) define formative assessment as "all those activities undertaken by teachers and by their students [that] provide information to be used as feedback to modify the teaching and learning activities in which they are engaged" (p. 140).

Several questions can guide the design and implementation of effective formative assessments into regular classroom practice:

- Where are you trying to go?
- Where are you now?
- How can you get there?

These questions—which essentially clarify goals, status, and opportunities—provide a framework for achieving effective formative assessment (Atkin, Black, and Coffey 2001).

Research supports the importance of formative assessment in ensuring student understanding of science. For example, in Black and Wiliam's (1998) review of more than 250 articles and books, they were able to conclude that formative assessment is an essential component of classroom instruction and, in fact, a necessary component when attempting to raise standards. The effectiveness of formative assessment depends on how teachers and students use the information from the assessment to inform teaching. The learning gains possible from systematic attention to the data from formative assessments have been documented to be larger than for any other type of education intervention. To make sure that formative assessment is an effective tool for increased learning, teachers must focus on the *quality* of their feedback. The use of descriptive, criterion-based feedback is much more useful and effective for the student contrasted with numerical scoring or letter grades that do not relate back to clear criteria (Black and Atkin 2014).

Summative Assessments

Summative evaluation is used most often by teachers and provides the basis for student grades and reports of achievement and abilities. Most frequently, these assessments are based on cognitive gains. For teachers, summative assessment is generally used at the end of an instructional unit to assess the final outcome of that unit in terms of student learning.

Effective summative assessment has a clear and valued target: It attends to many facets of learning, including content understanding, application, processes, and reasoning, and may include a role for student reflection (Black and Atkin 2014). Figure 17.1 summarizes the assessments for classroom instruction.

Figure 17.1. Assessments and Classroom Instruction

Diagnostic assessment refers to evaluations made prior to or very early in an instructional sequence. The results inform teacher decisions about students' knowledge, attitudes, and skills.

Formative assessment refers to assessments that provide information to students and teachers that is used to improve teaching and learning. These are often informal and ongoing.

Summative assessment refers to the cumulative assessments that usually occur at the end of a unit, chapter, or course of study. They are intended to capture what students have learned, or the quality of the learning, and judge performance against some standard. Although summative assessments are often viewed as traditional tests, this need not be the case. Summative assessments can be derived from an accumulation of evidence collected over time, as in a collection of student work.

Source: Adapted from NRC 2001.

Connecting Classroom Assessment and Teacher Accountability

Assessments in science education, whether in classrooms or international contexts, are based on assumptions about how people learn, what they know, and what they are able to do at different ages and stages of development. Assessments also include assumptions about the tasks, observations, data, and evidence that reveal individuals' knowledge and skills. Finally, there are assumptions about the evidence and inferences of learning (Pellegrino, Chudowasky, and Glaser 2001). Stated directly, assessments provide information that can and does influence policies, programs, and practices in science education.

Accountability entails the use of assessment information to confirm accomplishments or motivate changes in students' achievement and teachers' practices, among other issues, such as state policies for high school graduation.

Reporting assessment results represents a general form of accountability. How are students doing at district, state, or national levels? Who or what is accountable for gains or deficits? Assessments and accountability can result in policy changes, new research priorities, and an impetus for education reform. So, assessments and their connections to accountability can motivate change at different levels in the education system, from classrooms to national calls for reform.

LINKING THE *NGSS* AND ASSESSMENTS

The new standards for science education have presented a variety of opportunities and challenges for the science education community. Central among those are questions about assessment and accountability: Will assessments change? How will state and local assessments change? The brief answer to

these questions is yes, they will change. The more complicated questions are the following: *How* will the assessments change? *When* will they change? *What* will assessments based on the *NGSS* look like?

The Assessment Challenges of the *NGSS*

The *Next Generation Science Standards* (*NGSS*; NGSS Lead States 2013) present an opportunity to improve curriculum, teacher development, assessment and accountability, and ultimately student achievement in science. To bring this opportunity to reality, the science education community must recognize the challenge of addressing the innovations in the *NGSS*. While the innovations refer to the *NGSS*, in reality, the innovations have direct implications for assessment. The following brief description of innovations is based on *A Framework for K–12 Science Education* (NRC 2012) and the *NGSS*. We presented the innovations in earlier chapters. Here we highlight the implications for assessment.

Interconnect Three Dimensions

The *NGSS* have three dimensions: disciplinary core ideas, scientific and engineering practices, and crosscutting concepts. Most state and district assessments express these dimensions as separate entities, if at all. The integration of content and practices in assessments is a clear and direct implication of the *NGSS*. Expressing standards as performance expectations also implies the form of assessments—that is, they are performance based.

Recognize Learning Progressions

Science concepts in the *NGSS* build coherently from kindergarten through grade 12. Assessments also should present a progression, from grade band to grade band, that includes understanding and abilities throughout a student's K–12 science education.

Include Engineering and the Nature of Science

The *NGSS* includes both the nature of science and engineering design. Science and engineering practices and crosscutting concepts are designed as an integral component of the standards and by extension assessments.

Coordinate Science With Common Core State Standards (CCSS) *for English Language Arts (ELA) and Mathematics*

Implementing the *NGSS* presents an opportunity for science to become an integral part of students' comprehensive education. To do so implies the need to include appropriate components of ELA and math in assessments.

One clear implication is the need for integrated programs, including assessments. Furthermore, the incredible complexity of the *NGSS* innovations presents significant challenges for those with responsibilities for assessment. Among the most significant new features of the *NGSS* is a call for the integration—in instruction as well as assessment—across three dimensions of content: science and engineering practices, disciplinary core ideas, and crosscutting concepts. This is a new level of complexity for the assessment field. A National Research Council (NRC) report called *Developing*

Assessments for Next Generation Science Standards (Pellegrino et al. 2014) specifically addresses this challenge. The committee that prepared the report had to address challenges that centered on developing assessment tasks that

- integrate the three dimensions;
- assess where a student could be placed along a sequence of progressively more complex understandings of a given core idea, and successively more sophisticated applications of practices and crosscutting concepts; and
- measure the connections between the different strands of the disciplinary core ideas (e.g., using understandings about chemical interactions from physical science to explain phenomena in biological contexts).

The NRC report emphasizes an evidence-centered design for tasks and a systems approach to assessment of the *NGSS*. Essential components of the assessment system include classroom-based assessments designed to provide feedback on activities and instruction, assessments to monitor student learning, and assessments to track factors that influence learning outcomes (e.g., opportunities to learn, appropriate resources, teaching strategies).

Classroom assessments are an essential component of teaching and learning and should include both formative and summative tasks. Formative tasks are those that are specifically designed to guide instructional decision making and lesson planning. Summative tasks are those that are specifically designed to assign student grades. Curriculum developers, assessment developers, and others who create resource materials aligned to the *NGSS* or standards influenced by the *Framework* and *NGSS* should ensure that assessment activities included in materials (such as mid- and end-of-chapter activities, tasks for unit assessments, and online activities) engage students in science and engineering practices that demonstrate their understanding of core ideas and crosscutting concepts. These materials also should reflect multiple dimensions of diversity (e.g., by connecting with students' cultural and linguistic identities). In designing instructional materials, development teams should include experts in science, science learning, assessment design, equity, diversity, and science teaching (Pellegrino et al. 2014).

Assessment tasks must be designed to provide evidence of students' ability to use the practices, apply their knowledge of crosscutting concepts, and draw on their understanding of disciplinary core ideas, all in the context of addressing specific problems. Instruction and assessments must be designed to support and monitor students as they develop increasing sophistication in their ability to use practices, apply crosscutting concepts, and understand core ideas as they progress across the grade levels. Assessment developers should draw on the idea of developing understanding as they structure tasks for different levels and purposes and build this idea into the scoring rubrics for the tasks. Although understanding the language and terminology of science is fundamental and factual knowledge is very important, tasks that demand only declarative knowledge about practices or isolated facts would be insufficient to measure satisfaction of the performance expectations in the *NGSS* (Pellegrino et al. 2014).

Effective evaluation of three-dimensional science learning requires more than a one-to-one mapping between the *NGSS* performance expectations and assessment tasks. More than one assessment test may be required to adequately assess students' mastery of some performance expectations. For assessing both understanding of core ideas and facility with a practice, assessments may need to probe students' use of the practice in more than one disciplinary context. To adequately cover the three dimensions, assessment tasks will generally need to contain multiple components (e.g., a set of interrelated questions). Developers may focus on individual practices, core ideas, or crosscutting concepts in the various components of an assessment task, but together the components need to support inferences about students' three-dimensional science learning as described in a given performance expectation. Assessment tasks that attempt to test the science and engineering practices in strict isolation from one another likely will not be as meaningful as assessments of the three-dimensional science learning called for in the *NGSS* (Pellegrino et al. 2014).

Here are some key points regarding classroom assessments to support the *NGSS:*

- Assessments are aligned with the *NGSS*; are authentic; and include formative, summative, and self-assessment measures.
- Assessments are based on performance expectations and collect data on all three dimensions of the *NGSS.*
- Assessments have explicitly stated purposes and are consistent with the decisions they are designed to inform.
- Assessments are embedded throughout instructional materials as tools for students' learning and teachers' monitoring of instruction.
- Assessments reflect only knowledge and skills that have been covered adequately in the instructional materials.
- Assessments use varied methods, vocabulary, representations, and examples that are unbiased and accessible to all students and provide teachers with a range of data to inform instruction.

Translating the *NGSS* to Assessments

Producing an accurate translation requires a thorough understanding of both the language and the culture of the original product (e.g., a play or poem), as well as the contemporary culture. In addition, one must understand the trade-offs as well as the advantages of a translation. So it is with translating the *NGSS* to assessments. The *NGSS* have a language and innovations that are important. Likewise, there are constraints and concepts of assessment that should be honored.

Publications such as the NRC report give evidence that we have already moved beyond the challenges. The capacity to assess disciplinary core ideas and practices is clear. However, finding examples of assessing the three dimensions remains a challenge. Thinking in terms of assessment units (e.g., PISA) is an excellent way to move beyond old perceptions of a test with single items that emphasize content but do not integrate the three dimensions of contemporary standards.

Use of evidence-centered design (ECD; Mislevy, Almond, and Lukas 2003; Gorin and Mislevy 2013) will be essential for those at the state and local districts who have to develop and implement assessments. The ECD approach will facilitate the process of appropriately synthesizing the various examples into initial assessments of the *NGSS.*

In conclusion, translating the *NGSS* to assessments presents numerous challenges, particularly for the design of consequential assessments if current constraints of testing time, cost, and uses of data are followed. Science educators need to recognize and understand those challenges. As important as understanding the challenges may be, it is more important to move beyond the challenges and help those with the responsibility for developing the next generation of science assessments.

LARGE-SCALE NATIONAL ASSESSMENTS OF EDUCATIONAL PROGRESS

Assessments based on large, statistically selected samples, such as the National Assessment of Educational Progress (NAEP), provide a national perspective of student achievement across time. International assessments such as Trends in International Mathematics and Science Study (TIMSS) and the Program for International Student Assessment (PISA) extend the perspective of student achievement and abilities to international contexts (Britton and Schneider 2014). In this chapter, we introduce NAEP; in the next chapter, we discuss TIMSS and PISA.

National Assessment of Educational Progress (NAEP)

The National Assessment of Educational Progress (NAEP), known as the Nation's Report Card, is the only nationally representative and continuing assessment of what America's students know and can do in various academic subjects. Since 1969, the NAEP has been conducted on a national sample of students in the areas of reading, mathematics, science, writing, and other fields. In 2014, NAEP conducted a new assessment on technology and engineering literacy. By making objective information on student performance available to policy makers, educators, and the general public, NAEP is an integral part of our nation's evaluation of the condition and progress of education.

NAEP is a congressionally mandated project of the National Center for Education Statistics (NCES) at the U.S. Department of Education. Results are provided only for group performance. NAEP is forbidden by law to report data on individual students. As the ongoing national indicator of the academic achievement of American students, NAEP regularly collects information on representative samples of students in grades, 4, 8, and 12 and periodically reports on student achievement in reading, mathematics, science, and other subjects. NAEP reports results of student achievement in the categories of basic, proficient, and advanced.

In 1988, Congress created the National Assessment Governing Board (NAGB), whose purpose is to set policy for NAEP. The independent, 26-member board is composed of state and local policy

makers, teachers, curriculum specialists, testing experts, business representatives, and members of the general public.

The board is responsible for selecting subject areas to be assessed and determining appropriate achievement goals for each grade and subject tested. NAGB is responsible for developing test objectives and specifications, designing guidelines for reporting and disseminating results, and improving the form and use of the assessment. The board also is charged with ensuring that all items selected for use in NAEP are free from racial, cultural, and regional bias.

In 2009, NAGB released a new framework for NAEP science. This framework sets the design specifications for NAEP for the future, most likely until 2021. *Science Framework for the 2009 National Assessment of Educational Progress* dictates that the NAEP should

- be based on national standards,
- give considerations to TIMSS and PISA,
- include technological design, and
- address both knowledge and abilities.

Science Content

The science content for the 2009 NAEP is defined by a series of statements that describes key facts, concepts, principles, laws, and theories in three broad areas:

- Physical science
- Life science
- Earth and space science

Science Practices

The second dimension of the NAEP *Framework* is defined by four science practices:

- Identifying science principles
- Using science principles
- Using scientific inquiry
- Using technological design

Performance Expectations

The design of the NAEP science assessment is guided by the NAEP *Framework's* descriptions of the science content and practices to be assessed. Table 17.1 illustrates how content and practices are combined to generate performance expectations. The columns contain the science content (defined by content statements—propositions that express science principles—in physical, life, and Earth and space sciences), and the rows contain the four science practices. A double dashed line distinguishes Identifying Science Principles and Using Science Principles from Using Scientific Inquiry and Using Technological Design. The former two practices can be considered as "knowing science," and the

latter two practices can be considered as the application of that knowledge to "doing science" and "using science to solve real-world problems." The cells at the intersection of content (columns) and practices (rows) contain student performance expectations which specify the design of assessment experiences.

Table 17.1. Combining Content and Practices to Generate Performance Expectations

		Science Content		
		PHYSICAL SCIENCE	LIFE SCIENCE	EARTH AND SPACE SCIENCE
SCIENCE PRACTICES	**Identifying Science Principles**	Performance Expectations	Performance Expectations	Performance Expectations
	Using Science Principles	Performance Expectations	Performance Expectations	Performance Expectations
	Using Scientific Inquiry	Performance Expectations	Performance Expectations	Performance Expectations
	Using Technological Design	Performance Expectations	Performance Expectations	Performance Expectations

Source: NAGB 2008.

Table 17.1 represents a two-dimensional approach for which performance expectations play a central role. You have seen that performance expectations also are central to the *NGSS*. However, they add crosscutting concepts to present three dimensions in the performance expectations.

In 2014, the NAGB released the Technology and Engineering Literacy Framework, which was the basis for the NAEP in the same year. The Technology and Engineering Literacy Framework (TEL) begins with definitions that respond to the need for clarity in what is being assessed. Figure 17.2 presents the definitions.

Figure 17.2. Definitions From the Technology and Engineering Literacy Framework

- **Technology** is any modification of the natural world done to fulfill human needs or desires.
- **Engineering** is a systematic and often iterative approach to designing objects, processes, and systems to meet human needs and wants.
- **Technology and engineering literacy** is the capacity to use, understand, and evaluate technology as well as understand technological principles and strategies needed to develop solutions and achieve goals.

Source: NAGB 2014.

Major Areas of Technology and Engineering Literacy (TEL)

The TEL assessment framework centers on three major areas, each with a corresponding body of knowledge and skills with which students should be familiar:

- Technology and society
- Design and systems
- Information and communication technology

Each of the three areas has sub-areas in which students are assessed. Technology and Society, for example, includes Interaction of Technology and Humans; Design and Systems includes Systems Thinking; and Information and Communication Technology (ICT) includes Acknowledgement of Ideas and Information.

Technology and Engineering Literacy Practices

In all three areas of technology and engineering literacy, students are expected to be able to apply particular ways of thinking and reasoning when approaching a problem. These types of thinking and reasoning are referred to as *practices,* and the NAEP framework specifies three kinds in particular that students are expected to demonstrate when responding to test questions:

- Understanding technological principles
- Developing solutions and achieving goals
- Communicating and collaborating

The TEL Framework combined the three major areas and practices, creating two-dimensional sets of performance expectations for the assessment (see Table 17.2)

NAEP Achievement Levels

Congress authorized the NAGB to develop appropriate student achievement levels on the NAEP. The achievement level descriptions are statements of what students should know and be able to do on the NAEP in grades 4, 8, and 12. To fulfill its statutory responsibility, the NAGB developed a policy to guide the development of achievement levels for all NAEP subjects. Three levels of achievement were identified to provide the public, educators, and policy makers with information on student performance on NAEP. These levels—basic, proficient, and advanced—are used as a primary means of reporting NAEP results to describe "how good is good enough" at grades 4, 8, and 12.

Table 17.3 displays the generic policy definitions for basic, proficient, and advanced achievement that pertain to all NAEP subjects and grades.

Table 17.2. Examples of Types of Assessment Targets for Major Areas and Practices for Technology and Engineering Literacy

	Technology and Society	Design and Systems	Information and Communication Technology
Understanding Technological Principles	Explain costs and benefits	Analyze a need	Describe features and functions and ICT tools
Developing Solutions and Achieving Goals	Develop a plan to investigate an issue	Construct and test a model or prototype	Plan research and presentations
Communicating and Collaborating	Display positive and negative consequences using data and media	Represent data in graphs, tables, and models	Argue from an opposing point of view

Source: Adapted from NAGB 2014.
Note: The original figure had between five and nine examples in each cell. We have provided only one example in each.

Table 17.3. Generic Achievement-Level Policy Definitions for NAEP

Achievement Level	Policy Definition
Advanced	This level signifies superior performance.
Proficient	This level represents solid academic performance for each grade assessed. Students reaching this level have demonstrated competency over challenging subject matter, including subject-matter knowledge, application of such knowledge to real-world situations, and analytical skills appropriate to the subject matter.
Basic	This level denotes partial mastery of prerequisite knowledge and skills that are fundamental for proficient work at each grade.

NAEP: 2015 Science Results

In 2015, NAEP Science presented U.S. educators with some good news. Higher percentages of the nation's fourth- and eighth-grade students reached the proficient level. Compared to 2009, scores for both grade 4 and grade 8 increased by four points, from 150 in 2009 to 154 in 2015. Science scores in 2015 for grade 12 students were 150, the same as in 2009. (See Figure 17.3.) Although there was progress for students in fourth and eighth grades, significant percentages of all three grades remain below the proficient level. Only 38%, 34%, and 22% of students were at proficient levels or above for grades 4, 8, and 12, respectively (see Figure 17.4).

The 2015 NAEP results for science had some additional good news. African American and Hispanic students' science scores increased for eighth grade. These results continue to narrow the achievement gaps at both grades 4 and 8. Additionally, in 2015, grade 4 girls scored about the same as grade 4 boys, reducing the gender difference (NAGB 2016; Carr 2016; Hinton 2016).

On balance, we think these results indicate progress in our students' understanding of science. That said, it also seems clear that there is a need to address achievement in science at high school levels. Although some may make connections between these scores and the *Next Generation Science Standards* (*NGSS*), which were released in 2013, we believe it is too early to make any claim about the influence of *NGSS*. We hope new state adoptions of *NGSS* and the fact that many states have adapted *NGSS* will have a positive effect on NAEP science achievements, continue to narrow gaps, and improve grade 12 results.

Figure 17.3. Science Score Improvements in Grades 4 and 8 Since 2009

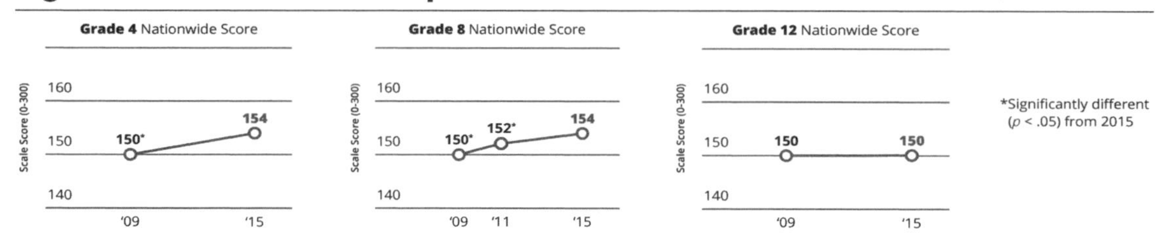

Source: NAGB 2016.

Figure 17.4. Percentage of Students at or Above Proficient in 2015

Source: NAGB 2016.

A PERSONAL PERSPECTIVE

ASSESSING THE *NGSS*—THE POTENTIAL, THE CHALLENGES

Stephen Pruitt

Assessment—a word that has become tied to negative feelings in today's society. Assessment has significantly changed since the No Child Left Behind (NCLB) legislation in 2001. It became synonymous with accountability, and for many teachers who did not teach math or English language arts, the idea of assessment left them feeling devalued. This is a tough place to be. I would say most educators have said, or at least heard, "If it is not tested, it is not taught." This is a despicable statement because one should not have to have a state test to do what is right by students, but it is an unfortunate reality. One promise of the *NGSS* is the hope that this will change if a state is to properly implement the *NGSS*. With the passage of the Every Student Succeeds Act (ESSA), there are new opportunities ahead for those who choose to embrace those challenges.

We were all tested as students, whether in grade school or graduate school. Assessment is far from a new idea in education. However, what is new is the degree to which assessment drives our system. In this essay, I intend to spend some time discussing the benefits and opportunities we have for science assessment in the near future. However, I will also spend some time talking about the challenges, some of which may not be to pleasant to discuss but which need to be discussed nonetheless.

A SYSTEM OF ASSESSMENTS

Before getting too far into some of the dynamics and lessons learned from the assessment development around the *NGSS*, I will take a little time to discuss a system of assessments. As educators in a post-NCLB era, we have become quite accustomed to one test at the end of the year being a be-all, end-all. We have seen many districts, at least in mathematics and reading, buying or developing a set of assessments that may or may not be aligned to either the standards or the summative assessment. What the *NGSS* call for is a system of assessments. This means it's not a simple matter of aligning key words; assessment must be a system in which each component relies on another. An assessment system does not predict the summative test score performance for a student; rather, it provides meaningful feedback to the student and teacher to determine next steps as they move toward the summative assessment. An assessment system has within its structure formative assessments that do not "feel" like a test. That is to say that a class does not stop to take the assessment; it is part of

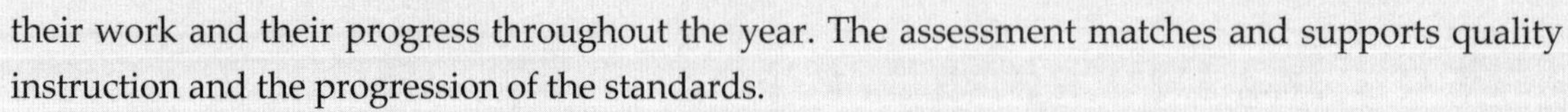

their work and their progress throughout the year. The assessment matches and supports quality instruction and the progression of the standards.

So, an assessment system is not one that a lot of educators have seen. I think we often fool ourselves into believing we have, but a true system builds on itself and, like a true system, each part affects the intent and behavior of other parts. It will take a lot of will to change to an assessment system. I am hopeful that states will do so. I am even more hopeful that our educators will do so.

Three-Dimensional Assessment

How can we discuss any aspect of the *NGSS* and not begin with the three-dimensional aspect? Well, we cannot. The design of the *NGSS* pushes us instructionally, but it also pushes assessment in a way never seen before, in my opinion. To discuss this properly, I need to share some of the discussions that occurred during the development of the *NGSS*.

A Framework for K–12 Science Education was very clear that to be scientifically literate, a student must be able to operate at the nexus of the three dimensions: scientific and engineering practices, disciplinary core ideas, and crosscutting concepts. As has been discussed many times before, these three dimensions in and of themselves were not new ideas. Even the idea that students should be able to use all three dimensions together was not completely new. What *was* new was the requirement that students must operate at this juncture to be considered scientifically literate. To this point, state agencies would develop a set of standards around inquiry (practices) and content (disciplinary core ideas). Some states, though few, would develop a set of unifying themes (crosscutting concepts) that would be placed in the science standards. The vision from the early sets of standards is that students would use these three dimensions to learn about the world. What happened—and I believe state testing must share some of the blame—was that large-scale assessments were developed using low-level content assessment items and nonmeaningful inquiry assessments items, and usually ignored anything dealing with the unifying themes. Additionally, the drive to get scores back quickly overwhelmed the practice of developing good items that require students to construct their answers as this took longer to score and longer to receive scores back from the state. These items were easy to write and score, so they became the norm.

If you ask most state-level science content experts, the idea (I believe) was that teachers would still do the things they were supposed to do with the deeper content and inquiry, but as the test required rote memorization and teachers and schools were being judged on student performance, less and less of that type of instruction was promoted. These types of tests promoted fear among teachers that if they taught using inquiry or project-based instruction, their test scores would suffer. Combine this with a large amount of content that was impossible to even cover in a year, and you have a recipe for poor science instruction and assessment. To be clear, this was not the intent of anyone in the science community. I do not believe that most science teachers wanted it either, but the test drove us this way. Knowing the four phases of mitosis became more important than explaining how the cell division process aids in genetic variation and the survival of a species. Knowing the math of gas laws became more important than students understanding how those laws are models of spacing and energy differences between molecules. In other words, the test tended to be more about memorization than was typically reflected in the standards.

At this point, you may be wondering what this has to do with the development of the *NGSS*. Well, it was a definite driver of the way the standards ultimately looked. When the project was beginning, there were many discussions, debates, and even arguments (and I do not mean the scientific practices kind) regarding how the standards were to be arranged. As with many new things, we began with our current understanding and reference, so the initial discussions were around how to write clear standards and yet also show the intent of the *Framework*. There were some who wanted to write standards for each of the dimensions separately. Most, however, wanted to see something significantly different. As I have said, something has to "perturb the system." Again, you may be saying, "This is not new information, so what does it have to do with assessment?" Well, a great deal, because much of what drove our decision was what test writers would do with the standards as they were used to developing large-scale assessments. Historically, because state science standards separated the practices (inquiry) from the content standards, test writers and states as a whole allowed the state assessment to be arranged as such. It is an unfortunate reality that we live in, one in which good instruction is overlooked because of the fear of the assessment from the state. This was not the intent of the standards in the 1990s. I believe this is why some believed those standards "failed."

True, they did not get science instruction to where we needed it to be, but they were far from a failure. Both the *National Science Education Standards* (*NSES*) and *Benchmarks for Science Literacy* (*Benchmarks*) presented inquiry, content, and unifying themes separately, but neither set of standards was developed to be adopted by states. The reality is that state standards are as much assessment standards as they are content standards. As such, the leadership and writing teams of the *NGSS* felt these had to embody the same reality. The *Framework* made very clear that to be scientifically literate, a student must be able to perform at the nexus of the three dimensions. The committee took this quite literally. If the students were meant to *perform*, then the large-scale assessment had to match that as well. As has been discussed earlier, assessment needs to be authentic and ascertain if students can use all three dimensions. The concern, which is a correct one, was that states would simply continue as they were, meaning large-scale assessment would not change, which meant that classrooms would not either.

So, the decision was made to follow the recommendations of the *Framework* to the letter by developing performance expectations that would drive the assessments. This was met with a fair amount of appreciation, but also some criticism. There is the unfortunate understanding from many that you should teach standards one at a time and exactly as written. This could not be further from the truth. I also think this is why some believe that content they believe is key is not included.

Standards-based education has become about structure, when it really should be about freedom. The *NGSS* allow instruction and assessment to go forward and grow. Once a teacher or test developer understands that the point of the *NGSS* is that students are able to explain phenomena, it becomes much easier to visualize how the standards are arranged. So, to properly assess student learning in a three-dimensional way, states and educators have to move past the current understanding and belief that each dimension must be done separately. Teaching the first chapter on the scientific method is not enough, so testing on it in isolation should not be either.

I do not want to mislead anyone here: Assessing science three dimensionally is not easy by any stretch. In general, we do not see a lot of assessment literacy being taught in preservice courses and often not even in our graduate courses. I believe most educators get a grounding in normative versus criterion assessment, some psychometrics, and maybe even some information on standard setting. What we miss, however, are alignment and quality of assessments beyond the statistics. What does it mean for an assessment item to be "aligned to the standards," and what does it mean to have a quality assessment? The problem becomes that the *NGSS* most definitely perturbed the system, and in doing so they will change our knowledge of assessment forever. The traditional psychometric calculations or observations are challenged by three-dimensional assessment. Instead of a question about each dimension, there is a series of questions around a scenario that elicit responses to observe whether students are able to use all three dimensions at once. Instead of a memorization question that a student needed to have been taught specifically to avoid bias or unpreparedness, the concepts are assessed in a way that focuses on the understanding of the content and the ability to attain it through the practices or crosscutting concepts. Instead of Webb's Depth of Knowledge, or DOK, we realize that a DOK 1 (the lowest level of knowing content) is not enough. As a good friend of mine (Peter McLaren) has said, "*NGSS* is where DOK 1s go to die."

NGSS Assessment: A Process

By the time I arrived in Kentucky, the work around assessment had been going on for a while. It began with a group of states coming together to start to tackle some of the tough issues around assessing the *NGSS*. The Council of Chief State School Officers (CCSSO), the organization for commissioners and state superintendents, convened all interested states and contracted with an assessment vendor to coordinate the development of an assessment plan that states could use in their own *NGSS* assessment Request for Proposals (RFP). This was difficult and grueling work. It was tough to get everyone on the same page, as it often is, but it was particularly tough with so many states having so many timelines and priorities. It is also tough in this sort of environment when you have science content experts and assessments experts working on something new to all. I believe the science experts were probably most in sync, but it was often hard to find "good enough." The assessment experts look at the realities of assessment, such as cost, psychometric requirements, and the general logistics of large-scale assessment. This work progressed nicely but sort of stalled as the various states began to make their own plans and tried to adhere to the timelines set by the state or the U.S. Department of Education.

Upon arriving in Kentucky, I immediately started taking a deep look at where we were on science assessment. I had worked pretty closely with staff from Kentucky before becoming commissioner, but now was time for me to get "into the weeds" of this new assessment. The original timeline had been set for a fully operational assessment in spring 2016. Additionally, Kentucky had made what I considered a brilliant move. The Kentucky Department of Education contracted with an assessment vendor to act as a thought partner in the development of their RFP and develop the full assessment system. I began meeting with the team each week, and it became clear we were not ready for the new assessment. So, I made the decision to delay assessment for two more years. Something else became clear as well. Just as with the CCSSO initiative, there came the point when arguing needed to end and a decision needed to be made. I have shared with other chiefs that we have to be involved with assessment to be the deciding vote or push forward. There is nothing wrong with our science experts continuing to look for perfection; in fact, it is admirable. However, at some point, it is time to move. We learned many things about our new system. We learned that "storyboards" needed to be created by an assessment developer and approved before moving forward. If a full set of items—or a cluster, as it has come to be known—gets too far in development without the science experts reviewing the items, it is almost impossible to correct course. We learned that starting with a phenomenon was key. In traditional tests, the standards were the starting point, but at least in the early going, while everyone is learning, it is a hindrance more than a help. We learned that we must develop clear guidelines and communications around the fact that we assess groups or "bundles" of performance expectations. This is also contrary to traditional thought, where one item is correlated to one standard. Once we were into development, it became clear this is the only way to assess the *NGSS* if we are to retain fidelity to the standards themselves. Finally, as we have moved forward with our pilots, we have discovered that the best compliment of an assessment item is when someone says, "Hey, this isn't a test. This is what we do in our classrooms." To follow through with the vision of the *NGSS* and its assessment, authenticity and reflection of good classroom practice are a must. Unfortunately, this is not what our teachers expect.

Test Is a Four-Letter Word. It Shouldn't Be, So Let's Say *Assessment Systems.*

As I mentioned earlier, our views on assessment have become skewed due to their nature and perhaps even their pervasiveness in our communities. *Assessment*, *standards*, and even *curriculum* are used so interchangeably that it is tough to figure out what some people are saying. Assessment has been such a big part of accountability and now has moved into teacher evaluation. I believe it is time to move science assessment out of compliance and into the powerful tool it was meant to be. To do this, states are going to have focus on systems of assessment and develop assessments that promote quality instruction rather than hamper it. It will be tough for sure, and probably expensive, but the benefit will be huge for our schools, our country, and of course, the most important people—our students. So, leadership is needed. Our teachers need to hold our states accountable for a quality test based on quality standards. They need to be engaged and willing to work on an assessment committee as the expert in the standards and an advocate for students. Assessment is one of the hardest parts of implementing the *NGSS*, but working through a fully integrated assessment system will pay off in unbelievable ways.

REFERENCES

Atkin, J. M., P. Black, and J. Coffey. 2001. *Classroom assessment and the* National Science Education Standards. Washington, DC: National Academies Press.

Black, P., and J. M. Atkin. 2014. The central role of assessment in pedagogy. In *Handbook of research on science education, volume II*, ed. N. Lederman and S. Abell, 775–790. New York: Routledge.

Black, P., and D. Wiliam. 1998. Inside the black box: Raising standards through classroom assessment. *Phi Delta Kappan* 80 (2): 139–148.

Bloom, B., J. Hastings, and G. Madaus. 1971. *Handbook of formative and summative evaluation of student learning.* New York: McGraw Hill.

Britton, E., and S. Schneider. 2014. Large-scale assessments in science education. In *Handbook of research on science education, volume II*, ed. N. Lederman and S. Abell, 791–808. New York: Routledge.

Carr, P. 2016. *National Assessment of Educational Progress 2015 science results.* Washington, DC: National Center for Education Statistics.

Gorin, J. S., and R. J. Mislevy. 2013. *Inherent measurement challenges in the* Next Generation Science Standards *for both formative and summative assessment.* Paper presented at the Invitational Research Symposium on Science Assessment, Washington, DC. *www.k12center.org/rsc/pdf/gorin-mislevy.pdf.*

Hinton, M. 2016. Science gains seen at 4th, 8th grades. *Education Week* 36 (11): 6.

Mislevy, R. J., R. Almond, and J. Lukas. 2003. *A brief introduction to evidence-centered design.* Research Report No. RR-03-16. Princeton, NJ: Educational Testing Service.

National Assessment Governing Board (NAGB). 2008. *Science framework for the 2009 National Assessment of Educational Progress.* Washington, DC: U.S. Government Printing Office.

National Assessment Governing Board (NAGB). 2014. *Technology and engineering literacy framework.* Washington, DC: U.S. Government Printing Office.

National Assessment Governing Board (NAGB). 2016. *The nation's report card: NAEP 2015 science results. https://www.nagb.org/newsroom/naep-releases/2015-science-release.html.*

National Research Council (NRC). 2001. *Knowing what students know: The science and design of educational assessments.* Washington, DC: National Academies Press.

National Research Council (NRC). 2012. *A framework for K–12 science education: Practices, crosscutting concepts, and core ideas.* Washington, DC: National Academies Press.

NGSS Lead States. 2013. *Next Generation Science Standards: For states, by states.* Washington, DC: National Academies Press. *www.nextgenstandards.org/next-generation-science-standards.*

Pellegrino, J., N. Chudowasky, and R. Glaser, eds. 2001. *Knowing what students know: The science and design of educational assessment.* Washington, DC: National Academies Press.

Pellegrino, J., M. Wilson, J. Koenig, and A. Beatty, eds. 2014. *Developing assessments for the* Next Generation Science Standards. Washington, DC: National Academies Press.

CHAPTER 18

INTERNATIONAL ASSESSMENTS OF SCIENCE

International assessments of science may seem quite distant and unimportant when contrasted to everyday practicalities of the classroom. In one sense, this perception is accurate. In another sense, the continued comparisons of U.S. students' achievement with the achievement of students from other countries, particularly countries with whom we compete economically, have increased our understanding of science education in other countries and elevated concerns about science teaching in this country. This heightened significance places a responsibility on all individuals in the science education community to understand what comparisons on international tests tell us about policies, programs, and practices.

PERSPECTIVES

In the early 1960s, we did not know very much about science education in other countries. Some academics maintained an interest, but for the most part educators concentrated on issues within U.S. borders. Later in that decade, there were initial efforts in what has become a tradition of international comparative studies of science and mathematics education. These nascent efforts have grown into major international assessments that the education community now knows as the Trends in International Mathematics and Science Study (TIMSS) and, beginning in the year 2000, the newest addition, known as the Program for International Student Assessment (PISA).

In this section, we provide introductions to TIMSS and PISA, discuss several insights based on results from the assessments, and conclude with reflections on international assessments. Portions of this discussion were adapted from Bybee and McCrae (2009). Additional information was retrieved from the National Center for Educational Statistics website under the title *Highlights From TIMSS*

and TIMSS Advanced 2015: Mathematics and Science Achievement of U.S. Students in Grades 4 and 8 and in Advanced Courses at the End of High School in an International Context (Provasnik et al. 2016).

TIMSS

TIMSS 2015 is the sixth comparison of mathematics and science achievement completed since 1995. TIMSS combines science and mathematics in one assessment and assesses student learning at different grades; in 2015, TIMSS evaluated grades 4 and 8 and advanced courses at the end of high school.

Since 1995, TIMSS has been coordinated by the International Association for the Evaluation of Educational Achievement (IEA), an international organization of national research institutions and governmental research agencies. TIMSS is funded by the U.S. Department of Education, the National Science Foundation, the World Bank, the United Nations Development Project, and participating countries. IEA is located in Boston, Massachusetts. In 2015, a total of 43 countries and education systems participated in TIMSS at the fourth-grade level, and 43 countries and education systems participated at the eighth-grade level.

TIMSS provides a perspective that links assessments to the curricula of cooperating countries. Thus, TIMSS provides an indication of the degree to which students have learned concepts in the mathematics and science they have had the opportunity to learn in school programs. TIMSS answers this question: Based on school curricula, what knowledge and skills have students attained by grade 4? By grade 8? The achievement scores from TIMSS represent the "learned" curriculum at different grade levels, specifically grades 4 and 8. Figures 18.1 and 18.2 (p. 280) present the science results for TIMSS 2015.

How did U.S. students do on TIMSS 2015? In 2015, U.S. fourth-graders' average score (546) was higher than the international TIMSS scale average, which is set at 500. See Figure 18.1 for a list of average scores of participating countries and education systems. For students in grade 4, the United States was among the top 11 education systems for science (7 education systems had higher averages and 3 were not measurably different) and scored higher, on average, than 47 education systems. The 7 education systems with average science scores above the U.S. score were Singapore, Korea, Japan, the Russian Federation, Hong Kong-CHN, Chinese Taipei-CHN, and Finland.

In 2015, the average science score of U.S. eighth-graders (530) was higher than the TIMSS scale average. At grade 8, the United States was among the top 17 education systems in science (7 education systems had higher averages and 9 were not measurably different) and scored higher, on average, than 32 education systems. The 7 education systems with average science scores measurably above the U.S. score were Singapore, Japan, Chinese Taipei-CHN, Korea, Slovenia, Hong Kong-CHN, and the Russian Federation. See Figure 18.2 (p. 280).

Are U.S. students making progress? On TIMSS, U.S. fourth graders have shown improvement for some periods (e.g., 2015 average scores were higher than 2003 and 2007) but not in others (e.g., there was no measurable difference between the average science score in 2015 and the averages in 1995 and 2011). We would suggest that the mixed results do not indicate a trend of improvement.

Figure 18.1. Average Science Scores of Fourth-Grade Students, by Education System: 2015

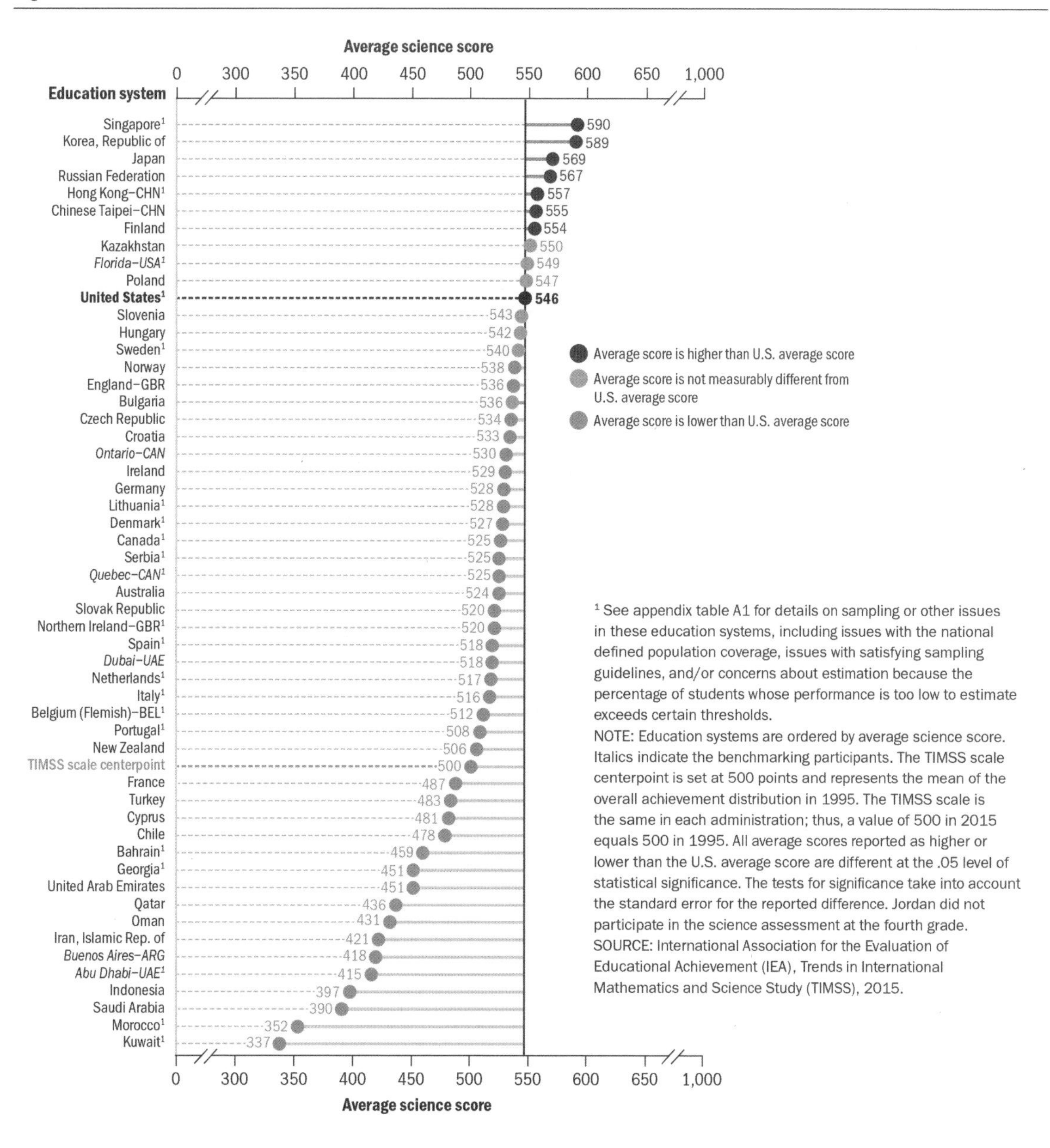

Source: Provasnik et al. 2016.

Figure 18.2. Average Science Scores of Eighth-Grade Students, by Education System: 2015

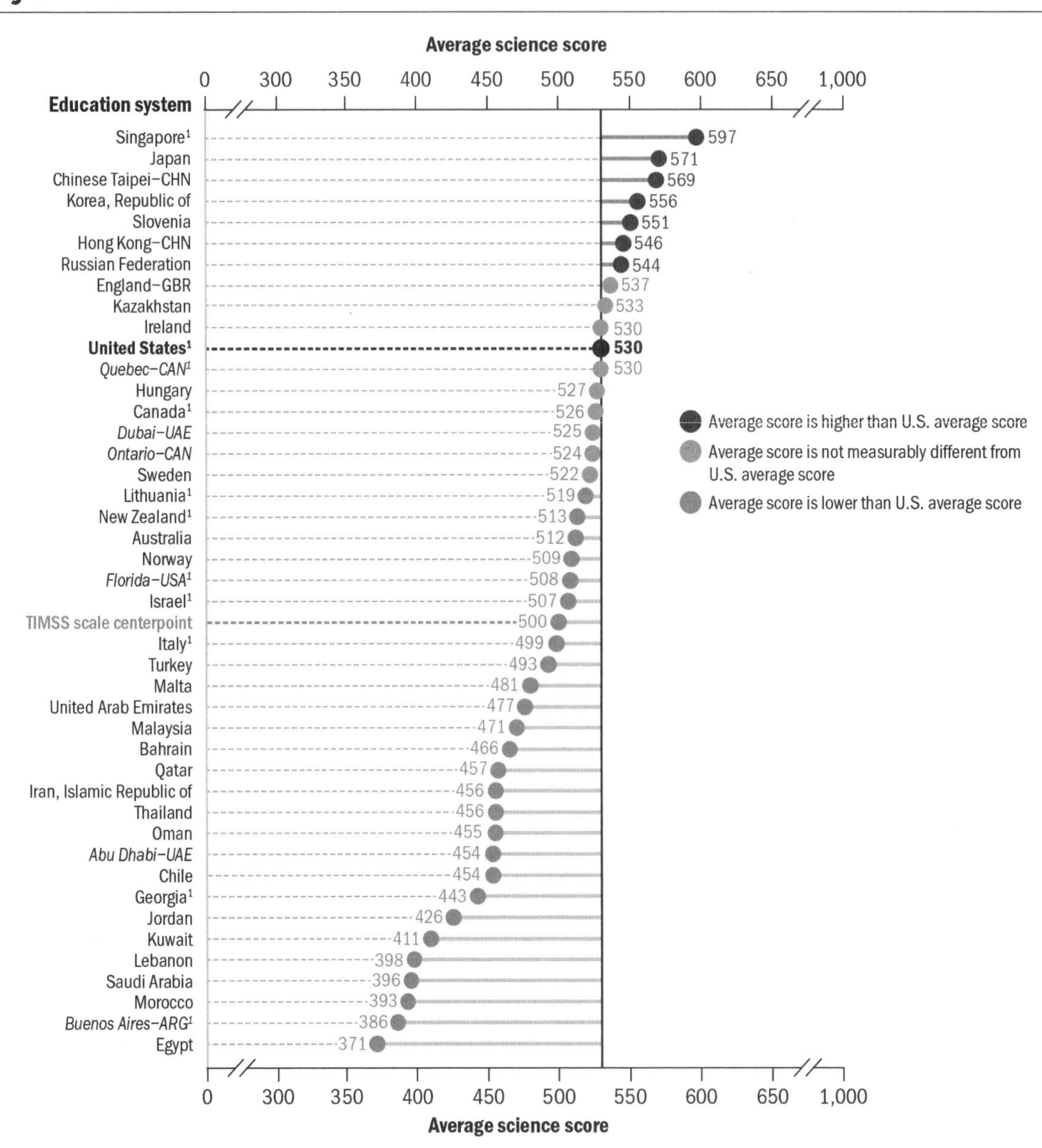

[1] See appendix table A1 for details on sampling or other issues in these education systems, including issues with the national defined population coverage, issues with satisfying sampling guidelines, and/or concerns about estimation because the percentage of students whose performance is too low to estimate exceeds certain thresholds.

NOTE: Education systems are ordered by average science score. Italics indicate the benchmarking participants. The TIMSS scale centerpoint is set at 500 points and represents the mean of the overall achievement distribution in 1995. The TIMSS scale is the same in each administration; thus, a value of 500 in 2015 equals 500 in 1995. All average scores reported as higher or lower than the U.S. average score are different at the .05 level of statistical significance. The tests for significance take into account the standard error for the reported difference.

SOURCE: International Association for the Evaluation of Educational Achievement (IEA), Trends in International Mathematics and Science Study (TIMSS), 2015.

Source: Provasnik et al. 2016.

What about any trends for 8th grade? Compared with 1995, the U.S. average science score in grade 8 was 12 score points higher in 2011 (525 v. 513). However, there was no measurable difference between the U.S. average science score in grade 8 in 2007 (520) and in 2011 (525).

PISA

PISA measures 15-year-olds' capabilities in reading literacy, mathematics literacy, and scientific literacy every three years. PISA was first implemented in 2000, and the most recent results are for the 2015 assessment.

Each three-year cycle assesses one subject in depth. The other two subjects also are assessed, but not in the same breadth and depth as the primary domain. In 2006 and 2015, science was the primary domain. PISA also measures cross-curricular competencies. In 2012, for example, PISA assessed problem solving (OECD 2014). Finally, each assessment includes questionnaires for students, school personnel, and parents.

PISA is sponsored by the Organisation for Economic Cooperation and Development (OECD), an intergovernmental organization of industrialized nations based in Paris, France. OECD and partner countries participate in each assessment cycle.

PISA uses the term *literacy* within each subject area to indicate a focus on the application of knowledge and abilities. Literacy refers to a continuum of knowledge and abilities; it is not a typological classification of a condition that one individual has or does not have. Originally, *scientific literacy* was defined as having the "capacity to use scientific knowledge, to identify questions, and to draw evidence-based conclusions in order to understand and help make decisions about the natural world and the changes made to it through human activity" (OECD 2003, p. 286). This definition was further clarified and elaborated for PISA Science 2006 (OECD 2006) and 2015.

Problem solving was originally defined as an individual's "capacity to use cognitive processes to confront and resolve real, cross-disciplinary situations where the solution is not immediately obvious and where the literacy domains or curricular areas that might be applicable are not isolated within the single domain of mathematics, science, or reading" (OECD 2003, p. 156).

Compared to the curricular orientation of TIMSS, PISA provides a unique and complementary perspective by focusing on the application of knowledge in reading, mathematics, and science for problems and issues in real-life contexts. PISA's goal is to answer this question: Considering schooling and other factors, what knowledge and skills do students have at age 15? The achievement scores from PISA represent the cumulate learning of students at age 15, rather than a measure of the attained curriculum at grades 4 or 8, as is the case with TIMSS. The PISA framework for assessment is based on content, processes, and life situations. For example, in 2003, the content for mathematical literacy consisted of major mathematical ideas such as space and shape, change and relationships, quantity, and uncertainty. The processes described what strategies students use to solve problems, and the situations consist of personal contexts in which students might encounter mathematical problems.

In PISA, a situation may be presented and several questions asked about it. Although some items are answered by selected response, the majority of items require a constructed response. The typical

PISA item makes more complex cognitive demands on the student than the typical item from TIMSS or the National Assessment of Educational Progress (NAEP; Neidorf et al. 2004). Table 18.1 summarizes TIMSS and PISA.

Table 18.1. PISA and TIMSS Overviews

	PISA	TIMSS
Organization Sponsor	Organisation for Economic Co-operation and Development (OECD)	International Association for the Evaluation of Educational Achievement (EA)
Location	Paris, France	Boston, Massachusetts, United States
Content	Reading, mathematics, and science	Mathematics and science
Emphasis	Knowledge and abilities as applied to real-world issues	Knowledge and abilities as attained based on participating countries' curricula
Age or Grade	15-year-olds (mostly grade 10)	Grade 4 (9-year-olds) and grade 8 (13-year-olds)
Assessment Cycle	Every three years, with one content area emphasized in each assessment—2003 emphasis: mathematics; 2006 emphasis: science.	Every four years, with variation of grades

How did U.S. students do on PISA 2015? The PISA 2015 survey had science as the major domain, with reading, mathematics, and collaborative problem solving as minor domains of assessment. The 2015 assessment was completed by approximately 540,000 students worldwide. This number represented about 29 million 15-year-olds in schools from 72 participating countries and economies. PISA scores are reported on a scale with a mean of 500. U.S. students had a score of 496, which placed our students 25th out of 72 countries and education systems. (See Figure 18.3.)

Reflections on International Assessments

Some educators question the fairness in the selection and testing of students in other countries. Also, science teachers rightfully ask, "What can we learn from international assessments?" Our reflections on the responses to these questions follow.

Regarding fairness, many educators perceive a bias in the selection and testing of students in other countries. A careful look clarifies the rigor and fairness of these international assessments. The international assessments use a random stratified sample of schools and of students within those schools. Students are selected in a way that represents the full population of students at the grade level (for TIMSS) or age (for PISA) in each country.

Figure 18.3. Snapshot of Performance in Science on PISA 2015

Countries/economies with a mean performance/share of top performers **above** the OECD average
Countries/economies with a share of low achievers **below** the OECD average

Countries/economies with a mean performance/share of top performers/share of low achievers not significantly different from the OECD average

Countries/economies with a mean performance/share of top performers **below** the OECD average
Countries/economies with a share of low achievers **above** the OECD average

	Science	
	Mean score in PISA 2015	Average three-year trend
	Mean	Score dif.
OECD average	493	-1
Singapore	556	**7**
Japan	538	3
Estonia	534	2
Chinese Taipei	532	0
Finland	531	**-11**
Macao (China)	529	**6**
Canada	528	-2
Viet Nam	525	-4
Hong Kong (China)	523	**-5**
B-S-J-G (China)	518	m
Korea	516	-2
New Zealand	513	**-7**
Slovenia	513	-2
Australia	510	**-6**
United Kingdom	509	-1
Germany	509	-2
Netherlands	509	**-5**
Switzerland	506	-2
Ireland	503	0
Belgium	502	-3
Denmark	502	2
Poland	501	3
Portugal	501	**8**
Norway	498	3
United States	496	2
Austria	495	**-5**
France	495	0
Sweden	493	**-4**
Czech Republic	493	**-5**
Spain	493	2
Latvia	490	1
Russia	487	3
Luxembourg	483	0
Italy	481	2
Hungary	477	**-9**

	Science	
	Mean score in PISA 2015	Average three-year trend
	Mean	Score dif.
OECD average	493	-1
Lithuania	475	-3
Croatia	475	**-5**
CABA (Argentina)	475	**51**
Iceland	473	**-7**
Israel	467	**5**
Malta	465	2
Slovak Republic	461	**-10**
Greece	455	**-6**
Chile	447	2
Bulgaria	446	4
United Arab Emirates	437	**-12**
Uruguay	435	1
Romania	435	6
Cyprus[1]	433	-5
Moldova	428	**9**
Albania	427	**18**
Turkey	425	2
Trinidad and Tobago	425	**7**
Thailand	421	2
Costa Rica	420	-7
Qatar	418	**21**
Colombia	416	**8**
Mexico	416	2
Montenegro	411	1
Georgia	411	**23**
Jordan	409	**-5**
Indonesia	403	3
Brazil	401	3
Peru	397	**14**
Lebanon	386	m
Tunisia	386	0
FYROM	384	m
Kosovo	378	m
Algeria	376	m
Dominican Republic	332	m

1. Note by Turkey: The information in this document with reference to "Cyprus" relates to the southern part of the Island. There is no single authority representing both Turkish and Greek Cypriot people on the Island. Turkey recognises the Turkish Republic of Northern Cyprus (TRNC). Until a lasting and equitable solution is found within the context of the United Nations, Turkey shall preserve its position concerning the "Cyprus issue".
Note by all the European Union Member States of the OECD and the European Union: The Republic of Cyprus is recognised by all members of the United Nations with the exception of Turkey. The information in this document relates to the area under the effective control of the Government of the Republic of Cyprus.

Notes: Values that are statistically significant are marked in bold.
The average trend is reported for the longest available period since PISA 2006 for science, PISA 2009 for reading, and PISA 2003 for mathematics.
Countries and economies are ranked in descending order of the mean science score in PISA 2015.
Source: OECD, PISA 2015 Database, Tables I.2.4a, I.2.6, I.2.7, I.4.4a and I.5.4a.

Source: OECD 2016.

Using PISA as the example, the population in each country consists of 15-year-olds attending both publicly and privately controlled educational systems in grades 7 and higher (Lemke et al. 2004). A minimum of 4,500 students from a minimum of 150 schools per country are required. Within schools, a sample of 35 students are selected in an equal-probability sample unless fewer than 35 students aged 15 were available (in which case all students were selected). There are strict requirements for school-response rates and exclusion guidelines for functionally disabled students, students with mental or emotional disabilities, and students with limited proficiency in the test language.

We describe these criteria because of the persistent dismissal of results from international assessments due to perceived issues of "who is tested." This brief summary indicates the procedures used to make the samples consistent for each country and serves to punctuate the need for U.S. science teachers to take these results seriously. (*Note:* For greater detail of the sampling, data collection, and response-rate requirements, see Lemke et al. [2004].)

So, what can we learn from international assessments? First, as mentioned, TIMSS is a curriculum-based assessment that educators can use to gain insights about the science curriculum in other countries. Especially helpful are perspectives from high-achieving countries and countries that are economic competitors. For example, U.S. teachers report that they teach many more topics during a school year than do teachers in Japan and Germany. On average, at the equivalent of U.S. eighth grade, Japan and Germany include between 6 and 8 topics in science textbooks. At the same grade level, U.S. science textbooks average about 60 to 70 topics per year (NRC 1999).

Second, PISA focuses on young people's ability to use their knowledge and apply their skills to life situations. PISA scores represent the product of learning experiences at age 15. As an example of PISA's benefit, the assessment provides specific information that pertains to the basic skills needed by U.S. students and our continuing interest in international competitiveness.

Third, information on students' problem-solving abilities should be of particular interest to science teachers. PISA included problem solving as a cross-disciplinary component of the 2012 program. PISA exercises assessed 15-year-olds' abilities to use reasoning processes to draw conclusions, make decisions, troubleshoot, and analyze procedures and structures of complex systems. The assessment required students to apply processes such as inductive and deductive reasoning, establishing cause-and-effect relationships, and combinatorial reasoning. Finally, the problem-solving assessment also related to other basic skills, namely working toward solutions and communicating the solution to others through appropriate representation. U.S. students scored 508, significantly higher than the OECD average of 500. U.S. students scored lower than their peers in 17 other countries or educational systems (e.g., countries such as Korea, Canada, and Germany and educational systems such as Macao-China, Shanghai-China, and Chinese Taipei). In short, our students' performance on problem solving compares moderately well with other industrialized democracies (OECD 2014).

Fourth, one might ask about the expectations of U.S. 15-year-olds as far as their future education and occupations. PISA did ask about these goals. In the United States, 64% of students reported that they expected to complete a bachelor's degree or higher. This was much higher than the OECD average (44%). Only South Korea reported a higher percentage (78%) than the United States. U.S.

students who expected to complete a bachelor's degree did score higher than their peers with lower educational expectations. However, compared with 37 countries reporting data, U.S. students with higher educational expectations were outperformed by their peers in 26 countries in mathematics literacy on PISA 2012.

Students also were asked about their job expectations by age 30. Responses were coded according to the International Standard Classification of Occupations (ISCO). The responses were then collapsed into three categories—high, medium, and low—based on skill level. In this study, 67% of U. S. students reported high job expectations and 32% reported medium expectations. The OECD average for high occupational expectations was 47%. As with degree expectations, U.S. students with high job expectations scored higher on the mathematics literacy scale than their U.S. peers with lower expectations. However, U.S. students with high job expectations scored lower than the OECD average in mathematics literacy.

Low scores on reading, mathematics, and problem solving contrasted with high educational and occupational expectations should be cause for concern, if not alarm. Not only do we have to increase performance in reading, mathematics, problem solving, and science, but we also need to establish accurate, realistic, and reasonable expectations for education and careers for our students.

In conclusion, perhaps the most educationally significant insight to be gained from PISA emerges from the difference between TIMSS and PISA. The difference is the orientation or emphasis of the respective assessment. TIMSS is grounded in the curriculum and provides feedback for how students are *attaining* what is intended and enacted vis-à-vis a country's curriculum. While not ignoring school curriculum, PISA asks how students can *apply* their knowledge in real-world situations. Lower scores on PISA suggest that our students do not do as well as the majority of our economic competitors when they have to demonstrate basic skills in contextual problems. This should be a concern for policy makers and educators alike.

CHAPTER 18

A PERSONAL PERSPECTIVE

PISA AND FINNISH EDUCATION

Rodger Bybee

As the science education community moves forward with the implementation of standards and the Every Student Succeeds Act (ESSA) of 2015, the time seems ideal to consider lessons from education systems in other countries. In this essay, I direct attention to international assessments, particularly PISA.

Like many, I have tried to find solutions to problems of American education, especially changes that will contribute to student achievement. One reasonable place to seek such solutions is in other countries, such as Finland, that have high levels of student achievement. In 2010, I had the opportunity to spend two weeks (one week scheduled and one additional week due to a volcanic eruption in Iceland) studying education in Finland. My time included visits with members of parliament, the ministry of education, members of the national board of education, university scientists, teacher educators, teachers, teachers in training, students, and parents. When I began reviewing notes, recalling discussions, and thinking about Finnish education, I confronted a number of observations contrary to my knowledge, experience, and beliefs about programs or practices in the U.S. education system.

Here are some facts you may have heard about Finland and some you likely have not heard:

- Finnish students had exceptionally high achievement in all three literacy domains—reading, mathematics, and science—in PISA studies of 2000, 2003, and 2006.
- Finnish students were among the best in terms of mean scale scores and in terms of a small variance. The latter refers to the small numbers performing below the level deemed necessary for full participation as citizens—that is, even the "low achievers" had higher scores than "low achievers" in most other countries.
- Finnish students, especially girls, do not find school science, careers, or occupations interesting. Furthermore, they do not think science is relevant for them.
- Finnish students do not begin formal schooling until age 7, and their daily school time is typically shorter than in many other countries. Furthermore, 66% of Finnish students indicate they have less than 2 hours of homework each week.
- Finish students have a basic comprehensive education from grade 1 through grade 9.
- Finnish students have highly qualified teachers. Selection to enter the profession is rigorous, teacher education programs are 6 years, and most teachers have master's degrees.

- Finnish students in the upper grades have class periods that are 75 minutes long. (This information was summarized from the following sources: Kupiainen, Hautamaki, and Karjalainen 2009; Pehkonen, Ahtee, and Lavonen 2007; Sahlberg 2013; Finnish National Board of Education 2004; Ministry of Education 2008.)

This combination of observations left me with a fundamental question: What is the explanation for Finnish students' reasonably consistent high achievement in reading, mathematics, and science? My search for an explanation was more than curiosity, as I had to discuss "Basic Education in Science: Innovations for the Year 2020" at a Finnish national forum on education reform. It seems self-evident that giving education advice to a country that is first in the world ought to be done with understanding and no small amount of humility, especially if you are from the United States, which in 2006 ranked 29th out of 57 countries participating in PISA.

Because Finnish students have consistently attained exceptionally high levels of achievement in reading, mathematics, and science, many education experts have investigated and written about the education system in Finland (Kupiainen, Hautamaki, and Karjalainen 2009; Pehkonen, Ahtee, and Lavonen 2007; Sahlberg 2013). My weeks of interviews, visits, and reviews revealed little new that could be stated. Most of my observations confirmed what others had observed—for example, teachers are well educated, and Finland is a society of fluent readers.

Upon further review, it struck me that there were several apparent contradictions with what I believed and assumed to be true about education in general and curriculum and instruction in particular. Here are several observations that originally went without special notice:

- The Finnish "standards" are fewer, and broader, than U.S. standards.
- Students attend local comprehensive schools.
- There are no national high-stakes assessments. Teachers create their own exams or use tests made by subject matter organizations.
- The teacher education program consists of lectures about pedagogy and associated school-based topics and few actual student teaching experiences.
- Teachers use lecture-discussion extensively.

Now, I can imagine you asking, "So, what accounts for the higher levels of achievement by Finnish students?" To be clear, variations of this question have been asked and answered by educators in Finland and elsewhere. Here is my answer.

First, there is remarkably high coherence across the different components of the Finnish educational system. In meetings at the Ministry of Education with teacher educators and classroom teachers, I heard the same vision of supporting a knowledge-based society and a comprehensive education for all students. This vision was clear, and it brought a unity of purpose from national policies to classroom practices.

Second, Finland places significant emphasis on recruiting the best individuals into teaching, training them well, then expecting great results. Teachers have a high level of public respect. Consequently, teaching, especially at the elementary level, is a popular profession. For example, only 10% to 15% of candidates for elementary school teaching are accepted into teacher education programs. For secondary teachers who select areas such as science and mathematics, their education allows them to continue into a variety of professions besides teaching. Most elementary and secondary school teachers attain a master's degree before they begin teaching. Their programs include the knowledge and skills needed for their respective levels—for instance, broad content for elementary subjects and specific content for secondary teachers.

Interestingly, the pedagogical strategies are not necessarily innovative. Teacher education programs emphasize a second subject matter major and a range of instructional methods. In a secondary classroom, for example, I observed an integrated sequence of instruction for 75 minutes. It was a mathematics class on the Pythagorean theorem. The teacher began by engaging students with several basic calculations; he continued with an introduction of the theorem and proceeded to introduce several problems. Students practiced calculations by selecting from easy or hard problems. During this time, the teacher talked to each student about his or her work, often directing students to try different problems based on their progress. Students took a short quiz, the effect of which was "non-negative"—that is, the result could help, but it could not lower the student's grade. This was one of the most effective lessons I have seen. The teacher was clearly competent and comfortable in the subject, demonstrated differentiated instruction across the class period, used several different technologies, was enthusiastic, and used questions, examples, direct instruction, and math problems at various levels in the course of the lesson.

Third, Finland has a core curriculum that describes objectives, core content, and final assessment criteria for grade level sets that vary across the disciplines. Each description of the "standards" for a domain (e.g., mathematics, physics and chemistry, history, music) is between three and five pages long. The document we would think of as the standards, *National Core Curriculum for Basic Education 2004,* is 319 pages. This document constitutes regulations used by educators to make decisions about the core curriculum. Compared to U.S. documents, this description really does present less and expresses expectations of more learning by all students in grades 1 through 9.

Here is a central point that was reiterated time and time again. Local authorities had freedom and autonomy to develop instructional programs to attain the content and skills described in the core curriculum document. There was, to state what seems to be a paradox, national regulation and local autonomy. These two apparent contradictions were connected and resolved by public trust in teachers and professional responsibility by educators.

Fourth, beginning in the mid 20th century, Finland initiated a radical education reform that resulted in a comprehensive nine-year education for all students. This education begins with grade 1 for 7-year-olds and ends at grade 9. After grade 9, the education system divides into general upper-secondary (i.e., high school) and vocational upper-secondary education and training. If students continue their education, the former enroll in universities and the latter go to polytechnic schools. Students can change "tracks" if they qualify.

There were new policy changes that decentralized education in the 1980s and 1990s and early in the 21st century. However, several important points stand out. A high consistency within the core curriculum continued. A requirement for classroom hours per week of instruction across all subject areas was established. The curriculum emphasized basic skills and knowledge with application to life situations. Finally, a central goal of Finnish educational policies was to provide all citizens with equal opportunities to obtain a basic education regardless of language, immigrant status, age, gender, or the family's financial situation.

Fifth, there is reduced emphasis on national assessments and no high-stakes tests in basic education. This said, it is still the case that students do have examinations during their school years. Assessments are left to schools and individual teachers. For example, teachers may develop their own examinations or use tests that accompany most textbooks. In addition, subject teachers' associations provide examinations with accumulated norms that can provide feedback to guide teaching and learning toward the comprehensive certificate used for entrance to upper-secondary schools. In short, testing is for learning and not consequential accountability.

The success for the Finnish education system, especially on PISA, has stimulated searches for key explanations, many of which are singular in nature. For example, some have suggested it is teacher education, the homogeneity of the society, or reading as a common pasttime. For sure, there is not a single key explanation; rather, there are several interrelated and interdependent explanations.

As the Finnish government realized the need to make a radical change from an agricultural society to an industrial and post-industrial, knowledge-based society, they also realized this involved education reform if Finland was to have the workers needed for the new society. It was serendipitous that education reform centered on a comprehensive, basic education accentuating math and science that apply to life and work. These foreshadow the orientation of PISA, as the assessment was developed 30 years later. I suggest this was not a coincidence of history. The orientation of both Finnish education and PISA reflects the fundamental purposes of a general education for all citizens in a free society.

While any review and search for explanations must acknowledge the role of teacher education, the coherence across the core curriculum, and reduced emphasis on high-stakes assessments, it is equally important to at least acknowledge the underlying common national goal, remarkable consistency from national purposes to local practices, positive perceptions of those in the education system, a culture of professional trust, and autonomy of school personnel. These attitudes and orientations are not commonly heard in other education systems. In the end, those may well contribute to less really being more.

Before discussing some implications of Finnish education for the U.S. educational system, it is important to make two points. First, Congress reauthorized the Elementary and Secondary Education Act (ESEA), the Every Student Succeeds Act (ESSA) of 2015. We have the *NGSS* that states have adopted and other states and school districts have adapted (see Table 18.2, p. 292). We also have the

U.S. National Assessments of Educational Progress (NAEP) for science, technology, engineering, and mathematics. These will continue to influence the U.S. system from national to local levels. They cannot be ignored in any discussion of education reform. In fact, current policies such as ESSA and assessments such as NAEP and TEL can be viewed as the basis of, and support for, some changes implied by high-achieving countries such as Finland.

Second, I avoid the error of recommending we "transplant" education in Finland to the United States. There is not donor-recipient compatibility between the two cultures; we do not know how to adequately suppress the educational "immune system" and those defensive political and professional antibodies that would be generated to reject such a large-scale reform based exclusively on the Finnish system. The Finnish education system was developed in historical, social, and cultural contexts, and the likelihood of replication by other countries is almost zero. There may be, however, some key insights and generalizations that apply to U.S. education at state, district, and classroom levels. Here are insights that may have implications for science education.

In Finland, there is a consistency from national purpose to classroom practice. From members of parliament to classroom teachers, establishing a "knowledge-based economy" was a consistent theme. As this purpose was translated to national educational policies in the *National Core Curriculum for Basic Education 2004* (Finnish National Board of Education 2004) and programs for teacher education, teacher educators, and teaching, I recognized the orientation to the major purposes of education. Associated with this consistency across different components of the system was a distribution of authority and recognition of the form and function of that authority's place and responsibilities in the education system. For example, the ministry of education had the responsibility for establishing polices while the local municipality and classroom teacher had responsibility for developing the actual curriculum and tailoring the materials to the local, unique needs of students. The glue that held all of this together was another theme heard everywhere—a culture of professional trust.

In the United States, it would be quite possible for the science education community in a state to establish a vision of "every student succeeds in science" and use the *NGSS* or new state standards as the basis for consistent policies, programs, and practices. In my view, this is not impossible. You undoubtedly notice the vision of "every student succeeds in science" is a reflection of the ESSA 2015 legislation. As states adopt or adapt new standards, these could certainly be the basis for the form and function of new policies, curriculum programs, and classroom practices.

Concurrent with the initial reform of Finnish education was a curriculum framework—that is, standards for an academically demanding curriculum. At that time, during the early to mid-1970s, there was a centralized system guiding the reform. That guidance was provided by the National Board of Education and included such things as school inspectors and approval of textbooks. In the 1980s, however, the political climate changed and the educational system became more decentralized. This change included the 1985 curriculum framework allowing greater freedom—and, I would note, responsibility—at the local level for development of the actual school program. The freedom was not absolute and total; rather, it was limited by the academic content and time requirements

across the basic academic subjects. The common curriculum in Finland includes all subjects—such as language, mathematics, sciences, social studies, arts, and physical education—and, very important, requirements for the distribution of lesson hours expressed in terms of weekly lessons per year for all subjects.

The U.S. does not have common standards for all academic subjects, so coherence across all subjects seems unlikely. However, due to ESSA 2015, the U.S. climate currently supports increased decentralization for decisions concerning the science curriculum and assessment. This insight does not imply absolute freedom with no accountability. Adapting the model from Finnish education suggests that the states adopt new content standards for science and make adaptations based on regional or district interests. Schools and teachers could exercise freedom in adopting textbooks, software, and other curriculum materials while assuming responsibility for students attaining the learning outcomes in state standards.

In Finland, teacher education was restructured in the 1970s along with the reform of basic education. Teacher education transferred from teacher colleges to universities and the requirement for teachers was set at a master's degree. For classroom (i.e., elementary) teachers, the degree is in education, and for subject teachers (i.e., upper secondary), the degree is in their respective subject (e.g., history, science). Their studies include both content and pedagogy.

Shifting teacher education to universities underscored a research-based foundation and prepares teachers for continued academic study. The academic status of classroom teachers likely contributes to the status, popularity, and trust of teachers in Finland. Because the teaching profession cares about prestige, the acceptance rate into university programs is only between 10% and 15% of candidates. Only the most motivated and qualified are accepted into teacher education programs. I should note that the same situation does not exist for subject teachers such as mathematics and science.

I have difficulty imagining a nationwide reform of teacher education in the United States. Increasing alignment of teacher education with state standards and providing professional development based on new state standards does seem possible.

There is a trend toward reducing the time spent on testing. This, too, is supported by ESSA 2015. The detrimental effects of excessive testing and consequential accountability for teachers now seem clear. The unintended consequences of emphasizing literacy and mathematics at the expense of science also are clear. I note that Finland does have a national test somewhat like the U.S. National Assessment of Educational Progress (NAEP). That test is primarily used for feedback and not accountability within the system.

In conclusion, my examination of the Finnish educational system reveals some possible changes that could result in improved science education in the U.S. I have suggested changes within a contemporary context and ones that relate to U.S. states, districts, and classrooms. Some of the changes, such as adopting new standards, have been initiated, and others will follow. There also is the responsibility of educators within states to express the new vision in their various policies, programs, and practices.

Table 18.2. Summary of States Adopting or Adapting the *Next Generation Science Standards*

States Adopting *NGSS*	States Adapting the *Framework* and *NGSS*
• Arkansas • California • Connecticut • Delaware • District of Columbia • Hawaii • Illinois • Iowa • Kansas • Kentucky • Maryland • Michigan • Nevada • New Jersey • Oregon • Rhode Island • Vermont • Washington	• Alabama • Georgia • Idaho • Indiana • Massachusetts • Missouri • Montana • Oklahoma • South Carolina • South Dakota • Utah • West Virginia • Wyoming
States With Standards in Development or Not Formally Adopted	**States That Have Not Revised Their Science Standards**
• Colorado • Louisiana • Minnesota • Nebraska • New Hampshire • New Mexico • New York • North Dakota • Pennsylvania • Tennessee • Virginia	• Alaska • Arizona • Florida • Maine • Mississippi • North Carolina • Ohio • Texas • Wisconsin

Note: Updated list as of September 2016.

OUR COMMON PERSPECTIVE AND LEADERSHIP OPPORTUNITIES

One of the first and most fundamental ways to increase coherence in components of the science education system is through valid and reliable classroom assessments. Leaders in science education can support teachers in this effort through professional development.

Linking contemporary state standards and assessments is a challenge, but it must be done. There are several critical points that leaders should recognize. One critical point is that the form and function of state assessments must be consistent with state standards.

To be fair, students should have had opportunities to learn the science content and abilities on assessments. This is another critical point. Indeed, this is essential. If the students do not have adequate and appropriate opportunities to learn the content and practices described in the standards, it is not fair to assess them on those outcomes.

Assessments beyond the classroom should align with K-12 learning progressions in science. This perspective, when implemented, will bring coherence to K–12 school science programs.

To the degree possible, science assessments should be coordinated with appropriate elements of standards for English language arts and mathematics. Leaders can implement connections between the experiences in science and other important learning outcomes for our students. This is one important example.

Contemporary assessments of science should include the nature of science, technology, and engineering. These aims are included in the standards for science and engineering practices and crosscutting concepts. Now it is important to include them on assessments.

Leaders at the school, district, state, and national levels should be knowledgeable about the general characteristics of and recent results of state, national, and international assessments. When presenting the ideas behind reform and improvement, leaders should have a depth and breadth of knowledge about the standards and assessments. In addition, they should be clear about the costs and benefits of the proposed changes.

ISSUES AND QUESTIONS FOR DISCUSSION

1. How would you complete this statement?
   Classroom assessments measure ____________________, provide information about science teachers' ______________________, and help them _________________.
2. "Teaching to the test" is generally a negative statement. What would have to change to make "Teaching to the test" a positive statement?
3. Do assessments and accountability represent a grand challenge for science education? If so, exactly what is the challenge? Is it the design of tests? Accountability or other issues?
4. How do you think the Every Student Succeeds Act of 2015 will affect assessment and accountability?

5. Do you think the National Assessment of Educational Progress (NAEP) tests are worth the time and expense? Provide an argument for your position.
6. Do you think the international assessments provide helpful information for U.S. educators? Provide an argument for your position.
7. What would you propose as a means for improving the U.S. position on international assessments?

REFERENCES

Bybee, R., and B. McCrae. 2009. *PISA science 2006: Implications for science teachers and teaching.* Arlington, VA: NSTA Press.

Finnish National Board of Education. 2004. *National core curriculum for basic education 2004.* Helsinki: Ministry of Education.

Kupiainen, S., J. Hautamaki, and T. Karjalainen. 2009. *The Finnish education system and PISA.* Helsinki: Ministry of Education Publications.

Lemke, M., A. Sen, L. Partelow, D. Miller, T. Williams, D. Kastberg, and L. Jocelyn. 2004. *International outcomes of learning in mathematics literacy and problem solving: PISA 2003 results from the U.S. perspective.* Washington, DC: U.S. Department of Education, National Center for Education Statistics.

Ministry of Education. 2008. *PISA06 Finland.* Helsinki: Ministry of Education.

National Research Council (NRC). 1999. *Global perspectives for local action: Using TIMSS to improve U.S. mathematics and science education.* Washington, DC: National Academy Press.

Neidorf, T. S., M. Binkley, K. Gattis, and D. Nohara. 2004. *A content comparison of the NAEP, TIMSS, PISA 2003 mathematics assessments.* Washington, DC: U.S. Department of Education, National Center for Education Statistics.

Organisation for Economic Cooperation and Development (OECD). 2003. *The PISA 2003 assessment framework: Mathematics, reading, science, and problem solving knowledge and skills.* Paris, France: OECD.

Organisation for Economic Cooperation and Development (OECD). 2006. *Assessing scientific, reading, and mathematical literacy: A framework for PISA 2006.* Paris, France: OECD.

Organisation for Economic Cooperation and Development (OECD). 2014. *PISA 2012 results: Creative problem solving. Students' skills in tackling real-life problems. Volume V.* Paris, France: OECD Publishing.

Organisation for Economic Cooperation and Development (OECD). 2016. *PISA 2015: Results in focus.* Paris, France: OECD.

Pehkonen, E., M. Ahtee, and J. Lavonen, eds. 2007. *How Finns learn mathematics and science.* Rotterdam, The Netherlands: Sense Publishers

Provasnik, S., L. Malley, M. Stephens, K. Landeros, R. Perkins, and J. H. Tang. 2016. *Highlights from TIMSS and TIMSS Advanced 2015: Mathematics and science achievement of U.S. students in grades 4 and 8 and in advanced courses at the end of high school in an international context* (NCES 2017-002). Washington, DC: U.S. Department of Education, National Center for Education Statistics.

Sahlberg, P. 2013. Teachers as leaders in Finland. *Educational Leadership* 71 (2): 36–40.

SECTION X

REFORMS, POLICIES, AND POLITICS IN SCIENCE EDUCATION

If it seems that American science education has a long history of reform, relatively frequent changes in policy, and fairly few substantial changes in school programs and classroom practices, your perception is mostly accurate. Why, one must ask, do so few reforms result in real changes in schools and classrooms? What is the role of politics in science education reform?

Depending on your perspective, you can view education reforms as a long and steady process of evolution or a series of independent events and issues. Lacking an understanding of history, especially in a larger social and political context, a perception of largely independent events and isolated calls for change is understandable but, in our view, inaccurate.

The fact that education functions within society and is subject to political and economic changes supports an evolutionary perspective, one that may have recognizable trends and cycles. From an evolutionary point of view, you should be able to identify changes in the larger social environment that result in selective pressures on education in general and science education in particular. Social changes may be perceived as, for example, threats to national security, health scares, economic recessions, and needed support for business and industry. The extension of societal issues such as these to education has been characterized by themes such as "life adjustment," "back to basics," and preparing for "college and career."

In Table X.1, we have listed examples of important political events that have influenced American education. This overview should provide insights and a larger perspective for Chapters 19 and 20.

Table X.1. Post–World War II Political Events Influencing Education Reform

1948	*Harry Truman is elected president.* President Truman supports general aid for public schools.
1952	*Dwight D. Eisenhower is elected president.*
1954	*Brown v. Board of Education* The U.S. Supreme Court rules that segregated schools are unconstitutional.
1956	*President Eisenhower is re-elected.*
1957	Union of Soviet Socialist Republics (USSR) launches Sputnik.
1958	The National Defense Education Act (NDEA) becomes law.
1960	*John F. Kennedy is elected president.*
1963	President Kennedy is assassinated, and Lyndon B. Johnson assumes the presidency.
1964	*Lyndon B. Johnson is elected president.* The Civil Rights Act becomes law.
1965	The Elementary and Secondary Education Act (ESEA) becomes law.
1968	*Richard M. Nixon is elected president.* The Bilingual Education Act becomes law.
1972	*President Nixon is re-elected.* Title I is administered as targeted assistance. Title IX, affecting girls and women, is adopted.

Table X.1. *(continued)*

1973	Section 504 of the Rehabilitation Act, which affects people with disabilities, is adopted.
1974	Amendments make Title I a targeted assistance program. In *Lau v. Nichols*, the Supreme Court rules that children must be given assistance to learn English. Congress broadens the Bilingual Education Act to provide more services and passes a provision to remove language as a barrier to education as part of the Equal Education Act.
1975	The Individuals with Disabilities Education Act becomes law.
1976	*Jimmy Carter is elected president.*
1980	*Ronald Reagan is elected president.* The U.S. Department of Education replaces the U.S. Office of Education.
1981	Reagan succeeds in cutting back on federal programs and funding. The Bilingual Education Act allows some funding for English-only programs.
1982	The *Sustaining Effects Study* finds modest academic improvement through Title I.
1983	*A Nation at Risk* is released.
1984	*President Reagan is re-elected.*
1987	The Civil Rights Restoration Act restores the broad effects of Title IX.
1988	*George H. W. Bush is elected president.* Title I is amended to emphasize the need to increase students' academic achievement. Congress increases funds available for English-only programs under the Bilingual Education Act.
1989	President Bush convenes the nation's governors at a Charlottesville summit on education. AAAS Project 2061 releases *Science for All Americans.*
1990	President Bush signs the Americans with Disabilities Act.
1992	*Bill Clinton is elected president.* President Bush's legislation related to national goals and statewide reform is impeded by the Senate.
1993	AAAS Project 2061 releases *Benchmarks for Science Literacy.*
1994	Goals 2000 and the ESEA amendments create a national standards-based program. The Bilingual Education Act is amended to permit more English-only programs.
1995	The Senate votes against President George H. W. Bush's national history standards. The National Research Council releases *National Science Education Standards* (with 1996 copyright).
1996	*President Clinton is re-elected.*

Table X.1. *(continued)*

2000	*George W. Bush is elected president.*
2002	The No Child Left Behind Act (NCLB) becomes law. The Bilingual Education Act becomes the English Acquisition Act.
2004	*President George W. Bush is re-elected.*
2008	*Barack Obama is elected president.*
2009	The American Recovery and Reinvestment Act (ARRA) becomes law.
2010	The *Common Core State Standards* are released by the governors and the Council of Chief State School Officers.
2011	The NCLB waiver program begins.
2012	*President Obama is re-elected.* The National Research Council releases *A Framework for K–12 Science Education* (*Framework*).
2013	The National Academies and Achieve, both non-federal agencies, release the *Next Generation Science Standards* (*NGSS*).
2017	The *NGSS* have been adopted by 17 states and the District of Columbia, and 13 states have new standards influenced by the *Framework* and *NGSS.*

SUGGESTED READINGS

Atkin, J. M., and P. Black. 2003. *Inside science education reform: A history of curriular and policy change.* New York: Teachers College Press, Columbia University. Personal histories of reform from two leaders in the United States and United Kingdom.

Colvin, R. L. 2013. *Tilting at windmills: School reform, San Diego, and America's race to renew public education.* Cambridge, MA: Harvard Education Press. An insightful case of reform and what the constructing and countervailing political forces can accomplish.

Cross, C. 2014. *Political education: Setting the course for state and federal policy.* New York: Teachers College Press, Columbia University. An in-depth review of politics and policies that have influenced education reforms. The emphasis is on federal policies.

DeBoer, G. E. 2006. History of the science standards movement in the United States. In *The impact of state and national standards on K–12 science teaching,* ed. D. D. Sunal and E. Wright, 7–49. Greenwich, CT: Information Age Publishing. A detailed history of *Benchmarks for Science Literacy* and the *National Science Education Standards* by an author with experience and knowledge of the standards movements.

Dow, P. B. 1991. *Schoolhouse politics.* Cambridge, MA: Harvard University Press. This is one insightful set of political lessons based on education reform in the Sputnik era.

Jennings, J. 2015. *Presidents, Congress, and the public schools: The politics of education reform.* Cambridge, MA: Harvard University Press. A 50-year history of congressional decisions and policies by a person who worked on Capitol Hill. The history is both well informed and unbiased.

Tycak, T., and L. Cuban. 1995. *Tinkering toward utopia: A century of public school reform.* Cambridge, MA: Harvard University Press. The title and subtitle tell the story. This is a classic historical discussion of educational reform.

CHAPTER 19

EDUCATION REFORM

HISTORICAL PERSPECTIVE

American education has a long history of making the curriculum available to diverse populations of students. Historian Lawrence Cremin referred to this theme as popularization (Cremin 1990). In science education, popularization is represented by the contemporary vision of scientific and technological literacy for all students.

American education has another enduring characteristic: its politicization, and specifically the continuing effort to solve social problems indirectly through education instead of directly through political processes (Cremin 1990). Examples of the intersection of education and politics abound: desegregation, the war on poverty, economic competition with other countries, and the war on drugs. The point here is the perspective that education reform is seen as the means to solve social and economic problems and not as one of multiple means to reduce the problem and improve society. Examining federal policies provides one perspective on politics and education reform. Both Christopher Cross in *Political Education: Setting the Course for State and Federal Policy* (Cross 2014) and Jack Jennings in *Presidents, Congress, and the Public Schools: The Politics of Education reform* (Jennings 2015) have provided insightful discussions of politics and education. We briefly look at some examples of education reform.

Periodically throughout history, schools were called on to reform so they would more accurately reflect the realities of the developing American society. Committee reports often reveal the nature of these reformations. Several committee reports from these periods of reform will serve as examples of the politics and policies influencing changes in education, including science education.

The Great Depression of the 1930s brought doubts and questions concerning education in science. There were two important publications in this period. The first, released by the National Society for the Study of Education (1932), was *A Program for Science Teaching*. This report emphasized the

importance of broad scientific principles that help students gain a fundamental understanding of nature. The second publication was from the Progressive Education Association (1938) and titled *Science in General Education.* This book stressed progressive goals such as personal-social relationships, personal living, economic relations, and reflective thinking. The general orientation of science programs was toward the more immediate needs of students; the content was of personal and social significance; and there were suggested programs in health, vocation, and consumerism. The greatest changes were in biology and general science; physics and chemistry remained largely as they had been.

The period after the Depression was relatively calm concerning science education. The National Society for the Study of Education (1947) published *Science Education in American Schools* with general objectives for science teaching; the objectives included functional information, concepts, and principles; skills and attitudes aligned with the scientific method; and recreational and social values of science. World War II gave the country a new sense of national purpose, which was reflected in the literature of science education.

After the World War II, policies stressed life-adjustment education. Examples of reports included the Educational Policies Commission's *Education for All American Youth* (1952); the National Association of Secondary School Principals' *Planning for American Youth* (1944); and the U.S. Office of Education's *Life Adjustment Education for Every Youth* (1951).

In the early 1950s, the United States entered a period of crises brought on by the rapid growth of science and technology after World War II and the simultaneous failure of science education to reflect these changes in curriculum, instruction, and career opportunities. The crises led to massive changes in the curriculum, predominately funded by the National Science Foundation (NSF). Generally, the emphasis shifted to knowledge-oriented aims and used the processes of science as a means of developing the primary objectives.

In the 1950s, during another economic recession, the schools were criticized by politicians and policy makers for lacking adequate academic goals (Bestor 1953; Rickover 1959). The cry was for a return to the basics, emphasis on traditional subjects, and special attention to the gifted. This tide of criticism rose with the October 1957 launch of Sputnik. The reform movement was then heading toward a period of curriculum revision and teacher training that was the largest in our history. By the middle of the 1960s, however, a new wave of education critics appealed for a greater understanding of student alienation and identity. The focus in education had shifted from the space race to urban disgrace. By the late 1970s, the public learned there was a "crisis in the classroom" (Siberman 1979).

In this historical survey, we have tried to clarify science education reforms resulting from various needs and demands of society. We describe the ongoing relationship between changed society and the type of curriculum and instruction that science educators are called on to provide in schools. The major social pressures have included the growth of an industrial-technological society, the demands of a depressed economy, and the emergence of an atomic age.

Reform in science education is not new; we have been continually reforming for more than 200 years. The process of reform takes place within our community of science educators, but it seems

clear from our history that reform is brought about and directed by the larger contextual forces in our society.

It should be clear that education as a social institution has a role in solving problems and contributing to the progress of America. This perspective was shared from Thomas Jefferson to John Dewey and contemporary reformers. That is not to say that education reform has been easy or that reforms have resulted in the changes presented in various reports and policies. The next section uses the 4*P*s perspective described in Chapter 1 to clarify some of the difficulties of education reform, particularly in science education.

PERSPECTIVES ON EDUCATION REFORM

We think it is important to look at various reforms from perspectives characterized in the introductory discussion of purposes, policies, programs, and practices and the parallel rhetoric of philosophical proposals, political talk, programmatic initiatives, and practical educations results. Indeed, we recommend taking this perspective one step further. Looking from left to right, or up and down the continuum of purpose, policy, program, and practice, how have various reform efforts addressed these different dimensions, and how have the efforts especially supported changes at the instructional core (i.e., classroom practices)?

Before continuing this discussion of education reform, we will digress to make a point about the four perspectives. We elaborate on a theme presented by Mark Windschitl (2006). The theme of Windschitl's article is the inability of individuals with different perspectives—traditionalist and reformist—to talk to one another due to the propensity to take on a position and belittle other positions. While we take a position that reform in science education ultimately must center on classrooms, it also is the case that individuals write, speak, and represent perspectives on purposes, policies, and programs, and these individuals should be recognized as part of the educational system and not rejected by sarcastic and dismissive comments. Yes, there is one's "real world" of the classroom; there are also real worlds of curriculum developers, assessment specialists, policy makers, and even philosophers.

From Purposes to Practices: A Brief Review

To address the issue of improving science education, we use a framework that identifies different domains of education reform. The framework has been described in other publications (Bybee 1993, 1997) and in Chapter 1 of this book. The framework proposes four connected and interdependent domains: purposes, policies, programs, and practices (i.e., the 4*P*s). Each domain has its advocates, audiences, problems, solutions, and politics in the processes of educational reform.

Establishing the Purposes of Science Education

Science educators directly or implicitly express purposes in statements of aims and goals in philosophical articles, national standards, and preambles to state frameworks and district syllabi. The term *purpose* refers to abstract and universal statements that apply to all and refer to what science education should achieve. "Achieving scientific literacy" serves as an example of a purpose statement. The strength lies in wide acceptance of the purpose. The weakness lies in the translation to specific situations, such as college and university teacher education programs, state assessments, high school chemistry teachers, middle school Earth science curriculum, or a second-grade teacher. So, there is a need for more statements and guidance that address different components of education systems.

Describing Policies for Science Education

Policy statements are translations of the purpose for different aspects of the science education system. Documents that give direction and guidance, but are not actual programs, serve as policies. Examples of policy documents include course plans for high school science classes, district syllabi for K–12 science, and state frameworks. Likewise, college or university requirements for undergraduate teacher education and state and national frameworks for assessing scientific literacy also fall into the category of policies for science educators. In the United States, and for this book, policy documents include the *Next Generation Science Standards* (*NGSS*; NGSS Lead States 2013).

Developing Programs for Science Education

Science education programs include the actual curriculum materials, textbooks, and electronic media based on the policies. Programs are unique to grade levels, disciplines, and aspects of science education, such as teacher education or secondary school science.

School science programs may be developed by independent national organizations and marketed commercially, or they may be developed by states or school districts. Who develops the materials is not the defining characteristic; the fact that schools, colleges, state agencies, and professional organizations have programs aligned with policies such as national standards is the important feature of this domain of reform.

Improving Practices of Science Education

Practice here refers to the specific actions, methods, strategies, and processes of teaching science in schools, colleges, or universities. The practices of science education include the personal interactions between teachers and students and among students, as well as the roles and uses of assessment, educational technologies, laboratories, and myriad other methods of teaching science. In the 4*P* organization described here, implementing a new innovative science curriculum implies that it would generally be consistent with state standards and designed to enhance learning, which would enable students to achieve higher levels of scientific literacy.

One can assume that greater coherence and consistency exists *within* a domain, such as policy, than exists *between* domains, such as policy and program. The explanation for this situation rests on the observation that rewards, incentives, and resources tend not to emphasize the translation of policies to programs and purposes to practices. Policy makers mostly tend to communicate with other policy makers, and practitioners mostly tend to communicate with other practitioners. If one is truly interested in widespread, large-scale, and long-term education reform, considerable time and effort

must focus on the interface between the domains. All those in the science education community must recognize that reform will occur through coordinated, consistent, and coherent translation of purpose into policies, programs, and practices. We note this issue as one significant factor that influences the impact of national standards on the science curriculum, a discussion we pursue in Chapter 20.

Throughout our history, the primary emphasis on reform has been at the policy level. The Committee of Ten is an example. In science education, reports such as those of the National Society for the Study of Education—including *A Program for Science Teaching* (NSSE 1932, *Science Education in American Schools* (NSSE 1947), and *Rethinking Science Education* (1960)—have stated broad objectives and plans for improving science education. The actual development of programs and changes to practices were left to others in the science education community. In the next section, we turn to the difficulties of education reform.

Dimensions and Difficulties of Education Reform

Here is a truthful note for present and future leaders: Education reform is not easy. We suspect most leaders already know this and potential leaders are recognizing the reality. Our intention here is to introduce a number of perspectives that may help leaders navigate the difficult seas of reform. We use the 4*P*s model—purpose, policy, program, and practice—to identify various ways of characterizing education reform.

One of the first things to point out is the need to translate the legislation (e.g., Every Student Succeeds Act [ESSA]), framework (e.g., *A Framework for K–12 Science Education*), standards (e.g., new state standards), or local guidelines (e.g., district syllabus) into appropriate and usable curriculum materials and classroom practices. This process of translation resides with educators at national, state, and district levels who interact with leaders that represent different perspectives. These education leaders have titles such as state science supervisor, district coordinator, or department chair, and they are responsible for introducing and clarifying the innovations of new standards, selecting or adapting instructional materials, and providing professional development.

The most critical and essential arena of education reform is the classroom. After all, the point of education reform is to improve student learning and provide a different emphasis for that learning. Yet, the classroom-level efforts and changing practices present the most complex issues of reform. We characterize some important dimensions and reform in Table 19.1 (p. 304). Table 19.2 (p. 305) summarizes some of the difficulties of education reform for the different perspectives. We have used a series of questions to capture the dimensions and difficulties. We point out that the numbers and years provided are estimates that may vary.

How long does it take to produce the changes? What is the scale in terms of people who must be addressed? Where does the reform occur? Once the reform has occurred (if it does), how long will it last? What is the product of the particular perspective? How difficult is it to get agreement by various constituents that will be directly influenced by the reform? Reviewing the answers to these questions reveals a pattern that leaders should understand. In general, the descriptions are smaller, simpler,

Table 19.1. Dimensions of Education Reform

Perspectives	Time: How Long Does It Take for Changes?	Scale: How Many Individuals Are Involved?	Location: What Is the Scope and Location of the Change Activities?	Duration: How Long Does the Innovation Last Once Change Has Occurred?	Materials: What Products Are Produced for the Reform?	Agreement: How Difficult Is It Reaching Agreement Among Constituents?
PURPOSE • Reforming goals • Providing justification for goals	**1–2 years** To publish document	**Hundreds** Philosophers and educators who write about the aims and goals of education	**National/global** Publications and reports disseminated widely	**1 year** New problems, new goals, and priorities proposed	**Articles/reports** Relatively short publications, reports, and articles	**Easy** Small number of reviewers and referees
POLICY • Establishing design criteria for programs • Developing frameworks for curriculum and instruction	**3–4 years** To develop frameworks and legislation	**Thousands** Policy analysts, legislators, supervisors, and reviewers	**National/state** Policies focus on specific areas	**Several years** Once in place, policies not easily changed	**Book/ monograph** Longer statements of rationale, content, and other aspects of reform	**Difficult** Political negotiations, trade-offs, and revisions
PROGRAM • Developing materials or adapting a program • Implementing the program	**3–6 years** To develop a complete curriculum program	**Tens of thousands** Developers, field-test teachers, students, textbook publishers, and software developers	**Local/school** Adoption committees or teams adapting current programs or developing curriculum materials	**Decades** Programs, once developed or adapted, used for extended periods	**Books/ courseware** Usually several books for students and teachers	**Very difficult** Many factions, barriers, and requirements
PRACTICES • Changing teaching strategies • Adapting materials to unique needs of schools and students	**7–10 years** To complete implementation and staff development	**Millions** School personnel and classroom teachers	**Classrooms** Individual teachers	**Several decades** Individual teaching practices for a professional lifetime	**Complete system** Books plus materials, equipment, and support	**Extraordinarily difficult** Unique needs, practices, and beliefs of individuals, schools, and communities

Table 19.2. Some Difficulties of Education Reform

Perspectives	What Is the Risk to School Personnel?	What Is the Cost to Schools in Financial Terms?	What Are the Constraints Against Reform for Schools?	What Is the Responsibility of School Personnel for Reform?	What Are the Benefits to School Personnel and Students?
PURPOSE • Reforming goals • Establishing priorities for goals	Minimal	Minimal	Minimal	Minimal	Minimal
POLICY • Establishing design criteria • Developing frameworks for curriculum and instruction	Moderate	Moderate	Moderate	Moderate	Moderate
PROGRAM • Developing materials or adapting a program • Implementing the program	High	High	High	High	High
PRACTICES • Changing teaching strategies • Adapting materials to unique needs of schools and students	Extremely high	Extremely high	Extremely high	Extremely high	Extremely high

and easier for purposes and policies, and they are larger, more complex, and more difficult when considering programs and practices.

Table 19.2 summarizes some of the difficulties of education reform. Here, the categories address questions about risks, costs, constraints, responsibilities, and benefits. Examination of the table reveals another pattern. At the levels of purpose and policy, the difficulties are minimal and moderate, respectively. For programs and practices, the difficulties are high and extremely high. We note one important point: When considering the benefits of education reform for practices, the response is extremely high.

The tables covering dimensions and difficulties of education reform provide general, but useful, ways of describing changes and progress in transforming science education. They also can be used to identify the political content and conflicts. The political issues of education reform range from national and abstract to local and concrete. The reminder to pay attention to the politics sets the stage for Stephen's personal perspective, one that uses the *NGSS* and state standards as the basis of a discussion about politics and science education.

A PERSONAL PERSPECTIVE

EDUCATION AND POLITICS: SOLUBLE OR INSOLUBLE?

Stephen Pruitt

For my portion of this chapter, I am choosing to focus on politics. I realize I spoke pretty extensively about politics and the interplay between politics and policy in a previous chapter. However, I would like to share some of the decisions made during the development of the *NGSS*. These decisions were not necessarily political decisions, but they were decisions that would both affect and be affected by politics. First, I should point out that for developing the *NGSS*, we had guiding principles that shaped the decisions we made. According to these guiding principles, the *NGSS* must (1) be a state initiative, (2) be transparent in all things, (3) engage anyone and everyone who has a stake in science education, (4) have a focus on communications at all times, and (5) meet the needs of all students represented in the *NGSS*. You may ask, "What about research? Why is that not a guiding principle?" Simply put, that was in the mix at all times, but the work done on the front end by the National Research Council (NRC) was critical to ensuring we were

tied to research. There were actual political and practical implications for each of these principles. As a science education leader, one should take time with every decision to ensure the sustainability and ramifications of those decisions. Policy and politics are soluble in reality, but only if the leader chooses to make them so. If they remain insoluble, the initiative is certain to fail. So, as in any chemical solution, the chemist must find the right conditions for successful solubility. As an example, I will walk through the guiding principles to illustrate the thinking.

GETTING THE SCIENCE RIGHT

Before spending time on the five principles, I need to talk about the very first phase of the process: getting the science right. Long before I came to Achieve to lead the development of the standards, a two-step process was put into place that had some of the best scientific and education minds charged with determining what science students should know in each grade band to be successful. As has been discussed earlier, standards had been around for a while—well, a type of standard, anyway. The *National Science Education Standards* (*NSES*) and *Benchmarks for Science Literacy* (*Benchmarks*) had been around since the 1990s. I say "a type of standard" because both of those documents were not intended to be state standards, as there was more work that had to be done to develop assessments for those standards. So, for state science supervisors and teachers, these were the guides to developing state standards. Frankly, the fact that the two strongest voices for science education, and science in general, had their own documents made state standards development a challenge. Generally, states would take the overlap between the two and then add the content that differed between the two, or they would pick one as a base document and fill it in with the other document. There was a great deal of overlap, but when extras were added, it also added content.

I would like to pause here for a moment to make a somewhat separate point. I have been told many times by people that the standards from the 1990s failed. I could not disagree more. Every state—well, every state that had science standards—used these documents and, I think, had the intent to move forward with inquiry as it was intended. I believe that the classroom dynamics were influenced by the enactment of No Child Left Behind (NCLB) and the rise in prominence of the practice of testing science. I will come back to this point later. Still, you could not attend a conference sponsored by the National Science Teachers Association (NSTA) or any other large science educator gathering without discussing inquiry, so I do believe the 1990s standards changed the face of science education. Perhaps the changes did not go as far as intended, but this time period did see a change.

At this point, you may be asking, "What does a discussion of the 1990s standards have to do with the politics and policy of the *NGSS*?" There was a tremendous effect on both because of how the development process for the *NGSS* began. First, the fact that two scientific voices (i.e., NRC and American Association for the Advancement of Science [AAAS]) came together to develop the *Framework* was essential. This combined effort gave authority to the document that would result in a set of standards (*NGSS*) that was bound to be heavily debated. The scientists set the science, so it took much of the argument about science out of the equation from the beginning. Here, I do not mean just

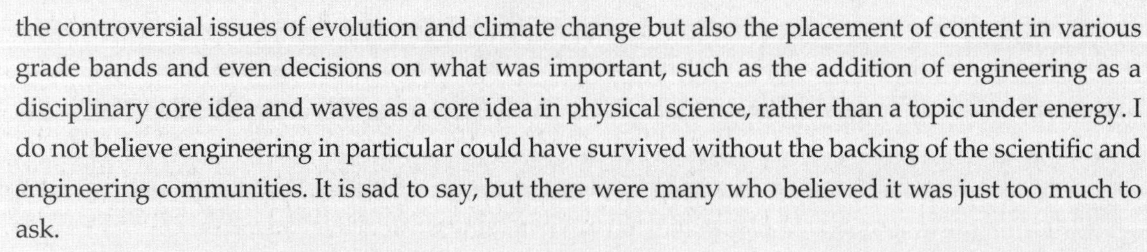

the controversial issues of evolution and climate change but also the placement of content in various grade bands and even decisions on what was important, such as the addition of engineering as a disciplinary core idea and waves as a core idea in physical science, rather than a topic under energy. I do not believe engineering in particular could have survived without the backing of the scientific and engineering communities. It is sad to say, but there were many who believed it was just too much to ask.

I was still in Georgia at the time. I was honored to have been a part of this process from the beginning. It was a group of real science education rock stars. The presence of two Nobel Laureates and people like my good friend and colleague Rodger Bybee, among others, made me feel like I should have walked around the room with an autograph book. They were, and are, a group of incredibly passionate and brilliant people who care about students being prepared for the 21st century. It was not about just the content; they wanted the standards done differently and done right. Many long debates later, the *Framework* was finished. After it had gone through many revisions and even a public review (my understanding is that it was only the second time that had ever happened since the NRC was established), the scientists had spoken. There would be no debate, as the charge to the group working on the *NGSS* was to develop a set of standards that had fidelity to the *Framework*. Politically, this was tremendous, as previous efforts, including the *Common Core State Standards* (*CCSS*), did not have the strong unified voice on the content areas to establish the direction. I think this brilliant decision set the environment that helped make the *NGSS* successful. This decision made the policy and politics soluble, as it gave authority to the next step—developing the guiding principles I mentioned.

STATE INITIATIVE

For the *NGSS* to be accepted at the state level, it had to be a state initiative. I realize that sounds obvious, but it is actually a fairly complicated matter. First, it needed to have a stamp of ownership by all the states willing to work on it. Having just a few select states would not have had the necessary impact. Additionally, if the goal is for the standards to have an effect across the country, states who did not directly work on the standards needed to feel they had, at the very least, knowledge of the process and some opportunity to affect the document as well. There is also the challenge of finding a way to leverage all the important groups in and around science education that need to be tapped for both their brainpower and their support. As a general rule, I believe, states do keep these groups in mind when developing standards, but there needed to be some consistency to be successful.

As the process began, all the previously mentioned issues needed to be considered. At that time, the *Common Core State Standards* (*CCSS*) had been released and states were enjoying a bit of a honeymoon period with those standards. However, as we now know, that did not last. Additionally, science needed to be treated differently. There was far more to consider regarding the standards than simply having multiple states adopt them. There are those inherently controversial issues that were discussed in earlier chapters. This is another factor that led to the need for as many states as possible to join the development effort. As we assumed the need for state involvement, we had to also realize

the states needed support in that process. What I mean by that is that the state science supervisor could not be the only individual from the state to work on the standards. There needed to be buy-in from the chief state school officer and the chair of the state board of education. From there, the state needed to expand its sphere of influence to provide the best feedback and direction possible. The staff at Achieve and members of the leadership team developed a list of roles that each state should have represented on their teams. As such, each state that chose to work with the development had to identify a lead, typically the state science supervisor, and a broad-based team of science education representatives from across K–12, postsecondary education, industry, informal science, and third-party STEM interest groups. A process was developed to provide these teams with drafts for review, capture the feedback, ask for direction, and leverage the work in the states. The leads for each state would work with these teams to answer specific challenges and direct the writing teams regarding the words of the standards as well as format and handling of specific topics such as engineering.

Well, that was the easy part. It is a lot of work to develop a plan or process, but it pales in comparison to implementation. How would we identify states? Given the challenges of implementing the *CCSS*, would any states want to come along for a two-year, in-depth process? How many should we have if we are given the choice? Finally, how could we tell if the states were really invested? It was one thing to say, "Sign on and give us some feedback every so often." It was quite another to say, "We need your time, your expertise, your public acknowledgement that you were part of this huge movement." We needed to know the states that chose to work with us had the commitment to work on this process. There was no funding beyond our ability to bring the leads together, so the states had to foot the bill. We decided to develop a proposal process that would both elicit details of states' commitment to the standards and push them to think about the future if they were to adopt them. To be considered, states had to submit a proposal on how they would engage their communities and shareholders (those invested in a solid science education). They also had to agree to the following set of assurances (NGSS Lead States 2013):

- Give serious consideration to adopting the resulting *Next Generation Science Standards* as presented.
- Identify a state science lead who will attend meetings with writers to provide direction and work toward agreement on issues around the standards, adoption, and implementation.
- Participate in Multi-State Action Committee meetings (committee of the chief state school officers) to discuss issues regarding adoption and implementation of the new standards.
- Publicly announce that the state is part of the effort to draft new science standards and make transparent the state's process for outreach and receiving feedback during the process.
- Form a broad-based committee that considers issues regarding adoption and provides input and reactions to drafts of the standards.

- Publicly identify a timeline for adopting science standards (though not necessarily the *NGSS*).
- Utilize the collective experiences of the states to develop implementation and transition plans while the standards are being developed that can be used as models for all states.

When the proposal for the lead states went out, we had no idea how many proposals we would receive. The original thought was that we would probably need at least 6. I actually had some people say to me, "Do you think you will even get 6 proposals? I mean, it's science, and there is not always a lot of support for standards given the controversy." Well, it turned out we received 20 proposals. This became a different issue unto itself. How do you tell 14 states you do not need their help? Well, you don't. They had all put together superb proposals, and we knew the more state involvement there was, the more likely it would be to keep the politics and policy soluble. So, we announced all 20 states as lead states. Some who did not apply then stated they would have applied had they known we would take everyone, so we reopened the applications. After sending a couple of proposals back for deficiencies in quality, we took 6 more states. This brought the total number of states to 26, representing well more than half of the nation's students. The following map and content reflect the state applications for Lead State Partners in 2011.

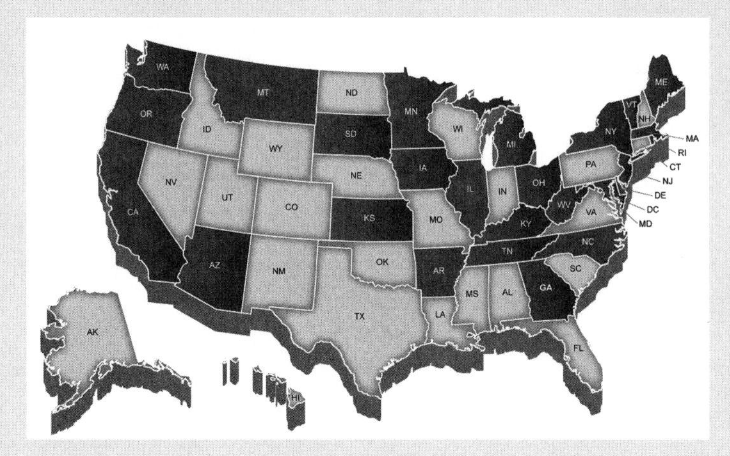

Source: NGSS Lead States 2013.

For this process to remain soluble, the states could not have a cursory role. The states had to have a meaningful role, and they had to direct the process. At the same time, the standards needed to be written by experts in the field with fidelity to the *Framework* as their guidepost. A writing team was established to translate the *Framework* into the standards. I was asked early in the process, "Who holds the pen?" My response was, "The writers hold the pen, and the states hold the hands of the writers." That is to say, the writers made sure we followed the *Framework*; the states made sure the standards met the needs of their states. So, everything from the architecture of the standards to placing standards into grade levels at the elementary grades to the language of the standards themselves were decision points for the states. Lead states would receive a draft, meet with their committees,

and come together to discuss the feedback from all states and give the writers direction on the next round of revisions. Each state got one vote; they were all treated as equals. This is important, as there are obviously some states that have better histories with science achievement than others. However, with the *Framework* to guide the process, it was imperative that each state had an equal voice. It was an incredible experience, albeit painful at times. At the end, states were able to proclaim they owned the standards because they did not just play a role—they led the process.

TRANSPARENCY

I will not spend a great deal of time on this as it is somewhat self-explanatory. There were a few key decisions that made the process more transparent. As leaders, we should always seek transparency on the front end of any project. It saves many headaches later, and it often affects the success or failure of the project. First, we made everything public. Everything you wanted to know was on the internet, and we sent the information out in newsletters as well. When I say *we*, I do not mean just the staff at Achieve; I also mean the states. The development of the *NGSS* was a topic at every state science conference, so we wanted to push information out to the Lead States as well as to every other state and teacher. The website homepage was developed such that each visit automatically showed the timeline and where we were in the process at that moment. We announced changes, proactively tried to send communications about the process, and published the results of each public review. This brings me to my second point, regarding the public reviews. From the outset, we made the decision to do two public releases. I believed this to be important on several fronts. There are the obvious ones of getting public input, but it is also important we chose to do two releases. That is important because we felt strongly the public needed two opportunities to see the full draft to hold us accountable for changes made or not made. We also published the results of each release so that people could see the type of feedback we received. Finally, I should be clear that no leader is successful if they are not transparent. The more it appears you do things behind closed doors, the more you degrade trust. It is never completely comfortable to be transparent, but it is a must if you are to keep the solubility between politics and policy.

ENGAGING ALL

As critical as it was for the states to lead the effort, it was equally important for the science, science education, and general education communities to be engaged in the process. We needed to find ways to build relationships with as many of these groups as possible. After all, even with the states leading the effort, if the science community did not support the standards, they would fail. Groups want to see their membership as having a role in any initiative they support. This is a key understanding, I think. I have joked in the past that partnerships would be so much easier if it were not for all these partners, but the reality is that if you do not have them, your success will be limited. You can choose to be uncomfortable during a process or after—trust me, during the process is far more preferable.

To keep the initiative soluble between policy and politics, you need all your external partners to be at the table. Otherwise, the politics will prevent the policy.

FOCUS ON COMMUNICATIONS

This is not something that comes easy to educators, but it is critical. Communications cannot be an afterthought. On the front end of planning, communications strategies need to be planned and replanned. Just because something may seem to be going well or "flying under the radar" doesn't mean that it is. Bottom line: Communication plans are always a priority, whether it is with a supervisor, students, or the media.

ALL STANDARDS, ALL STUDENTS—REALLY

This was a keystone to the development of the *NGSS*. If we are going to make policy and politics soluble, I felt this was a key feature. So many of our students today still live with the misguided view that science is for a few, the elite. However, what we know is that science opens a world for all students if they are engaged the right way. For that reason, the state teams and the writing team were designed to have representation from all groups of students to ensure we considered all students during development, as opposed to an afterthought. We needed expertise in educating students with disabilities, English language learners, low-performing students, students living in poverty, students in alternative education situations, underserved minorities, and gifted and talented students. To be clear, this was more than just a ploy to keep policy and politics soluble, even though it did help. This was a leadership decision to meet the real promise of standards. Standards are meant to be an equalizer for all students. We cannot hope to make that claim if we do not keep every child in our minds from the beginning. I am very proud of the *NGSS*, but I actually think I am most proud of the fact that we developed *NGSS* with the intent that every student can and should have a quality science education. The team came up with the title *All Standards, All Students*. This was not because we only think that all kids need all content. It stems from the fact that all students can learn at high levels if given the opportunity and the support.

In closing, I would like to briefly reflect on the idea of solubility between policy and politics. They both affect one another, whether we like to admit it or not. Again, let me reinforce it is not always about the big *P* politics, but many of the things discussed in this essay are little *p*s. These are just a few things that leaders must do to ensure viability of their initiatives and to affect children's lives for the better. Unfortunately, as leaders, it is easy to get wrapped up in our content. I want to encourage you to think about the big picture and find ways to make education better for all students. Science education will do that, but all things must be considered in the bigger picture of education if policy and politics are to be soluble.

REFERENCES

Bestor, A. 1953. *Educational wastelands*. Urbana, IL: University of Illinois Press.

Bybee, R. 1993. *Reforming science education: Social perspectives and personal reflections.* New York: Teachers College Press, Columbia University.

Bybee, R. 1997. *Achieving scientific literacy: From purposes to practices.* Portsmouth, NH: Heinemann.

Cremin, L. 1990. *Popular education and its discontents.* New York: Harper & Row.

Cross, C. 2014. *Political education: Setting the course for state and federal policy.* New York: Teachers College Press, Columbia University.

Educational Policies Commission. 1952. *Education for all American youth: A further look.* Washington, DC: National Education Association and the American Association of School Administrators.

Jennings, J. 2015. *Presidents, Congress, and the public schools: The politics of education reform.* Cambridge, MA: Harvard University Press.

National Association of Secondary School Principals. 1944. *Planning for American youth: An American educational program for youth of secondary age.* Washington, DC: National Education Association.

National Society for the Study of Education (NSSE). 1932. *A program for science teaching.* Chicago, IL: University of Chicago Press.

National Society for the Study of Education (NSSE). 1947. *Science education in American schools.* Chicago, IL: University of Chicago Press.

National Society for the Study of Education (NSSE). 1960. *Rethinking science education.* Chicago, IL: University of Chicago Press.

NGSS Lead States. 2013. *Next Generation Science Standards: For states, by states* Washington, DC: National Academies Press. *www.nextgenscience.org/next-generation-science-standards.*

Progressive Education Association. 1938. *Science in general education.* New York: Applton-Century Crofts.

Rickover, H. 1959. *Education and freedom.* New York: E. P. Dutton.

Silberman, C. 1979. *Crisis in the classroom: The remaking of American education.* New York: Random House.

U.S. Office of Education. 1951. *Life adjustment education for every youth.* Washington, DC: U.S. Government Printing Office.

Windschitl, M. 2006. Why we can't talk to one another about science education reform. *Phi Delta Kappan* 87 (5): 349–355.

CHAPTER 20

STANDARDS-BASED REFORM

This chapter will focus on the contemporary era of standards-based reform, in particular the emergence of national standards for science education. After two decades of experience with standards, we recognize the long-term positive influence of national standards for science education. First, national standards can influence all of the fundamental components of the education system. Second, standards clarify the most basic goals—the learning outcomes—for all students. Third, standards at the national level are necessary for equality of educational opportunity. Finally, standards have the potential to reduce significant variations between state standards and district programs.

The fundamental idea underlying national standards is that they should describe clear, consistent, and challenging goals for science education. Then, based on the standards, educators and policy makers should reform school science programs and classroom practices to enhance student learning. The effect that the standards have on three channels of influence—curriculum programs, teachers' professional development, and assessment and accountability—will determine whether the standards have provided adequate implementation as a basis for reform. These channels, in turn, influence teachers, teaching practices, and, ultimately, student learning (NRC 2002). While the process just described may be the ideal, history shows that numerous and varied reform efforts in the United States have had little effect on teaching and learning in classrooms.

We hope the discussions in this chapter shed some light on factors that can set a different course for the early 21st century and standards-based reform. How did we arrive at an era of national standards for science education? The following discussion answers this question.

CHAPTER 20

A NATION AT RISK AND THE EMERGENCE OF NATIONAL STANDARDS

In 1983, the National Commission on Excellence in Education (NCEE) released *A Nation at Risk: The Imperative for Educational Reform.* Thanks to scientists on the commission, the report stimulated the development of new political perspectives on education for the remainder of the 20th century. The report's rhetoric was strong and included statements such as, "The educational foundations of our society are presently being eroded by a rising tide of mediocrity that threatens our very future as a Nation and people" (p. 5). This language caught the American public's attention, as it stood out from numerous other reports that declared the need for education reform.

Terrell Bell, President Ronald Reagan's Secretary of Education, convened the commission at least in part to divert the president from his intention to eliminate the U.S. Department of Education. With the release of *A Nation at Risk,* Reagan declared the need to improve education through school prayer and vouchers for private school tuition; neither recommendation was in the report. The report expressed a general theme of going "back to the basics," including three years of science and mathematics and longer school days and school years. Four components of education had to change: the content of programs, the expectations of students, time in school, and teaching practices.

Science and Technology in *A Nation at Risk*

The subjects of science and technology were prominent in the report. The report warned of a generation of scientifically and technologically illiterate Americans and a growing chasm between a small scientific elite and a citizenry both ill-informed and uninformed in matters of science and technology. *A Nation at Risk* made content recommendations for science:

> *The teaching of science in high school should provide graduates with an introduction to (a) the concepts, laws, and processes of the physical and biological sciences; (b) the methods of inquiry and reasoning; (c) the application of scientific knowledge to everyday life; and (d) the social and environmental implications of scientific and technological development. Science courses must be revised and updated for the college bound and those not intending to go to college. (NCEE 1983, p. 25)*

This set of recommendations was consistent with those of other education reports of the period. However, the report quickly became a milestone in the literature on education reform. *A Nation at Risk* expressed the need for reform in language that captured the public's attention and stimulated the drive for reform. The latter is a positive thing about powerful reports such as *A Nation at Risk.* On the negative side, a report is only words and recommendations, which lack the power and resources to change fundamental components of the education system, namely programs and teaching practices. Stated another way, this was a political report without support for curriculum, professional development, or assessment programs.

As noted above, the commission recommended "the new basics," which for this discussion involved requiring three years of high school science. This requirement implied a policy change that was to be implemented at the state level. The resulting reform in graduation requirements resulted in dramatic changes for high school course-taking in the sciences. For example, between 1982 and 1992, many students—white, African American, Hispanic, Asian, Native American, male, and female—were taking more science courses. During this period, the statistics for all students indicate changes: The percentage of students taking high school biology increased from 78.7% to 93.0%; those taking high school chemistry increased from 31.6% to 56.5%; and those taking high school physics increased from 13.5% to 24.7% (NCES 1994, pp. 242–243). While these changes in course-taking were dramatic, it is not clear how much the school programs and teaching practices actually changed. Furthermore, the increase in the number of courses taken does not necessarily translate to higher levels of academic achievement in the sciences.

A Nation at Risk in 1989 set the stage for President George H. W. Bush and the governors to develop and adopt six comprehensive goals for education. We discussed this meeting and the goals in Chapter 7. One goal—that by the year 2000, U.S. students would be first in the world in mathematics and science achievement—was bold but unachievable. However, another goal became the impetus for an era of standards-based educational reform (NCEE 1983, p. 25):

> *By the year 2000 American students will leave grades 4, 8, and 12 having demonstrated competency in challenging subject matter, including English, mathematics, science, history and geography; and every school in America will ensure that all students learn to use their minds well, so that they may be prepared for responsible citizenship, further learning, and productive employment in our modern economy.*

This goal presented achievable outcomes only if there were concrete statements of content for grades 4, 8, and 12 and processes that would help students "use their minds well." For science, this goal stimulated the development of *Benchmarks for Science Literacy* (AAAS 1993), the *National Science Education Standards* (NRC 1996), and, nearly two decades later, *A Framework for K–12 Science Education* (NRC 2012) and the *Next Generation Science Standards* (*NGSS*; NGSS Lead States 2013).

Standards-Based Reform and Two Generations of Standards

As mentioned in the prior section, national goals emerged from the governors' 1989 education summit. Two of the goals emphasized academic achievement for all students. We described one goal in detail and indicated how the second goal made political sense but was an unrealistic educational goal; being first in the world in math and science by the year 2000 was not even close to being achieved. To sustain the pressure for higher academic achievement for all students, the U.S. Department of Education and other federal agencies (e.g., National Science Foundation [NSF]) supported the development of voluntary national standards. The National Research Council (NRC) was awarded a grant to develop standards for science education.

At this time in the early 1990s, with the change of presidents, both the George H. W. Bush and Bill Clinton administrations recognized the need for national standards and assessments that described what students should learn and evidence of the degree to which they learned what the standards described. Policy makers subsequently supported the development of national standards. The reasons for their support included the measured decline of student achievement on national and international assessments; the gap in achievement between students of different racial and ethnic groups; the increasing number of reports from organizations, businesses, and industries proclaiming the need for education reform; the public's concern about economic and international competitiveness; and the general concern of public officials, educators, business leaders, and citizens that American education had no clear definition of what students should know and be able to do after 13 years of public education.

Relative to this discussion, the result was the development of the *National Science Education Standards* (*NSES*; NRC 1996) and later the *NGSS* (NGSS Lead States 2013). Support for science education standards came from the scientific, business, and education communities as well as the public. American public schools needed to establish the priority of educating students in the knowledge and skills associated with core disciplines, including science. In *National Standards in American Education,* Diane Ravitch (1995) provided this excellent summary:

> *. . . the purpose of establishing standards and assessments is to raise the academic achievement of all or nearly all children, to signal students and teachers about the kind of achievement that is possible with hard work, to emphasize the value of education for future success in college and careers, to encourage improvement of instruction and collaboration among teachers, and to motivate students to have higher aspirations in their school work. (Ravitch 1995, p. 5)*

The purpose of standards for science education is to clearly define the knowledge and skills that all students should develop. Once defined, these learning outcomes will result in aligned assessments, curriculum, and instruction. Accordingly, teacher preparation programs, graduation requirements, and ongoing professional development will be reformed.

While this may have represented the common sense and good judgment of those supporting and working on standards, the multiple and diverse factors influencing education seldom can be characterized as "common sense and good judgment." Colleges and universities design teacher preparation programs the way they think best. Within the states' guidelines for certification, textbook publishers prepare programs based on input that aligns with the results of marketing, and assessment (at the local, state, and national levels) is developed within the budgets and expertise of specialists. Standards should be influential forces, even if the forces are weak, to increase the alignment among the primary channels of influence between the standards and classroom practices. Those channels include curriculum, teacher development, and assessment and accountability (NRC 2002).

In reflecting on the implementation of *NSES* (NRC 1996), we believe it is safe to say that the reform efforts avoided any significant political conflicts experienced in other fields—for example, the "Math Wars" (Phillips 2015) and debates about the history standards. We should note that the *NSES* did include biological evolution in the life science standards, which resulted in political efforts by some religious groups to include nonscientific perspectives in school science programs. The National Science Teachers Association (NSTA) supported the *NSES* by requiring articles in its journals to address and cite these standards. The National Science Foundation (NSF) supported development of curriculum materials based on the *NSES*. Upon reflection, it is clear that implementation of the *NSES* did not pay adequate attention to the role states play in education reform. This omission was clearly remedied in the development and continuing implementation of the *NGSS* (NGSS Lead States 2013).

The *NGSS* presented new political issues in the processes of education reform. Recalling congressional leader Thomas "Tip" O'Neill's aphorism that "All politics is local," we can report that the political issues related to the *NGSS* are at the state and district levels and primarily center on evolution and climate change. Several examples make the point about local politics. For example, the Wyoming legislature initially blocked any consideration of the *NGSS* because of the document's treatment of climate change. This position was subsequently rescinded, and the state board has recommended new standards for this state. After a long series of objections, Oklahoma passed the Oklahoma Academic Standards for Science. West Virginia also objected to the inclusion of climate change; subsequently, the standards were approved with a minimal change (i.e., change of one word) and the state adopted the modified *NGSS*. As another contemporary example, the Alabama House of Representatives introduced a bill that would, if enacted, encourage teachers to introduce their unique, idiosyncratic opinions about topics such as biological evolution. Missouri introduced a similar bill, but it died in committee without a hearing.

Some states, such as Oklahoma and South Dakota, have adapted the *NGSS* and developed acceptable new state standards for science education. This is the political influence of the *NGSS* that should be recognized beyond the states that have adopted the *NGSS* as state standards.

Table 20.1 (pp. 320–321) presents a summary of major reforms since World War II.

Implementing Contemporary Standards for Science Education

Developing new state standards or adopting the *NGSS* presents a variety of questions for the science education community. Constructive responses to the questions of science teachers, science coordinators, providers of professional development, and curriculum developers and their related concerns will be critical to a successful implementation of the *NGSS*. Questions such as these will be central to science educators:

- How can the new standards be used to create curriculum programs and improve instructional practices?
- Will the national, state, and district assessments change?
- How do I translate the standards to classroom practices?
- How will the *NGSS* affect my teaching?

Leaders must provide constructive, appropriate, and practical answers, especially for classroom teachers. The questions are appropriate, as the daily task of science teaching centers on teachers

Table 20.1. Summary of Science Education Reforms

	Sputnik Era	*A Nation at Risk*	National Science Education Standards	Next Generation Science Standards
CALL TO ACTION **Why reform/ change?**	To win the space race with the USSR	To improve global competitiveness and the workforce in the United States	To respond to *A Nation at Risk*	• To focus on college and career • To improve the U.S. economy
GOALS **What has to change?**	School science curriculum	Renewed emphasis on basic science education	• Scientific literacy • National goals for science education	• Update and improve national standards for science education • Focus on states
FUNDING **Who paid?**	• Federal support (e.g., National Science Foundation [NSF]) • Some private foundations	Federal government	• U.S. Department of Education • NSF, NASA, National Institutes of Health (NIH)	Private foundations (i.e., no federal funds)
LEADERS **Who did the work—on what?** **Who was involved?**	Elite scientists	Educators, including scientists	• Scientists • Science educators • Science teachers	• State leaders supported by scientists • Science educators • Science teachers • Business and industry
INNOVATIONS **What's new?**	• New curriculum materials based on the structure of science disciplines • Inquiry investigations to enhance learning	Policies for improvement	Content standards for traditional disciplines of life, Earth, and physical science, plus standards for inquiry, technology, history and nature of science, and science in personal and social perspectives	• Performance standards • Combine disciplinary core ideas, science and engineering practices, and crosscutting concepts in standards

Table 20.1. *(continued)*

	Sputnik Era	*A Nation at Risk*	*National Science Education Standards*	*Next Generation Science Standards*
RESULTS **What was produced?** **Reform?** **What changed?**	• Wide adoption of new curriculum materials • Physical Sciences Study Committee (PSSC), Biological Sciences Curriculum Study (BSCS), introductory physical science (IPS), Earth Science Curriculum Project (ESCP), Science Curriculum Improvement Study (SCIS), Elementary Science Study (ESS), Science—A Process Approach (SAPA)	• Provided stimulus for standards • Translated goals to policies	Led to development of first generation of standards for science—*National Science Education Standards* (NRC 1996)	Developed *Next Generation Science Standards* (NGSS Lead States 2013)

providing opportunities for students to learn. Briefly, teachers' concerns are about curriculum, instruction, and assessment.

There are several critical points about this period of *NGSS*-influenced reform. First, we should not expect funding for instructional materials and professional development from federal agencies such as the NSF. Second, in the years since release of the *Framework* and *NGSS*, the assessment community has responded with meetings that addressed new *NGSS*-based assessments (see, e.g., Pellegrino et al. 2014; Bybee 2013a). Third, based on issues of "local control," states and districts have significant responsibility for addressing the reform of science education.

CHAPTER 20

Recommendations for Implementing State Standards and the *NGSS*

What follows are five initiatives that center on reforming curriculum programs and classroom practices in K–12 science education.

Recommendation 1: Focus on the Instructional Core

Richard Elmore (2009) introduced the term *instructional core* as a response to issues of improving student learning at scales that make a difference. Here is Elmore's fundamental point:

> *There are only three ways to improve student learning at scale: You can raise the level of content that students are taught. You can increase the skill and knowledge that teachers bring to the teaching of that content. And you can increase the level of students' active learning of the content. (Elmore 2009, p. 24)*

Focusing on the instructional core accommodates the complex and difficult work of science teaching and student learning. Put simply, the role of science teaching is too important to avoid and too critical to misrepresent. Figure 20.1 presents the instructional core with adaptations for *NGSS*-based reform.

The *NGSS* and new state standards change the rigor, focus, and depth of science content. There is a need to increase students' active learning of the science content, and this, for me, directly implies changes in curriculum, instruction, and assessments in science classrooms (Bybee 2010). Finally, there is the need for professional development for science teachers with the *NGSS* in mind.

Think of a three-legged stool. In this case, the three legs are science content described in the *NGSS*, school science programs, and science teachers' knowledge and skills for classroom practices. In the world of three-legged stools, there is a law—not a hypothesis or theory—that once you change one leg, you must change the other two if your goal is stability. The *NGSS* and new state standards have changed one leg of the instructional core. Now, the science education community must address changes in curriculum programs, classroom practices, and teachers' knowledge and skills.

Recommendation 2: Develop a New Generation of Curriculum Models

Design specifications for these instructional materials include (1) integrating scientific and engineering practices, disciplinary core ideas, and crosscutting concepts; (2) providing adequate time and opportunities to learn; (3) using appropriate and varied instructional strategies; (4) aligning classroom curriculum, instruction, and assessments; and (5) making connections to *Common Core State Standards* for English language arts and mathematic.

The models and tools for new instructional materials should include several programs for each grade-level set (e.g., elementary school); incorporate educational technologies (e.g., use of e-books, games, and simulations); and address the needs of all students (i.e., for both college and career).

Figure 20.1. Instructional Core and *NGSS*-Based Reform

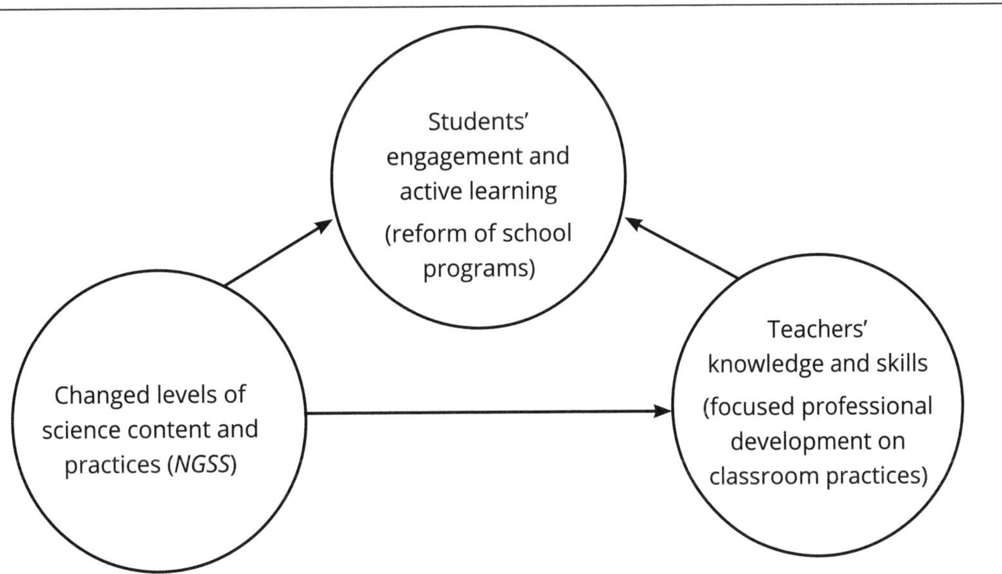

Recommendation 3: Support Professional Development for Teachers of Science

The professional development programs should provide enough initial time to establish a clear foundation for teaching and learning. The program should be a multiyear experience and include continuous work on improving curriculum, instruction, and classroom assessments. The educational context for the programs may include curriculum models based on the *NGSS*—that is, content and pedagogy with direct classroom application. Core disciplinary ideas, crosscutting concepts, and the practices of science and engineering must be the programs' focus. Finally, the programs require the establishment of professional learning communities with teams of teachers analyzing teaching, engaging in lesson study, reviewing content, and working on the implementation of instructional materials.

The initial work of the institutes and professional development experiences should concentrate on model instructional units as concrete examples of the innovations that form the basis of *NGSS*-based reform.

Recommendation 4: Build States' and Districts' Capacity for NGSS*-Based Reform*

This recommendation uses the critical leverage of teacher certification to facilitate reform of undergraduate teacher education programs. No discussion of improving science education escapes acknowledging the need to change teacher education. This includes changes in states' certification and national accreditation (e.g., Teacher Education Accreditation Council [TEAC]). In addition, federal support for colleges and universities that prepare significant numbers of future science teachers will be a major impetus to reform. To this recommendation, we would add that there should be special support for colleges and universities with significant populations of Hispanic, African American, and Native American students so those institutions can recruit and prepare STEM teachers from diverse backgrounds.

Recommendation 5: Explain to the Public the Basis for NGSS *Curriculum Programs and Classroom Practices and Why the Reform Will Benefit Their Children and the Nation*

One of the great insights from the Sputnik era was the fact that national leaders provided clear and compelling explanations of what the reform was and why it was important. Furthermore, there was continued support for science teachers and a national enthusiasm for reform.

This is a low-cost, high-value initiative, one that has not been clear since the Sputnik era. The importance of a nonmilitary national mission and positive aspirations for the future could be invaluable to the American public. It also is the case that science education could make clear the direct connection to the public through topics such as health, environmental quality, resource use, and energy efficiency.

A PERSONAL PERSPECTIVE

REFORMING SCIENCE TEACHING: ANSWERING THE MONDAY MORNING QUESTIONS

Rodger Bybee

"Where are the curriculum materials I can use Monday morning?" "Will there be professional development for the curriculum materials?" The "Monday morning questions" mentioned in this essay's subtitle include variations on these questions. These questions are reasonable for science teachers to ask as they respond to calls for reform. Let me begin by placing the "Monday morning question" in the context of reform using the 4*P*s.

As you have likely noticed in prior chapters, there is continuous talk of improving science teaching and enhancing student learning. With the 4*P*s—purpose, policy, programs, and practices—as context, I point out that the majority of talk in the numerous reports takes place at the policy level. In general, those preparing reports and making recommendations have stimulated but not answered science teachers' Monday morning questions: Where are the curriculum materials I can use on Monday morning? What knowledge and skills do I need to use those materials?

The prestige and authority of various groups of policy makers, business leaders, and some educators get the public's attention through statements about the dire consequences of low student achievement and poor teaching. Historically, the public heard how the Russians were winning the space race, the nation was at risk, and we must "rise above the gathering storm." Unfortunately, while these reports make great political rhetoric, they have seldom influenced the school science programs or classroom practices. Why, you should ask, is there so much talk and so little change? My answer: All things considered, preparing a report is easy and not too costly. Developing the report includes few restraints and minimal responsibilities, and the benefits for the actual reform of science teaching are moderate at best. Much more difficult, costly, and time consuming are the next steps: developing curriculum materials and providing professional development.

Because we are in an era of standards-based reform, one has to ask, "Are the possibilities of improving science teaching and student learning any better?" At this point, I would say the possibilities are good, but the science education community must find ways and means to influence the enduring and essential features of science teaching. We must respond to science teachers' concerns and answer the Monday morning questions.

In November 2015, *NSTA Reports* published the results of a survey that asked science educators about their experiences preparing for implementation of the *NGSS*. Most of respondents (83%) were from states or districts adopting the *NGSS*. A majority of the teachers (59%) said they had modified lesson plans to align with the *NGSS*. When asked to name the one thing they would request for help in transitioning to the *NGSS*, 35% of respondents said they needed access to curricular materials and resources aligned to the *NGSS*; 30% wanted high-quality professional learning opportunities; and 22% indicated they wanted support for science from school and district leaders. Educators reported the most common challenge to implementing the *NGSS* is determining how to assess whether students have met the standards (57%). In the survey, 44% of respondents cited adapting lesson plans and finding and selecting instructional resources as challenges.

I think these responses are especially informative. To summarize responses to the Monday morning questions: Science teachers need curricular materials and resources; they want high-quality professional development; they want support from school administrators and school boards; and they require assessments aligned with the *NGSS*.

The *NSTA Reports* article also included selected statements about the effect of the *NGSS*. These statements revealed common perspectives and concerns. The statements expressed frustration about the arrangements of standards versus curricular themes; reservations about the reduced content and common assessments versus the place of honors and AP classes; an understanding of how the *NGSS* and implied changes in teaching; recognition of the time needed to implement the standards; interpretations of the *NGSS* and seeing how they align with current practices; and failure to recognize that the innovations in the *NGSS* imply changing what and how science is currently taught.

This era of standards-based reform has the possibility to make the essential transition from policies to programs and practices. There are, however, several fundamental issues that must be addressed, and those issues are fairly clear. The issues were identified in the NSTA survey mentioned earlier. First, curricular materials and resources aligned to the *NGSS* must be developed and made available. Second, high-quality professional development centering on the knowledge and skills needed to teach *NGSS*-aligned programs must be provided. Third, district leaders, administrators, school boards, and parents have to support teachers and the teaching of standards-based science. Finally, there is the need for assessments, both formative and summative, that exemplify the standards-based reforms described in policy documents such as the *NGSS* and recently revised state standards. Let me elaborate on these basic points.

The *NGSS* and revised state standards influenced by *NGSS* present innovations that do not exist in current instructional materials. For example, current materials do not include three-dimensional teaching—that is, how to address science and engineering practices, crosscutting concepts, and disciplinary core ideas in an instructional sequence or unit. A secondary question to "What do I do Monday morning?" is "How do I incorporate the three dimensions so students will learn them?" I assume science teachers recognize and accept the new state standards. As they ask how the new standards align with their current curriculum (putting aside the observation that this question subtly shifts their current program as the controlling entity), they will find significant discrepancies; Hence the need for exemplary lessons, units, and programs. I do know that some curriculum developers and publishers (e.g., Full Option Science System [FOSS] and and Science and Technology Concepts [STC] at Carolina Biological Supply) are preparing *NGSS*-aligned editions of their products. There are other examples, such as units prepared by the American Museum of Natural History and the Lawrence Hall of Science. NSTA journals also include lessons aligned to the *NGSS*.

Changing the strategies and practices of teaching is difficult and daunting. Those at the national, state, and district levels and in academic positions need to recognize the challenge of changing teaching behaviors. The need for professional development cannot be stressed enough. Ideally, the professional development will take place within the context of instructional materials and concentrate on the knowledge and skills required to implement the new standards effectively. I would also note that one workshop or session is not enough. The experiences also must move beyond the introduction of standards to the management and evaluation concerns of teachers.

Implementing new standards has personal, financial, and political costs. I alluded to the personal costs in the prior paragraphs. The changes we are proposing are personal for teachers. Science teachers in kindergarten through grade 12 are the ones with the most personal changes, and their concerns are real. The financial costs for materials, equipment, and professional development will also be significant. Finally, political issues related to content (e.g., biological evolution, climate change), local control, and accountability require administrative and public support.

One concern that teachers have is often expressed in the question, "Will the assessments change?" The answer is yes, they will. But pursuing the question often reveals the real concern—accountability. Administrators at local and state levels need to address the accountability question, and professional developers need to include assessment in their programs for teachers.

The issues just discussed echo many prior discussions of standards-based reform. I would like to take the discussion to deeper practical and political issues. Here are the basic questions: How do educators make standards responsive to the diversity of schools, teachers, and students? Standards, whether national or state, typically present one set of statements, yet answering the Monday morning question has to accommodate different schools within a state or district; variation in science teachers' experience, knowledge, and skills; and diversity in students' backgrounds, current scientific knowledge and abilities, and prior experience with science. Standards by their very nature include statements of content and performance and cannot address all of the unique circumstances that may underlie the Monday morning question. I could continue with the different approaches among elementary, middle, and high school teachers but will leave that to the readers' understanding. I can add the fact that new curriculum frameworks, textbooks, kit materials, and combinations of activities, texts, and other resources do not provide answers to the many singular challenges of classroom teachers. So, what is the response? Standards require translation to school programs and classroom practices. I have written about this in *Translating the* NGSS *for Classroom Instruction* (Bybee 2013b). Some interpretations will be completed by curriculum developers and commercial publishers, but the ultimate translation is by individual teachers.

There is a practical reality embedded in the process of translating standards to classroom practices. As curriculum developers and teachers identify topics and themes and design lessons, units, and full programs, some concepts and information will be added out of necessity. Now comes the hard part. Those responsible for translating will be required to focus on the fundamental ideas, practices, and concepts and replace or eliminate the unnecessary, repetitive, and distracting details that are in many current curricula or may be added under the political policy of local control. There is a need for parsimony. These are the trade-offs of translating, and the challenge is finding the adequate and appropriate balance in the cost, risk, and benefit analysis.

This discussion directs our attention to fundamental structures of the education system—the instructional core—an idea introduced by Richard Elmore in *School Reform From the Inside Out: Policy, Practice, and Performance* (Elmore 2004). I have discussed Elmore's idea of instructional core in prior publications, such as *The Teaching of Science: 21st-Century Perspectives* (Bybee 2010), and the instructional core is a theme in Elmore's book:

> *By "the core of educational practice," I mean how teachers understand the nature of knowledge and their students' role in learning, and how these ideas about knowledge and learning are manifest in teaching and class work. The "core" also includes structural arrangements of schools, such as the physical layouts of classrooms, student grouping practices, teachers' responsibilities for groups of students, and relations among teachers in their work with students, as well as processes for assessing student learning and communicating it to students, teachers, parents, administrators, and other interested parties. (Elmore 2004, p. 8)*

In this quotation, Elmore implies the importance of teachers' understanding of the nature of scientific knowledge and their students' role in learning science. This feature has direct implications for professional development. Second, he underscores how ideas about scientific knowledge and students' learning of science are realized in the classroom. For me, this translates to the traditional categories of curriculum and instruction. Third, he recognizes broader programmatic and systemic factors such as classroom student grouping, teachers' responsibilities and collegiality, and, finally, the process of assessing student learning. Assessment certainly will be on science educators' agenda for the foreseeable future. The basic categories of the instructional core can be identified in traditional terms of curriculum, instruction, and assessment with the underlying foundation of student learning and the continuous professional development of science teachers.

The instructional core can be summarized as involving three components: the content of science, the science curriculum and assessments, and the knowledge and skills of teachers as they teach. In light of this discussion, the *NGSS* has defined the content and practices for science, so the need for reform centers on the curriculum, assessment, and knowledge and skills of teachers.

Answering teachers' Monday morning questions gives direction to issues associated with translating policies of standards-based reform to curriculum materials and teaching strategies. This is the challenge of making standards responsive to the diversity of students, schools, and districts within states and the nation. Using the instructional core as a framework recognizes the critical role of teachers' knowledge and skills and the implied necessary professional development. This view presents a reciprocal obligation between science teachers, school administrators, and policy makers. As the pressure for changing classroom teaching and student performance increases, there is an obligation of leadership to provide time, budgets, and expertise so teachers have the curriculum materials, assessments, and knowledge and skills to implement the new science education standards. This, it seems to me, is a basic and appropriate response to the Monday morning questions.

In this personal perspective, I addressed one question science teachers frequently ask: "Where are the curriculum materials that will help me implement the new standards in my classroom?" This is a critically important question. Although there are several initiatives focusing on assessment, there are few discussions of new instructional materials. One purpose in this essay was to help the science education community answer the critical question of making curriculum reforms based on contemporary standards.

The instructional materials may be adapted from current programs, provided by states, or developed by professional organizations. They may come as print books, e-books, or other electronic forms, but they must be available. At a minimum, model units are needed as soon as possible. Arguing for a coherent curriculum based on the standards is not new. Indeed, there is a long history of curriculum serving an essential role in science teaching. If there is no curriculum for teachers, I predict the standards will be implemented with far less integrity than intended by the National Research Council's (NRC) *A Framework for K–12 Science Education* and those who developed the *NGSS*.

I conclude by emphasizing a key feature of implementation—the need for clear and coherent curriculum and instruction that connect the standards and assessments. Curriculum materials will be the missing link if they are not developed and implemented. The absence of a curriculum based on the standards will be a major failure in this era of standards-based reform and assessment-dominated changes. When science teachers at all levels in grades K–12 ask, "Where are the materials that help me teach to the standards?" the education system must have a concrete answer.

OUR COMMON PERSPECTIVE AND LEADERSHIP OPPORTUNITIES

Improving science education will result from sustained, systematic efforts of leaders, especially teachers.

The larger science education community must recognize the unique and central position of teachers in the improvement of science teaching and student learning. Reciprocally, teachers must accept the responsibility to change programs and practices in ways that align with policies (i.e., standards). All members of the community have some responsibility for improving teaching and learning.

Initiating reform of science education at any level—local, state, or national—must do more than revise policies (i.e., standards, frameworks); reforms must attend to the changes in programs and practices implied by reform of policies. We want to emphasize the fact that teachers and teaching make the difference in student learning. Standards set goals and give direction, but they are only part of education reform.

Focusing on the instructional core will enhance the possibility of a reform reaching a scale that makes a difference. Focusing on the instructional core has been a continuing recommendation in this seminar.

ISSUES AND QUESTIONS FOR DISCUSSION

1. Think of reform initiatives with which you have been associated. They may be local, state, or national. How would you describe the call to action? Goals? Funding? Leaders? Innovations? Products? Results?
2. How would you propose bridging the gap between research-based theory and teacher practice?
3. While education reforms may vary, can you identify any common features and pitfalls of reforms?
4. How have the policies and politics of science education reform changed across the decades?
5. Given what you have learned about educational change, how would you approach reform in your science department, school, district, or state?

REFERENCES

American Association for the Advancement of Science (AAAS). 1993. *Benchmarks for science literacy.* Washington, DC: AAAS.

Bybee, R. 2010. *The teaching of science: 21st-century perspectives.* Arlington, VA: NSTA Press.

Bybee, R. 2013a. *Invitational research symposium on science assessment: Measurement challenges and opportunities.* Princeton, NJ: Educational Testing Services.

Bybee, R. 2013b. *Translating the* NGSS *for classroom instruction.* Arlington, VA: NSTA Press.

Elmore, R. 2004. *School reform from the inside out: Policy, practice, and performance.* Cambridge, MA: Harvard University Press.

Elmore, R. 2009. Improving the instructional core. In *Instructional rounds in education: A network approach to improving teaching and learning*, ed. E. City, R. Elmore, S. Fiarman, and L. Teite, 21–38. Cambridge, MA: Harvard Education Press.

National Center for Education Statistics (NCES). 1994. *The condition of education, 1994.* Washington, DC: U.S. Department of Education.

National Commission on Excellence in Education (NCEE). 1983. *A nation at risk: The imperative for educational reform.* Washington, DC: U.S. Government Printing Office.

National Research Council (NRC). 2002. *Investigating the influence of standards: A framework for research in mathematics, science, and technology education.* Washington, DC: National Academies Press.

National Research Council (NRC). 1996. *National science education standards.* Washington, DC: National Academies Press.

National Research Council (NRC). 2012. *A framework for K–12 science education: Practices, crosscutting concepts, and core ideas.* Washington, DC: National Academies Press.

National Society for the Study of Education (NSSE). 1932. *A program for science teaching.* Chicago, IL: University of Chicago Press.

NGSS Lead States. 2013. *Next Generation Science Standards: For states, by states* Washington, DC: National Academies Press. *www.nextgenscience.org/next-generation-science-standards.*

NSTA. 2015. Science educators are preparing for the *NGSS* 27 (4): 16–17.

Pellegrino, J., M. Wilson, J. Koenig, and A. Beatty. 2014. *Developing assessments for the* Next Generation Science Standards. Washington, DC: National Academies Press.

Phillips, C. J. 2015. *The new math: A political history.* Chicago, IL: The University of Chicago Press.

Ravitch, D. 1995. *National standards in American education.* Washington, DC: Brookings Institution.

SECTION XI

LEADERSHIP AND EDUCATION

Larger-than-life educational leaders are rare. We must recognize that a majority of individuals in education have leadership responsibilities.

Beginning in Chapter 1, we developed a theme of challenges. That theme was in part influenced by articles on "Grand Challenges in Science Education," introduced in the April 19, 2013, issue of *Science*. The articles in that special issue of *Science* identified an initial set of problems that also represent leadership opportunities for those in the science education community.

The community's search for solutions to the complicated problems of contemporary reform directs us toward the topic of leadership. We certainly recognize the problems, because hundreds of reports document the multiple reasons and many directions for reform of science education. We realize there are understandable and acceptable solutions to our problems, but science education needs widespread leadership to mobilize the available resources to achieve the reform.

In Chapter 21, we develop the themes of leadership and responsibility in education and consider the role of teacher educators, state and district science supervisors, science education researchers, and science teachers. We realize science teachers have the greatest burden and heaviest responsibility for reform. Our comments are grounded in recognition of the essential position of teachers in contemporary reform, the need for change, and compassion for their difficult task.

Chapter 22 addresses the challenge of leadership and power. The chapter also includes a personal perspective on leadership from Stephen.

SUGGESTED READINGS

American Association for the Advancement of Science (AAAS). 2013. Grand challenges in science education. Special issue, *Science* 340 (6130). This theme issue of *Science* addresses a variety of challenges that require leadership in science education.

Burns, J. M. 1978. *Leadership*. New York: Harper and Row. A classic in the literature on leadership.

Gardner, J. 1990. *On leadership*. New York: The Free Press. Another classic on leadership.

Rhoton, J., ed. 2010. *Science education leadership: Best practices for the new century*. Arlington, VA: NSTA Press. A compilation of chapters by many science education leaders. Themes include challenges, school and district leadership, school improvement, and public understanding of science.

CHAPTER 21

PERSPECTIVES ON LEADERSHIP

Because of the immense and diverse structure of science education and the scale and complicated nature of the contemporary reform, we are convinced that *distributed leadership* is one key to a successful reform. That is, all individuals associated with science education must contribute to the vision of scientific literacy for all students. This chapter describes perspectives on leadership, beginning with a definition.

DEFINING LEADERSHIP

Leadership defies easy definition. Figure 21.1 lists a number of definitions drawn from a variety of sources. Two common themes connect the definitions of leadership: relationships with people and achieving a group purpose. From our perspective, we define leadership as an individual's ability to work with others to improve science teaching and learning to enhance the scientific literacy of

Figure 21.1. Definitions of Leadership

LEADERSHIP IS

- **making** things happen, or not happen;
- **including** others to act for the wants and needs, aspirations, and goals of both leaders and followers;
- **inspiring** hope and confidence in others to accomplish purposes they think are impossible;
- **perceiving** what is needed and right and knowing how to mobilize people and resources to accomplish these goals;
- **creating** options and opportunities, clarifying problems and choices, building morale and coalitions, and providing a vision and possibilities of something better than currently exists; and
- **empowering** and liberating people to become leaders in their own right.

all students. Our definition implies the involvement of a majority of individuals within the science education community.

Status and power do not define leadership. A person who has the status may or may not be an effective leader. Status, however, certainly enhances the possibility of leadership. Being in a position with some status conveys expectations of leadership to others, and their expectations can contribute to one's ability to lead. Leadership is not only power, however. Individuals can have power for a variety of reasons, such as money, authority, or position. These can result in power, but they do not necessarily result in an individual being able to lead a group toward a common purpose.

Assuming you agree with the need for leadership, you may ask, "What else should I know about leadership? What type of leadership is appropriate for my situation?" In this section, our discussion of leadership centers on three classic works: James MacGregor Burns's *Leadership* (1978), Bernard Bass's *Leadership and Performance Beyond Expectations* (1985), and John Gardner's *On Leadership* (1990). We begin with a description of several different perspectives on leadership.

Transactional Leadership

Burns (1978) described transactional leadership as a relationship between leaders and followers based on exchanges of, for example, jobs for votes and favors for campaign contributions. Transactional leaders let their followers know what is expected and what they will receive for meeting those expectations. This type of leadership focuses on services and rewards; according to Burns, such transactions represent the majority of relationships between leaders and followers. Burns primarily describes political relationships.

In a later work, Bass (1985) extended the definition to supervisor-subordinate relationships. In these relationships, the transactional leader first identifies what subordinates want from their work and sees that they get the rewards, if the performance warrants it, and thus exchanges rewards for efforts.

In science education, the transactional approach requires the leader to identify outcomes and clarify the role of subordinates (e.g., students, teachers, colleagues). At the same time, the leader must identify what rewards will be bestowed when the outcomes are achieved. The subordinates must have confidence that they can achieve the desired outcomes and believe the reward is worth the effort.

Transformational Leadership

Burns (1978) also described a type of leadership that is both more complex and more powerful than transactional. Transformational leaders also have goals, but these goals focus on the personal needs or demands of their followers. Here, the leader identifies the motives and aspirations of followers and seeks to engage individual followers on a personal level. Burns also suggests that transformational leadership may result in a relationship of mutual stimulation that elevates the followers to leadership activities and agents that recognize moral outcomes.

Burns based his conception of transformational leaders on Abraham Maslow's hierarchy of needs (1968, 1970, 1971), a theory that is important for the theme of leadership, particularly transformational leadership. The hierarchy of needs begins with the most basic physiological needs, such as food, water, air, and sleep, which have the greatest motivational force. As physiological needs are fulfilled, there emerge needs for order, structure, stability, and freedom from chaos and fear. The terms *safety* and *security* describe this level. Next, individuals need to belong, to give and receive affection, to have a friend, and to belong to a group. The next level is the need for self-esteem—to have a stable, firmly based, positive, and realistic perception of one's self. Finally, there is a need for self-actualization. When all other needs are largely met, there emerges the need to develop and use one's capabilities, talents, and potential. The ways in which individuals actualize their potential vary, but having a mission such as improving science education is certainly one such way—and one that transformational leaders understand.

Again, Bass (1985) slightly modified Burns's (1978) original conception of transformational leadership. Bass suggested that transformative leaders influence followers first by raising their level of awareness about goals or aspirations; second, by helping them transcend their self-interests for the sake of the larger mission; and third, by altering their level on Maslow's hierarchy from lower to higher levels, or horizontally, by expansion at their current level.

Burns (1978) originally characterized transactional and transformational as two types of leadership at opposite ends of a continuum. Most leaders actually exhibit both types of leadership in varying amounts. Observation or review of leadership supports this refinement of the conceptually distinct types (Bass 1985).

Moral Leadership

We cannot avoid mention of morals and ethics because many of the characteristics discussed so far are equally applicable to Adolf Hitler, Winston Churchill, Idi Amin, and Martin Luther King, Jr. But real differences do exist among leaders when one considers the moral dimensions of the relationship between these leaders, their followers, and the purposes these leaders wanted to achieve.

Burns (1978) clearly described the moral dimensions of leadership. He indicated first that the leader and the led have mutual needs, aspirations, and values; second, that followers should have knowledge of alternative leaders and programs; and third, that leaders should take responsibility for their commitments. Burns writes,

> *Moral leadership emerges from and always returns to the fundamental wants and needs, aspirations and values, of the followers. I mean the kind of leadership that can produce social change that will satisfy followers' authentic needs. I mean less the Ten Commandments than the Golden Rule. But even the Golden Rule is inadequate, for it measures the wants and needs of others simply by our own. (1978, p. 4)*

In discussing the moral dimension of leadership, Gardner (1990) appealed to the categorical imperative (i.e., act only as you would have others act in the same situation) when he stated, "We believe, with Immanuel Kant, that individuals should be treated as ends in themselves, not as a means to the leader's end, not as objects to be manipulated" (p. 73). Although leadership by district coordinators, state science supervisors, teacher educators, professional developers, and classroom teachers does not involve the influences and power that national or international leaders may have, it is still important to have high and clear ethical aspirations and to assume responsibility for those aspirations.

Science educators can use the transactional and transformational models in understanding approaches to leadership. Teacher-student, supervisor-committee, and professor–graduate student relationships provide opportunities to apply both transactional and transformational leadership. The lack of connection between these ideas about leadership and educational change can be discerned in the understanding of the concerns of beginning teachers and those of experienced teachers implementing innovative programs, as elaborated in the original work of Frances Fuller (1969), Gene Hall and Shirley Hord (1987), and Michael Fullan (1982). These individuals take a perspective paralleling that of transformational leadership in that they focus on the needs and concerns of individuals in the process of educational change and professional development.

Reform Leadership

Burns (1978) devoted a chapter of *Leadership* to reform leadership. Burns argued that the leadership of a reform movement, as opposed to that of a revolution, is particularly difficult and exacting. Reform leaders usually work within extant institutions, coordinating individuals with a reform when they have no reform goals of their own, and reform leadership implies moral leadership because proper means must be used to achieve the goals. This is not encouraging news for contemporary education leaders.

Regarding political reform, Burns points out the tendency of members of the aristocracy to assume leadership. Burns (1978) described another troublesome problem of reform leadership:

> *Because reform leaders typically accept the political and social structures within which they act, their reform efforts are inevitably compromised, and usually inhibited, by the tenacious inertia of existing institutions. (p. 200)*

He concluded with this insight about reform leadership: "Reform is ever poised between the transforming and transactional—transforming in spirit and posture, transactional in process and results" (p. 200).

Given this perspective on reform leadership and the history of education change, is there any hope of achieving reform in science education? We believe there is. According to Bass (1985), in times of dissatisfaction, distress, and pressure, transformational leadership is more likely to emerge and be effective. We suggest the potential for real reform in science education is high. In the early 21st century, the United States has increasingly recognized the connection between education and the critical issues of economic growth, environmental stability, and national security. New national standards in English language arts, math, and science have been accepted with clear goals of attaining higher levels of college and career readiness for all students. That is, the social support and policies for change

are strong, and the reform includes significant components of education. It is not solely a reform of science education. However, our optimistic view of the system assumes that science educators, and especially science teachers, will provide their share of leadership.

So, one reasonably asks, "Where do I begin?" Our answer centers on a fundamental challenge of leadership. Here we will review a point made in Chapter 1.

ESTABLISHING A VISION AND PROVIDING A PLAN

All leaders provide a vision. Leading in our context means establishing a vision; translating that vision for students, committees, organizations, and schools; and sustaining the vision while adapting it to school science programs and teaching practices. Expressions of each leader's vision must be in the context of that individual's own environment, organization, or institution. Although visions need not be complex or elaborate, they must be different from current programs and practices. Visions are not statements of the status quo; they are new, they are substantial, and they look to the future.

The countervailing force for any leader's vision of the future is the institution's or individual's current perspective. Variations on statements such as, "We already have standards," "We have tried that," "We can change the curriculum, but what about assessments?" and "Where is a particular topic in this new program?" are all examples of evaluations of the new in terms of old perspectives. Many variations of memory can act as a force against vision. Teachers, parents, and even students have past experiences that they use to evaluate anything new. We cannot tell you what the unique examples will be, but we can tell you that your vision of science education will be compared to past and present programs. The effective leader has internalized his or her vision and conveys it at every opportunity. In most instances, providing leadership means persuading constituents of their potential to achieve new goals. A vision with substance meets the criticisms that will be offered; mere slogans do not.

After vision, leadership requires a plan. Following a leader's initial question—"What are we going to do to improve science education?"—the second question is, "How are we going to do it?" This seems a simple enough question, but some individuals have a vision without a plan. Others may have plans but no vision. Providing leadership requires a plan of both directed action and flexibility. One must give a sense of direction and be responsive to those with whom one is working. Being too rigid and authoritarian is ineffective; being too flexible and *laissez-faire* is equally ineffective. Providing leadership means continually reaffirming your vision while simultaneously adjusting your plans.

Consider the challenge of implementing new national or state standards, examples we know well. There is little question about the district science coordinator's need to communicate the standards' vision and clarify the plans for school curricula, classroom instruction, district assessments, and professional development. One has to have both the vision and plan.

The form and function of leadership in science education will vary with contexts and situations. The requisite qualities for effective leadership in science education—whether in science classrooms, school districts, colleges, universities, state departments of education, national organizations, or

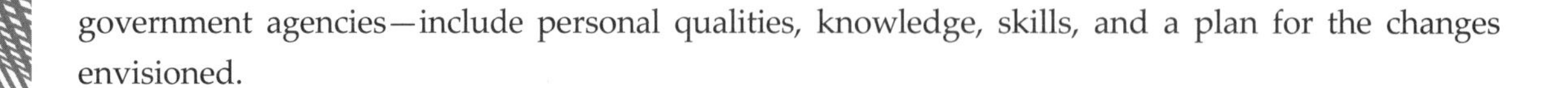

government agencies—include personal qualities, knowledge, skills, and a plan for the changes envisioned.

A PERSONAL PERSPECTIVE

RECOGNIZING AND RESOLVING THE PARADOXES OF LEADERSHIP

Rodger Bybee

Science educators who assume the responsibilities of leadership will inevitably confront people, situations, and actions that contain apparently contradictory aspects. My interest in paradox and its relationship to leadership was formed several decades ago. In *The State of the Presidency* (1980), Thomas Cronin presented the general idea that there are paradoxes associated with leadership. For example, Americans demand powerful, popular presidential leadership, yet they are suspicious of strong, centralized leadership and the potential abuse of power. Educators assuming responsibility for leadership will have to address such paradoxes. Indeed, I would argue that a leader's effectiveness can be assessed by the way he or she resolves paradoxes. What follows are paradoxes that are basic to leadership in science education.

THINKING ABSTRACTLY AND ACTING CONCRETELY

Our discussion of leadership emphasized the importance of a vision. Leaders must develop for themselves and convey to their followers a new and unique image of science education. The leader's vision, by its very nature, is abstract and complex. In science education, this vision must incorporate many ideas about science, students, schools, teachers, and society.

Our discussion of leadership also stressed the vision's implementation. The leader must translate the abstract qualities of the vision into concrete plans for action. The process includes such things as identifying policies and procedures, clarifying the process of reform, assembling teams, providing professional development, creating support for the plan, and revising the plan based on feedback.

I place this paradox first because having a vision and a plan is critical to resolving the other paradoxes. Consider, for example, the numerous workshops that school districts provide each year. The relationship of many such workshops to some larger vision for the school district is only tangential. Or consider the lofty ideas of policy makers who have no plan for implementing the ideas. First and foremost, effective leadership in science education requires that individuals think abstractly but act concretely.

A contemporary example of this paradox might involve recognition of the role of the *Next Generation Science Standards* (*NGSS*) and addressing changes in science classrooms. Another example might be thinking of new science standards and curriculum reform in a school district and subsequently selecting instructional materials for the elementary, middle, and high schools.

HAVING DIRECTION AND RETAINING FLEXIBILITY

As mentioned, effective leadership requires a vision and a plan. Still, the leader cannot ignore suggestions and feedback from those with whom he or she is working. This paradox is like science teaching. One must have a goal for the lesson and unit, yet also be flexible in responding to the various unexpected situations that occur in classrooms. I have seen beginning science teachers lock themselves into a lesson plan and continue teaching long after they have lost control of the students. On the other hand, there are the science teachers who are so flexible that little learning occurs. The effective leader simultaneously has a direction and demonstrates flexibility. A variation on this paradox is planning for a new school science curriculum while also handling each teacher's questions and concerns. Examples might include the process of teaching a science class or conducting a professional development workshop.

INITIATING CHANGE AND MAINTAINING CONTINUITY

Science educators must respond to the calls for reform based on new national and state standards. Such responses have accompanying demands for change. Yet a school system or a particular school also needs continuity with the past. The enduring purposes of science education—contributing to scientific literacy—bring continuity to the changing nature of assessment, teacher education, curriculum, and science teaching. Some science educators will inevitably perceive new standards and the implied innovations in science education as disruptive; the inevitable protectors of the status quo usually articulate positions that resist change. We need only to remind them that the status quo is the reason for a decade of reports and standards that demand education reform. Balancing change with continuity results in a successful transformation of science education. You cannot have progress without change, and you cannot have change without disruption.

ENCOURAGING INNOVATION AND SUSTAINING TRADITION

This paradox is similar to the one just discussed, except that it centers on innovative curriculum programs and classroom practices. There is a need for bold, innovative curricula that also accommodate old policies, mandates, and obsolete syllabi. If you present a new curriculum program, you can immediately expect a question grounded in an old topic. For example, if you present an integrated

science program, someone may ask where the unit on rocks and minerals is. A variation on this paradox is promoting risk-taking while providing security.

FULFILLING A NATIONAL AGENDA AND INCORPORATING LOCAL MANDATES

Every leader in science education should be aware of *A Framework for K–12 Science Education* (NRC 2012) and the *NGSS* (NGSS Lead States 2013). Even if your state is not adopting the *NGSS*, those standards are influencing the standards and assessments in non-adopting states; thus, they do represent a national agenda in science education. At the same time, leaders at the local level must incorporate state requirements and local mandates into school science programs. Whether you are adopting a national program or developing materials locally, there is a need to balance these priorities.

ACHIEVING YOUR GOALS AND ENDURING CRITICISM

An irony exists in successful leadership: The more successful you are in achieving your vision and implementing your plan for better science education, the more you will be subject to criticism, some of which will be unjust. Whether it is from deans and colleagues in the school of education, superintendents and parents in school districts, or science teachers within your school, with success comes criticism. Abraham Lincoln, a great leader, was severely criticized throughout his presidency. Lincoln's words from his Cooper Union Address of February 27, 1860, provide encouragement. Lincoln concluded with a challenge for people to hold their beliefs even under criticism:

> *Neither let us be slandered from our duty by false accusations against us, nor frightened from it by menaces of destruction to the Government, nor of dungeons to ourselves. Let us have faith that right makes might, and in that faith let us to the end dare to do our duty as we understand it. (Phillips 1992, p. 149)*

Science educators have neither the weight of fighting a civil war nor the power of a president, but we do have an important mission, and doubtless there will be judgments, reviews, and claims of catastrophe. Leadership requires self-confidence, fortitude, and a way to deal with this paradox. Perhaps we could use Lincoln's example—ignore most of the criticism and respond with vigor to the important attacks. More important, have courage and maintain a sense of humor (Phillips 1992).

ASSUMING THE RESPONSIBILITIES OF LEADERSHIP

For the person willing to assume responsibilities of leadership, the first question is, "Where do I begin?" Begin with some introspection and clarification of your personal goals. If personal goals center on increasing a salary, obtaining prestige, evading stress, and avoiding criticism, then perhaps the leadership role is not for you. But if you understand the problems of leadership and still see your mission as contributing to a cause that transcends material welfare and contributes to personal and professional growth for yourself and others and to society's aspirations for education, you have probably already assumed a leadership role.

The second question is, "Whom am I leading?" Obviously, you have a first answer: your students, the science teachers in your district or state, your colleagues in science education, your community, and so on. However, based on our descriptions of leadership, there is another answer that may be more subtle and elusive. I refer to the motives, needs, aspirations, and goals of potential followers, as individuals and as a group. There should be some congruence between your motives and goals and those of your followers. As pointed out earlier, leadership will probably be a combination of the transactional and transformational, but you should base it on an accurate view of your followers' motives and goals. In the beginning, you may have to use persuasion and rewards, but these should give way to mutuality of goals based on higher levels of motivation and values.

The third question is, "Where am I going?" Articulation of your personal vision is a prerequisite to all that follows. You should be able to describe your vision for the classroom, school, state, or nation in a paragraph of clear, ordinary language. Writing the statement may provide you opportunities to elaborate on the details of your vision. Being able to articulate your vision and elaborate on its details defines the difference between having a vision and stating a slogan.

"How will I achieve my vision?" The fourth question returns to the need for a plan. You should consider immediate, short-term (between six months and one year) and long-term (between one and three years) objectives and specific actions. In the course of implementing your plan, these objectives and actions will undoubtedly change. This is acceptable. Remember that improving science education is your goal, and changing the system will result in modified plans and actions. In a sense, the more effective your leadership, the more you will have to modify your plans.

"What are the obstacles and barriers?" This question is difficult to answer because I cannot anticipate all the various obstacles you might encounter. Having a clear vision and plan and using a systems model should help overcome the barriers.

You may see these concluding questions and discussions as a process you already use. I hope so. In a very real sense, that observation confirms my point of distributed leadership and supports the widespread assumption of the responsibility for leadership by science educators.

REFERENCES

Bass, B. 1985. *Leadership and performance beyond expectations.* New York: Free Press.

Burns, J. M. 1978. *Leadership*. New York: Harper and Row Publishers.

Cronin, T. 1980. *The state of the presidency.* Boston, Little, Brown.

Fullan, M. 1982. *The meaning of educational change.* New York: Teachers College Press, Columbia University.

Fuller, F. 1969. Concerns of teachers: A developmental model. *American Educational Research Journal* 6 (2): 207–226.

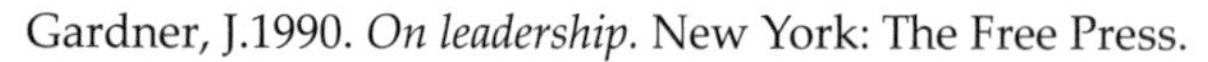

Gardner, J.1990. *On leadership.* New York: The Free Press.

Hall, G., and S. Hord. 1987. *Change in schools: Facilitating the process.* Albany: State University of New York Press.

Maslow, A. H. 1968. *Toward a psychology of being.* New York: Van Nostrand Reinhold.

Maslow, A. H. 1970. *Motivation and personality.* New York: Harper and Row.

Maslow, A. H. 1971. *The farther reaches of human nature.* New York: Harper and Row.

National Research Council (NRC). 2012. *A framework for K–12 science education: Practices, crosscutting concepts, and core ideas.* Washington, DC: National Academies Press.

NGSS Lead States. 2013. *Next Generation Science Standards: For states, by states.* Washington, DC: National Academies Press. *www.nextgenscience.org/next-generation-science-standards.*

Phillips, D. T. 1992. *Lincoln on leadership.* New York: Warner Books.

CHAPTER 22

LEADERSHIP AND POWER

Bertrand Russell (1938) once said that "the fundamental concept in social sciences is power, in the same sense in which energy is the fundamental concept in physics" (p. 12). Providing leadership means giving energy, empowering the group or individuals (Ramey 1991). Individuals require power to make decisions, develop ideas, and influence tasks as they relate to overall vision. You likely have been in discussions or groups that lacked either energy in the form of an enthusiastic leader or power in terms of the capacity (e.g., budget) for implementing the recommendations. In either case, your participation may have been interesting, but it was an empty exercise.

PERSPECTIVE

Leadership requires power. The issue is not whether leaders have and use power—they do and they must—but rather the ethical uses of power. Is the leader using power to achieve morally sound and ethically defensible goals? Leaders need to understand power for at least two reasons. First, effective leadership in science education entails empowering others, especially science teachers. Second, we need to understand the power relationship in school systems if we hope to achieve the purposes of contemporary reform (Sarason 1991). In short, if the leaders in science education do not understand and change the power relationships, the current system will not change and reform initiatives will falter and fail.

Leadership and the Empowerment of Others

In *Power and Innocence* (1972), Rollo May described the types of power that are evident in human interactions and, by extension, in interactions between leaders and followers. *Exploititative power* is the destructive use of power to subject others to a leader's goals; slavery is an example. *Manipulative power* is the capacity to control another without the physical dimensions and force implied by exploitation; physical conditioning is an example of manipulative power. *Competitive power* is power against another person or group; the obvious example is sports. The final two kinds of power are most relevant to our discussion of leadership. According to May, there is *nutrient power*, which is power on behalf of another person; parents' care for their children and teachers' concern for students are both examples. Most discussions on empowering teachers exemplify nutrient power. Finally, there is *integrative power*, the power that results from joining with another person; cooperative learning is an example of integrative power. In reflecting on the issue of empowering people and the types of power just described, it becomes clear that a leader's use of power is another area in which the dimensions of leadership are evident.

In their book *Leaders: The Strategies for Taking Charge* (1985), Warren Bennis and Burt Nanus described several dimensions of empowerment. One of the first dimensions is developing a sense of *significance* in others. Effective leaders create a vision that makes others feel as though they make a difference. To be significant, this vision must have substance and transcend the superficiality of slogans. Individuals are, for example, translating the vision into innovative standards-based science programs and sustaining the new programs through difficult situations. In so doing, they are really making a difference in students' science education.

A second dimension of empowerment involves developing new *knowledge, skills,* and *attitudes*. This results in greater competence and a sense of mastery. Third, empowerment provides a sense of *community*. For example, when all the science teachers in a school system have the common purpose of implementing the *Next Generation Science Standards* (*NGSS*) and cooperate in achieving that goal, they develop a sense of community and collegiality.

Empowering people results in their experiencing greater *enjoyment* in their work. This is a fourth dimension. Outdated theories of motivation and leadership suggest that only rewarding or punishing individuals could achieve desired results. Those theories recognized only the lower levels of Maslow's hierarchy of needs. Contemporary theories of motivation and leadership recognize that individuals have higher needs, including needing to know and understand, engaging in meaningful work, and developing personal and professional efficacy.

A critical aspect of empowerment is assuming responsibility for achieving the tasks. So, with empowerment, science teachers have the responsibility to improve science teaching in their classrooms. Teachers also need to understand that every time they request how-to activities, require supplemental materials from publishers, resist understanding the purposes of science education, and refuse information because it is not relevant to their "real world," they are relinquishing power and avoiding the responsibilities they have for improving science education.

Empowering Science Teachers

Empowering science teachers is a prominent theme in the contemporary reform of science education. *Teacher empowerment* stands in contrast with the phrase popular in the 1960s, *teacher-proof programs*. A teacher-proof curriculum was a set of materials designed to enhance student learning independent of, or in spite of, the science teacher. As early as 1965, in *The Genius of American Education*, Lawrence

Cremin pointed out the misguided nature of a teacher-proof curriculum. He suggested that reformers had a legitimate concern about contemporary teachers and teaching, but he saw their solution of designing materials impervious to misuse as flawed. Cremin gave advice that is appropriate for any generation of leaders and reformers, especially those teacher educators and district science coordinators who are directly responsible for the professional development of science teachers:

> *But education is too significant and dynamic an enterprise to be left to mere technicians; and we might as well begin now the prodigious task of preparing men and women who understand not only the substance of what they are teaching but also the theories behind the particular strategies they employ to convey that substance. A society committed to the continuing intellectual, aesthetic, and moral growth of all its members can ill afford less on the part of those who undertake to teach. (Cremin 1965, p. 57)*

The teacher-proof approach to the curriculum did not work, in large part because of the science teacher's power once the classroom door was closed. That it did not work demonstrated the power of classroom teachers.

After teacher-proof materials, there emerged teacher-dependent materials and programs. Essentially, teacher-dependent programs are sets of how-to activities on which the teacher depends and which can be used without understanding the subject, pedagogy, or context of the lesson. We use the term *dependent* here in two respects. First, in a reaction to attempt to teacher-proof the curriculum, there was a pendulum swing toward designing curriculum materials that were entirely contingent on the teacher. Second, and most important, in the decades since the teacher-proof programs, many science teachers have developed a dependence on short-term, quick-fix lessons and a general intolerance of curriculum programs with a fully articulated scope and sequence. Just as the pendulum swung too far in the 1960s, the corrective reaction was also too extreme in the 1970s, 1980s, and 1990s. Empowering science teachers to adapt science materials, whether local or national, will enhance the teaching and learning of science.

CHAPTER 22

A PERSONAL PERSPECTIVE

LEADERSHIP: A CHOICE, A NEED

Stephen Pruitt

Leadership takes on many different forms. There are leaders who are loud, leaders who are quiet, leaders who are always out front, and leaders who like to work behind the scenes. What kind of leader are you? Have you actually considered the type of leader you are or the type you want to be? In Rodger's essay in Chapter 21, he does a wonderful job of laying out the paradoxes of leadership. I would like to spend a little time talking about leadership through some examples, and maybe providing a few anecdotes along the way.

I believe strongly that leadership is a choice in many respects, but not being a leader is a bigger choice. I know that sounds odd, so let me explain. As educators, we are meant to be leaders. We are not trained to be, and perhaps some of society expects us not to be. I think that the very nature that brought us to the field of teaching means we are leaders. I think the choice not to be a leader, while often unconscious, is a detriment to our students and our communities.

I am often dismayed, even angered, when I hear one of our wonderful educators say, "I am *just* a teacher." Many of you reading this book will have heard me discuss this. We do our colleagues, our students, and ourselves a tremendous disservice when we belittle ourselves this way. I think this has become a societal label about us, and if it is ever to stop, we educators must stop it. Even if the word *just* is what the world wants to use to label us, we perpetuate that by using it as well. Educators are the most passionate, compassionate, caring, and intelligent people I know. What is missing, unfortunately, is the aspect of leadership outside of our classrooms or buildings. It is not included in most preservice teaching discussions; in fact, I am not so sure it is not discouraged subliminally.

So much of this book is about leadership. You bought it thinking it was about science education in general and the *Next Generation Science Standards* (*NGSS*) in particular, but what I hope you are getting from it is the idea that leadership matters. How we lead is up to each of us individually, and our leadership style must fit our personalities. However, I would also say we may not know our style if we do not spend some time in study of other leaders and how they are looked at by those they lead. I would say I have made a career out of reading people. I have observed them, studied them, and tried to emulate the qualities that fit my personality. I am going to use my time to talk about the leaders that have affected me, what I learned from them, and how what I have learned influences my decisions and actions today. Some of these leaders are people who I have had the opportunity to know personally, some are historical figures, and, oddly enough, some are fictitious, but still I learned from them.

In her book *Harry Potter and the Deathly Hallows*, J. K. Rowling has Professor Dumbledore state, "It is a curious thing, Harry, but perhaps those who are best suited to power are those who have never sought it. Those who, like you, have leadership thrust upon them, and take up the mantle because they must, and find to their own surprise that they wear it well" (Rowling 2007, p. 718). Leadership is not something educators look for, but we are in a time when educators are having leadership thrust upon them. We do wear it well, but we must still choose to put it on. The traits that we use in our classrooms every day are the same that make us good leaders. In the next few pages, I am going to talk about leadership and the people who shaped me into the person I am today. I beg your indulgence for a bit as it may sound like a history lesson, but my point is to help you see that leadership is a choice. But, like good science, as we consider our leadership, we should have evidence of what good leadership is and what traits best suit us.

STRENGTH AND PRESENCE

The first trait of a good leader I would like to discuss is strength. Well, you could say, "Duh," but I mean strength in an unconventional way. When I say *strength*, I really mean presence. I mean that when a leader is in the room, people know it and respect the leader. This does not mean that they simply and blindly follow him or her, but that they recognize there is someone who may have achieved much, yet is eager to work with others so they achieve more together. This is a decision one makes. I have seen this trait in some of the greatest leaders I have known or read about. To be clear, this is not the person who says, "Here I am, look at me!" Rather, it is the person who need not say a word, yet people know he or she is there. I am a huge fan of George Washington. I love reading historical novels, not just because I am a history buff, but because of what I can learn about leadership. It is clear why Washington holds the station he does in our country. If you have not read about him, I encourage you to do so. In many historical novels and accounts, you will find that Washington was not giving fiery speeches, nor was he always addressing the crowd. More often than not, the books will state he was there in the room or talking to small groups. Everyone knew who he was and what he was about. He was this way his whole life. Yes, he jumped on his horse and charged the enemy, often to the dismay of his superiors and the concern of his men, but he also knew how to influence without having to be out front all the time. He knew the value of good relationships and how to hold people accountable. For me, I wanted to build my knowledge and my abilities as a leader and leverage my relationship skills to build presence. I do not have a need to be in front all the time, although I am clearly comfortable there. But I did and do want to lead by example and offer my thoughts on tough problems—not because of vanity, but because I believe I have something to offer. Washington had much to offer, but he did so in a manner that suited him.

Another great example is my other favorite president, Abraham Lincoln. I know Rodger mentioned him earlier, but it is worth bringing him up again. We know of his speeches and of course his achievements. What is often left out of his story is that he was a fairly soft-spoken man with a high voice. He, like Washington, did not feel the need to "control" a room. He did anyway, but in a very different manner. He would control meetings through his stories, for instance. If a discussion was off track or moving in a direction he did not like, rather than always redirecting it with a mandate, he did so through the use of one of his many stories. Another thing I loved about Lincoln that I will expand on later was that he was a master at positioning himself in a room full of people. His presence was felt, but it was often him deciding where he would stand or how he could empower or comfort someone based on where he was in the room. Like a good educator, he knew that where you stand makes a difference.

If educators build presence, they will no longer say, "I am *just* a teacher," they would simply say, "I am a teacher," and that will be enough to drive the conversation. In the development of the *NGSS*, I knew I needed to play a prominent role, but I also needed to not be seen as the leader. There were many smarter and more savvy people I had the chance to work with, but I like to think that there was some comfort in knowing I was there. Perhaps not, but influencing others with my presence is a goal I will continue to strive for. A leader is not the loudest or even the smartest person; it is the person who can influence others to be better than they thought they could be. As educators, we have the ability to have that presence, but we have to choose to use it with adults as well as with our students.

COURAGE

A second major trait is courage. This is another word that seems obvious, but I would stipulate it is not. As part of developing the *NGSS*, I have said often that we must have the courage to be patient. In education, we are all about doing things quickly, and that often fails us. After all, no great education initiative ever failed in the vision phase; failure happens in implementation. It is here we need courage. It is not only courageous to simply be out front of the development of an idea. To have the patience to plan, envision, build relationships, execute a difficult plan or vision, or be willing to think out of the box on a tough issue—that is courage. The *NGSS* are very different than any state standards we have ever had. Yes, it has taken some courageous people to develop and adopt them. It has taken some courageous teachers to implement them in classrooms when they realize the type of instruction required is very different than what they have done in the past. However, it has also taken courage to not rush into testing or accountability with the *NGSS*. Assessment of these standards is very different, as it should be, since the point of their architecture was to push change in our science classrooms. But rather than rushing, most educators have been patient despite criticism from those who do not understand the need to get it right. So I would say that courage is actually doing what you know to be right, even when others around you are telling you to go a different direction. This leads me to my next point about courage: doing what is right, not what is easy.

An unfortunate part of our profession is that it is easy to feel cut off from the outside world while we are in our classrooms working with our students. It is easy to become cynical and perhaps even downright obstinate. A courageous education leader realizes that we can always improve. Educational excellence is not a destination; it is a journey. Sorry for the analogy, but it is true. Educational excellence today looks very different than it did 20 years ago and different from what it will be 20 years from now. The world we prepare students for today is very different from the one we were prepared for, yet many still teach as they were taught. I am personally inspired by those educators courageous enough to take risks and do what they know they must to improve their instruction for the sake of their students. They do this knowing they are learning themselves and that any change or transition is not easy. I am sorry to quote him again, but Dumbledore also said, "Dark times lie ahead of us and there will be a time when we must choose between what is easy and what is right" (Heyman and Newell 2005). For some, the development of new standards were dark times. I am happy to say that I believe for most of us they have represented new hope. Still, it is easy to assume the *NGSS* are another fad, that if I wait long enough they will pass too, or the "state" will just not test them properly. I have seen some incredibly courageous educators who have said they simply will not stand for this. They push forward, they serve on state and local committees to get the implementation right, and they stand as models for the rest of us who struggle each day. They did not go the easy way; they went the right way.

PERSEVERANCE AND RELATIONSHIPS

Washington and Lincoln certainly went the right way, and it was never easy. What they showed us was that perseverance and relationships matter. Whether Washington was experiencing his many military battles or his early struggles with trying to move up the ranks, he stuck with it. He built relationships that eventually even led to him owning his beloved Mount Vernon. This is why he had presence; he had perseverance and relationships. Lincoln certainly had the same traits. The passage of the 13th Amendment, which was incredibly controversial, happened because of his perseverance, and in no small part because of relationships. He showed people he cared in small ways. A hand on the shoulder when his men were panicking, or taking the time to sit and talk about an issue with people regardless of their societal level—these were ways he built his relationships. He was also tenacious about his beliefs. The 13th Amendment was very contentious. He lost some relationships with the passage, but he also knew it was right. It may appear odd to place these two traits in the same section, but they are connected. A person can persevere through adversity, but an idea can only persevere if many support it. That comes through relationships. The development of the *NGSS* required perseverance and relationships. I was able to build many during that time, and it is possible I hurt some as well. It was important we persevered even in a time when other standards were under attack. I felt the future of education in our country was at risk. The relationships between key partners and educators were key to the standards themselves persevering, so it was important that leaders from across all fields came together to not only work on the standards but also advocate for them.

VISION AND CONTEXT

As an education leader, one must realize the great importance of vision. Again, I realize that Rodger has addressed this already, but it bears repeating. Without vision, there is nowhere to go. I have had opportunities to work with some great leaders, and they were great because they had vision. They were not only out-of-the-box thinkers; I am not sure they even knew where the box was at times. In particular, I think of the principal of the first school where I taught, Wayne Robinson. He was one of the greatest education leaders I have ever known. He had all the traits that both Rodger and I have mentioned above, and more. But what I remember most about him was his vision. He was committed to making our school the best in the county and probably the state. He did so with a passion and presence that inspires me still today. Many times, people will talk about great visionaries as being just that and not worrying about details. Wayne taught me to "sweat the small stuff" because that is what will undermine your work. So he kept his eye on the big vision while making sure nothing derailed it. He was flexible in how we got there but never in what he expected. I am still proud to have worked with him. In many ways, he embodied all the great leaders I have studied. And, at the end of the day, his vision for *why* he wanted the best school in the state was clear—our students. So, it is great for a leader to have their vision, but can they discuss the *why*? Wayne could do so at the drop of a hat. It was about kids for him—not about accolades or accountability scores, but about making lives better for the kids. Vision without context is a dream quickly forgotten; vision with context is a dream come true. As education leaders, we should always be an advocate for education. However, as soon as that advocacy becomes about our content more than our students or about preserving a way of life more than changing the lives of students, we have lost. Trust me, for years I have worked at the policy level of education. Those who come in not articulating why or the context for their advocacy become noise. So, to be an education leader, we really do need to know not only where we want to go, but why as well.

HUMBLE AND HONORABLE, YET STRONG

Of course, there are many traits of a quality leader, but I will end with one that I think is overlooked: humbleness. As I write this, I realize my highly respected co-author (and very good friend) will get a bit uneasy, because he is one of the most humble men with whom I have ever had the pleasure of working. Rodger is one of the smartest and most driven people I know. I could use pages up talking about his achievements. From his "discovery" of the 5 E Instructional Model to his work with PISA (Program for International Student Assessment) to his work leading the *National Science Education Standards* (*NSES*), he has done more to influence science education than perhaps anyone in the United States. I have known Rodger for several years now. I remember meeting him the first time and how in awe I was. What I also remember from that meeting was how comfortable he made me feel by caring about what I had to say. At that time, I really am not sure, at least in my mind, that I had earned the right to have a say, but he listened. I watched him with the *NGSS* writing team, and if there was anyone who had a right to have an ego, it was Rodger, but he worked hard and was extremely humble in working with everyone regardless of pedigree or résumé. Through this process, he was honorable. He honored all opinions, even when he knew the history behind an idea did not bode well for it. He honored all he worked with, even though they may not have had the experience or knowledge he did. In doing this, however, he was always strong in his beliefs. Because he was honorable and not because of his résumé, people listened to him. Because he is humble and honorable is why he has the résumé he has, not the other way around. So, as education leaders, I urge you to be humble and honorable. Yes, you may be able to attract more attention and have a short time of notoriety if you are loud and boisterous, but being humble and honorable builds a lifetime of leadership and achievement.

I would like to end with a short commentary on leadership for educators. I mentioned earlier that we cannot continue to say, "I am *just* a teacher." I strongly hold that we need to use our teacher voice for something more than getting students quiet. We need to proudly and boldly proclaim, "I *am* a teacher!" As education continues to become more political, our profession must stand ready to be a leader in our country. Through strength, presence, perseverance, relationships, vision, context, humbleness, and honor, we can be a voice for students. In fact, we know from research that the most important voices in our communities come from teachers—not politicians or celebrities, but teachers. My hope is that someday leadership for our educators will be the norm in preservice programs and that our teachers will stand with one voice to say we are the experts in education: "We *are* teachers."

OUR COMMON PERSPECTIVE AND LEADERSHIP OPPORTUNITIES

Distributed leadership may be the key to successful reform. The hope for greater coherence within the diverse components of the science education system lies in the increasing leadership within those components.

Leaders, from classrooms to state departments and national organizations, should use standards for science education as the foundation for reform. There is a need for a common vision, one with small variations, within science education. State standards provide that vision.

Empowering science teachers is a vital focus for improving leadership in science education. The role and importance of science teachers and their interactions with all students cannot be overstated.

ISSUES AND QUESTIONS FOR DISCUSSION

1. What would you list as challenges that require contemporary leadership in science education?
2. Considering your position and role in science education, what would you state as a vision and plan for reform?
3. Think of a person you would identify as a leader, and define that person's qualities of leadership.
4. Can you identify examples of transactional and transformational leadership in science education? If so, how would you describe them?
5. What is required for the empowerment of science teachers?
6. Can you think of another paradox of leadership? If so, how would you describe it?
7. Imagine that you are going to assume a leadership role for a component of science education. What is that component? What is your vision and plan for the change? What perspectives of leadership apply to your leadership? What paradoxes?

REFERENCES

Bennis, W., and B. Nanus. 1985. *Leaders: The strategies for taking charge.* New York: Harper and Row Publishers.

Cremin, L. 1965. *The genius of American education.* New York: Vintage Books.

Heyman, D. (Producer), and M. Newell (Director). 2005. *Harry Potter and the Goblet of Fire.* United States: Warner Bros.

May, R. 1972. *Power and innocence.* New York: W. W. Norton & Company.

NGSS Lead States. 2013. *Next Generation Science Standards: For states, by states.* Washington, DC: National Academies Press. *www.nextgenscience.org/next-generation-science-standards.*

Ramey, D.1991. *Empowering leaders.* Kansas City, MO: Sheed & Ward.

Rowling, J. K. 2007. *Harry Potter and the deathly hallows.* New York: Scholastic.

Russell, B. 1938. *Power: A new social analysis.* New York: W. W. Norton & Company.

Sarason, S. 1991. *The predictable failure of education reform.* San Francisco: Jossey-Bass Publishers.

SECTION XII

LEADERSHIP IN CONTEMPORARY SCIENCE EDUCATION

SECTION XII

The science education community must recognize the individuals and processes of leadership required in contemporary reform. We must realize that those with leadership responsibilities include classroom teachers, district administrators, state supervisors, teacher educators, state and national policy makers, professional organizations, and the American public; the responsibilities for leadership are distributed across the science education community.

The American education system is rooted in a network of hundreds of thousands of schools, colleges, and universities. In addition, there is a system of informal education that includes the media, museums, science and technology centers, botanical gardens, and a variety of outdoor and environmental education groups. In these institutions, educators face a difficult but achievable goal: improving the way they teach and what students learn about science, mathematics, engineering, and technology. How can we effect change, sustain reform, and improve education across such a diverse system? The answer lies in a common vision founded on a cohesive set of ideas, communicated broadly, and implemented with commitment.

The two chapters in this section address the themes that have been the foundation of this seminar and our closing perspectives, respectively. We are asking you to provide leadership in the various aspects of science education within which you have influence.

SUGGESTED READINGS

Fullan, F. 2005. *Leadership and sustainability: Systems thinkers in action.* Thousand Oaks, CA: Corwin. This is a very insightful book, with discussions of leadership at the school, district, and system levels.

Hitt, D. H., and P. D. Tacher. 2016. Systemic review of key leader practices found to influence student achievement: A unified framework. *Review of Educational Research* 86 (2): 531–569. The review article identifies and synthesizes empirical research on how leadership influences student achievement. It presents evidence with significant implications for school leadership.

Kaser, J., S. Mundry, K. Stiles, and S. Loucks-Horsley. 2013. *Leading every day: Actions for effective leadership.* Thousand Oaks, CA: Corwin. This book is based on the curriculum of the National Academy for Science and Mathematics Education Leadership. The clear, simple, and direct actions will be helpful to all science education leaders.

Malone, H. J., ed. 2013. *Leading educational change: Global issues, challenges, and lessons on whole-system reform.* New York: Columbia University, Teachers College Press. An insightful set of essays that address categories such as international comparisons, equity, accountability, and whole-system reform (e.g., Singapore, Finland).

Rhoton, J., ed. 2010. *Science education leadership: Best practices for a new century.* Arlington, VA: NSTA Press. Although published before release of *A Framework for K–12 Science Education* and the *Next Generation Science Standards,* this volume addresses the theme of science education leadership, provides different perspectives, and identifies challenges that must be addressed in the second decade of the 21st century. The authors are leaders and include Jeanne Century, Julie Gess-Newsome, Norman Lederman, Judith Lederman, Emily van Zee, and George DeBoer.

CHAPTER 23

LEADERSHIP IN SCIENCE EDUCATION

Release of the *Next Generation Science Standards* (*NGSS*; NGSS Lead States 2013), which were based on the vision and recommendations of *A Framework for K–12 Science Education* (*Framework*; NRC 2012), initiated 21st-century efforts to improve science education. Judgments about the degree to which the *NGSS* actually improved science education must be deferred to the future. However, there are preliminary indicators that contemporary standards may have some influence on science achievement (see, e.g., Carr 2016; Hinton 2016). Although preliminary and correlational, the increases in achievement in the fourth and eighth grades on the 2015 National Assessment of Educational Progress (NAEP) in science, as well as the closing of racial and gender gaps, are worth monitoring and analysis relative to the influence of standards.

It is reasonable to examine aspects of education reform in general and science education reform in particular. We are particularly interested in exploring the translation of policies to programs and practices.

CONTEMPORARY LEADERSHIP

Contemporary leaders in science education must recognize and address themes such as economic stability, basic skills for the 21st-century workforce, and resource use and environmental quality. Such themes differ from earlier justifications such as winning the space race and responding to *A Nation at Risk*. The economic rationale emerged from a significant recession, the realization that the U.S. economy is part of a global economy, and the acknowledgment that the education levels of the broad public influence the rate and direction of a country's economic progress.

Concretely, contemporary leadership for science education centers on responses to national and state standards and the implied reforms of curriculum, instruction, and assessments. Providing leadership requires, among many tasks, communicating the innovations and implications of new standards to those with responsibilities for change. To state the obvious, the leader's message should

be clear, simple, and positive, but it must convey a vision that challenges the status quo. The vision must be new and substantial and look to the future. *Science for All Americans* (Rutherford and Ahlgren 1989) is an excellent example of a vision statement by science education leaders in the late 1980s. In this era of standards-based reform, leaders might send a message such as, "Science Education by States for Every Student." Such a statement expresses a vision, subtly includes standards, identifies the importance of states, embraces equity, and addresses the ultimate point of science education reform—the students.

CONTEMPORARY CHALLENGES

To say the least, leaders will confront and must address challenges associated with efforts to improve science education in general and student learning in particular. Among the essentials of leadership, we would identify a challenge associated with a leader's ability to recognize and address the political realities of his or her work. Our insight here is that the leader has to recognize that initiating changes means addressing the politics. Not all issues are solely educational. Indeed, it may be the case that all education issues ultimately are political. The paradox embedded here can be stated as achieving educational goals while addressing political realities. We have found that *either/or* thinking often expresses the paradox, while *both/and* thinking provides insights into the leader's resolutions.

Experience has taught us one more lesson about challenges and leadership: If you are leading, you cannot avoid conflict and controversy. And the larger the system and greater the change, the more controversy you will experience. Indeed, this is another paradox. It can be thought of as the fact that achieving your goals requires enduring criticism, and the criticism often is unfair, constant, and personal.

Here are several insights we have learned about leadership and standards-based reform. These may seem like contradictions, but the effective leader is able to resolve them. The context of this discussion is the reform initiated by the *Framework* (NRC 2012) and the *NGSS* (NGSS Lead States 2013). Our reference to new standards may involve the national, state, or local standards.

Individuals will report concerns about the standards. Take the expression of these concerns seriously. The individuals, often teachers, are the ones confronted with significant change, and they may rightfully express concerns. It will help you if you understand the levels of concern and means of addressing those concerns (see, e.g., Hall and Hord 2014).

Another challenge of leadership is the need to fulfill a national or state agenda and respond to local requirements. This includes, for example, standards and the Every Student Succeeds Act (ESSA) legislation. This challenge requires paying attention to local mandates and addressing them in the translation of policies (i.e., standards) to school programs and classroom practices.

Although new standards may be abstract and general, you will have to act concretely. You should recognize the roles and goals of new standards and be able to describe changes in science classrooms at the elementary, middle, and high school levels.

By implementing new standards, you are initiating change in the educational system. This is one aspect of a paradox. The other part of the paradox is maintaining continuity in various components of the system. For example, the changes may be the content emphasis and organization, but there may still be life, Earth, and physical science courses. The latter feature will maintain continuity in science programs.

By its very nature, leadership in science education requires you to acknowledge, live with, and resolve challenges such as those described. Often the politics and realities of education reduce your aspirations. This should not deter you from leadership. It will be the students as future citizens who benefit from your leadership.

We conclude this chapter with our personal perspectives on challenges and their implications for your leadership.

A PERSONAL PERSPECTIVE

LEADERSHIP IN K-12 SCIENCE EDUCATION: MY RECOMMENDATIONS

Rodger Bybee

In 2006, I read an article titled "Grand Challenges and Great Opportunities in Science, Technology, and Public Policy" (Omenn 2006). The idea of grand challenges caught on and has been used as a theme in a variety of contexts and disciplines. With leadership from Bruce Alberts, then the editor of the American Association for the Advancement of Science (AAAS) magazine *Science*, the April 19, 2013, issue was devoted to the theme "Grand Challenges in Science Education." A distinguished group of educators addressed 20 challenges that included enabling students to build on their own enduring, science-related interests; adapting learning pathways to individual needs; and harnessing new technologies and social media to make high-quality science professional development available to all teachers.

Critically important features of grand challenges include the fact that they have confronted the education community for a considerable time; they also may have been identified and even addressed, but they have not been solved. Finally, these challenges are system wide. Addressing grand challenges is not, to use David Tyack and Larry Cuban's title, *Tinkering Toward Utopia* (1995). Addressing grand challenges requires more than a few marginal changes such as a new activity, a different test, or a longer school day. Here, I propose challenges for K–12 science education.

I asked myself, "What are the challenges faced by leaders in the K–12 science education community?" There are, to be clear, many challenges, but which are significantly large and affect multiple components of the system? My perspective on these challenges is based on nearly a half century of involvement in science education. For much of that time, I tried to synthesize ideas and communicate with K–12 teachers.

I must also say that the 2016 election caused me to review and revise my challenges. As I reflected on this situation, I recalled an editorial in the theme issue of *Science* on "Grand Challenges in Science Education" in which Alberts stated that a priority among his grand challenges would be "helping the business community promote new directions in precollege science education" (AAAS 2013, p. 249). The discussion included the need to be competitive in a global economy; workers who can "think for a living"; "employees who can apply abstract, conceptual thinking to "complex and real-world problems"; and how to function in environments that require communication skills. Alberts concluded the paragraph with this insightful statement: "Harness the influences of business organizations to strongly support the revolution in science education specified in the *Next Generation Science Standards*" (AAAS 2013, p. 249).

With the election of Donald J. Trump as the 45th president of the United States and his theme and experiences of business leadership, it seems that Alberts's perspective on grand challenges is especially appropriate. This is the case when science, technology, engineering, and mathematics are integral to the innovations that fuel business and industry.

To be clear, I am completing this essay late in 2016 and thus have no clear idea what a Trump administration will support relative to science or STEM education. As he makes the transition from campaigning to governing, I am sure he and his administration will find most things more complex and difficult than expected, and to be effective, they—like prior administrations, both conservative and liberal—will have to moderate extreme positions.

My advice to leaders in science education is to talk about STEM education, support state standards, and follow Bruce Alberts's themes of business and industry.

Finally, I think it wise to follow the advice made popular in the United Kingdom during World War II: Keep calm and carry on.

The recommendations I advance are the following:

- Develop curriculum materials that accommodate the innovations of standards for science education.
- Provide professional development for classroom teachers so they have the knowledge and abilities to implement the reform initiatives.
- Increase students' understanding of the nature of science.
- Emphasize personal and social perspectives in K–12 science education.
- Improve students' learning at scales that make a difference.

The following discussion provides more details on this list of challenges for science education leaders.

DEVELOP CURRICULUM MATERIALS THAT ACCOMMODATE THE INNOVATIONS OF STANDARDS FOR SCIENCE EDUCATION

Immediately after the release of the *NGSS*, teachers began asking for instructional materials they could use to implement the new standards in their classrooms. In fact, I conducted a workshop on *NGSS the day after* the release of *NGSS* and teachers asked, "Where are the instructional materials?" Teachers are still asking the same questions or attempting to adapt current materials or develop new instructional units. These are the right questions coming from those who have the ultimate responsibility for implementing *NGSS* and new state standards. Unfortunately, curriculum materials that one can propose as excellent examples of *NGSS*-based programs are few compared to the demand. Yes, commercial publishers have released textbooks, and there are some units that already exist—for example, *Disruptions of Ecosystems*, developed by the American Museum of Natural History and the Lawrence Hall of Science. There also is a recent project by the GLOBE Implementation Office to design and develop units on the theme Weather and Climate: Local Experiences for Global Understanding. Support for this project comes from the National Aeronautics and Space Administration (NASA). Both of these projects have a strong professional development component. I am also aware of efforts by several organizations to increase the alignment of their programs with the *NGSS*. For example, the Biological Sciences Curriculum Study (BSCS) is revising *BSCS Biology: A Human Approach*; Lawrence Hall of Science (LHS) is doing the same for *Full Options Science System*; and the Smithsonian Science Education Center is developing new science units. There are rays of hope that science education leaders will have exemplary curriculum materials for classroom teachers to see what contemporary, standards-based materials look like. But, in general, the demand for new materials is much greater than the current supply.

I think this presents a major challenge for the science education community. In the past, the National Science Foundation (NSF) funded proposals to curriculum development groups such as LHS, Education Development Center (EDC), and BSCS. The materials produced by these groups generally contributed to the reform efforts and specifically responded to teachers' concerns. In this era, the challenge is at least as great, and the federal response is at most marginal. States and local districts have the responsibility of answering the classroom teachers as they attempt to translate new standards to classroom instruction.

The assessment community has held conferences and produced reports on the general theme of the *NGSS* and state classroom testing. Unfortunately, there have been no conferences, reports, or meetings addressing themes associated with new science standards and curriculum and classroom instruction.

PROVIDE PROFESSIONAL DEVELOPMENT FOR CLASSROOM TEACHERS

Clearly related to the demand for instructional materials based on the *NGSS* and new state standards is the need for professional development. It would be unreasonable to expect classroom teachers to recognize the complexity of contemporary standards and develop the knowledge and abilities to design new materials for their science programs. Yes, some can and will do this, but for many the challenges of time and complexity will be too great. If we expect to succeed at any reasonable level, ongoing professional development will have to be provided by national, state, and district agencies as well as independent groups.

INCREASE STUDENTS' UNDERSTANDING OF THE NATURE OF SCIENCE

Most of us in the scientific and education communities have encountered variations on the following comments: "Climate change is a hoax." "I do not believe in biological evolution; it is only a theory." Disturbingly, such statements often are the basis for dismissing climate change as a constellation of short- and long-term problems that must be addressed. Assertions such as these also serve as justification for including creationism in school science programs as an equally valid explanation for the unity and diversity of life on Earth. I maintain that these examples highlight the need for a greater understanding of the nature of science.

A better illustration of the problem is found in interviews conducted as preparation for revising *Science, Evolution, and Creationism,* a National Academy of Sciences publication. College-educated adults were asked how they thought about science and creationism. The alarming message from these interviews was that those surveyed saw no difference between science as an evidence-based system and religion as a personal-belief system (Alberts 2012). So, in the case of science and creationism, the individuals could choose either because science and religion provided apparently equivalent explanations for natural phenomena. The individuals did not understand that science is a way of knowing, and, very important, that the way of knowing in science is a process based on accumulated empirical evidence explaining natural phenomena.

These observations are troublesome. Even more alarming is the research on students' and teachers' understanding of the nature of science. Although there is a long and continued support for teaching students about the nature of science (see, e.g., Driver et al. 1996), research on students' and teachers' understanding of the nature of science can be summarized in one word—*inadequate* (Lederman and Lederman 2014; Abd-El-Khalick 2014). I would describe this situation as worsening at a minimum and, most appropriately, alarming.

Positions that are grounded in nonscientific beliefs are not equal to the accumulated empirical evidence from multiple lines of research supporting explanations of, for example, climate change and biological evolution.

Often the response to such positions is that the deniers or creationists do not understand, so we must teach more about climate change or evolution to help individuals understand the scientific facts and concepts. While I certainly support the teaching of science concepts, I propose that understanding the nature of scientific knowledge may be of equal or greater importance for citizens. Here is the challenge.

Many scientists and science educators can support observations. They need only think about conversations they have had with seatmates on airline flights, for example, paraphrasing Winston Churchill, who was talking about democracy. I suggest the best evidence supporting an education about the nature of science is a five-minute conversation about science with the average citizen.

All citizens, college educated or not, confront science-related life situations in personal, national, and global contexts, and their understanding of the nature of scientific knowledge is limited, to say the least. To the degree citizens are making decisions without understanding the fundamentals of science and the nature of scientific knowledge, the scientific and education communities have both failed to fulfill their roles in democratic societies. Helping citizens make scientifically reasonable and prudent decisions within personal, social, and global situations is the challenge, and this is a significant one. An enlightened citizenry would be a significant countervailing force to those perpetuating nonscientific views. The beliefs of vocal minorities—such as creationists and climate change deniers—and others creating controversies based on nonscientific beliefs likely will continue. My argument is to complement knowledge based on the physical, life, and Earth sciences disciplines with equal emphasis on understanding the structure and function of the larger scientific enterprise and the nature of scientific explanations.

Will my recommendation for greater emphasis on the nature of science in education be a solution to problems such as those mentioned? Probably not. But it certainly will not hurt, and it may contribute to increased support for science among many citizens. Inclusion of the nature of science as part of the *NGSS* for grades K–12 serves as a positive step. The standards include nature of science themes such as science as a way of knowing; scientific knowledge is based on empirical evidence; science models, laws, mechanisms, and theories explain natural phenomena; and scientific knowledge is open to revision in light of new evidence.

My point in addressing this recommendation at length is that the majority of citizens need to understand the nature of science. They should understand how scientific progress is made and know enough about science to support it when nonscientific beliefs are presented as counterpoints to the ways science has contributed to their lives. Achieving these goals requires an explicit emphasis on the nature of science in the K–12 curriculum. (The content and learning progressions for the nature of science are in the *NGSS* Appendix H, p. 96.)

EMPHASIZE PERSONAL AND SOCIAL PERSPECTIVES IN K-12 SCIENCE EDUCATION

Historically, one purpose of science education has been to provide citizens with knowledge and abilities they can apply to life situations. Yes, the fundamental concepts and processes of science connect to life situations. But understanding the concepts of Newtonian mechanics, atomic structure, plate tectonics, the germ theory of disease, biological evolution, or molecular genetics does not necessarily mean the average citizen can apply those concepts when confronted with life situations.

In K–12 education, students should have experiences applying science concepts to everyday situations. For several decades, school science programs have emphasized the basic concepts and processes of science and assumed students could apply science knowledge and abilities. Evidence and experience indicate this assumption is questionable. I argue that citizens rarely appeal to the science taught in school as they make decisions concerning life situations. It is time to reclaim this diminished goal of K–12 science education. I note use of the term *diminished*. The goal still exists but has not been a clear and explicit goal of school science programs for some time. The purpose for which I argue does have historical precedent.

I point out the fact that the original purpose of public education as viewed by the founders of the American republic was to prepare all people for their roles as citizens. That is, public education has a civic mission to prepare informed, rational, responsible individuals who can participate in their duties in a democracy. Historically, the goal of literacy—teaching everyone to read and write—was not what we now refer to as college and career; it was so they could perform their duties as citizens. So this goal is not new; it has a long history in American education (see Butts 1980, 1991; Boyer 1990).

What are science educators doing to contribute to citizenship as a fundamental purpose of education? Should we connect our goal of scientific literacy to the historical reason for emphasizing literacy in public schools? My answer is yes. Science education should contribute to the preparation of students who, as citizens, will confront personal and social situations related to science topics such as health, resources, environments, and material hazards.

Our programs and practices emphasize the basic concepts and processes of inquiry with the assumption that students will apply this knowledge when confronted with personal and social issues. The evidence for citizens' ability to do this is weak, at best. The challenge is to provide some experiences where students apply the science concepts and processes to meaningful life situations.

IMPROVING STUDENT LEARNING IN SCIENCE AT SCALES THAT MAKE A DIFFERENCE

Over the decades the science education community has been witness to numerous innovations, most with the underlying aim of improving student learning in science. Some of the initiatives have

reached a scale that made a difference. For example, the reform of instructional materials in the Sputnik era did reach a scale with positive results, at least in my view. Unfortunately, most reform initiatives do not reach a scale that influences the system of science education. My recommendation is to improve student learning in science at scale—no small challenge indeed.

I first read about the idea of scale in 1996 when an article by Richard Elmore called "Getting to Scale With Good Educational Practice" caught my attention. Why is it that many good innovations do not get to scale? More important, what could the science education community do to improve student learning? As it turns out, Richard Elmore and colleagues answered the second question, which, by the way, also provides answers to the first (City et al. 2009).

A chapter called "The Instructional Core" addresses the theme of getting to scale by asserting that there are only three ways to improve student learning at scale: raising the level of content, developing teachers' knowledge and skills, and increasing students' active learning of the content. There it is—the constructive response simply stated. Well, it is a bit more complex, but still achievable.

Here is my summary response to this recommendation: Getting to scale involves changing the science content, providing professional development for present and future science teachers, and reforming curriculum and instruction to improve students' classroom experiences with the new science content. To this adaptation of the original statement, I would add that we need to provide assessment of student learning that is aligned with the content, embedded in curriculum, and included in science teachers' knowledge and skills. I have summarized my perspectives on the instructional core in Figures 23.1 and 23.2 (p. 364).

Figure 23.1. Principles for Reform Based on the Instructional Core

1. Increases in student learning occur only as a consequence of improvements in the content, teachers' knowledge and skills, and student engagement—the curriculum.
2. If you change one element of the instructional core (e.g., changes to content via new standards), you have to change the other two elements as well.
3. If you cannot see an element in the instructional core, it is not there.
4. Tasks for students and teachers predict performance.
5. The real accountability is in the performances students are asked to accomplish.
6. Students and teachers learn to do the work by doing the work.
7. Remember to have description before analysis, analysis before prediction, and prediction before evaluation.

Figure 23.2. The Instructional Core

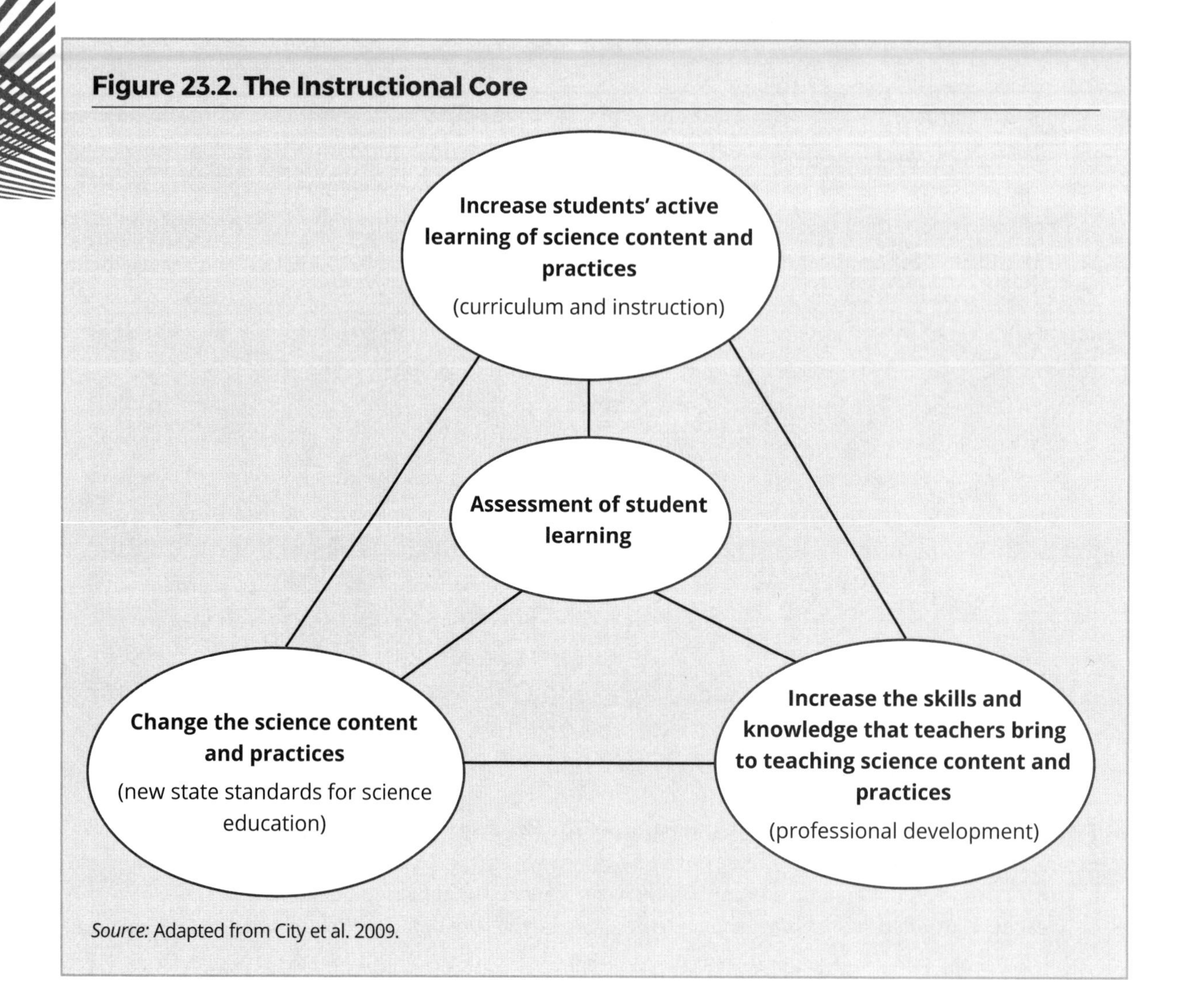

Source: Adapted from City et al. 2009.

A PERSONAL PERSPECTIVE

LEADERSHIP IN K-12 SCIENCE EDUCATION: MY RECOMMENDATIONS

Stephen Pruitt

As a classroom teacher in the 1990s, when the *Benchmarks for Science Literacy* and the *National Science Education Standards* (*NSES*) were released, I am not sure I fully understood the impact or even the purpose of those standards. I certainly did not understand their issues and importance when they were first released. As a third-generation teacher, I am ashamed to say I thought I knew as a young teacher what my students needed in my class. I actually thought I knew because of what, and how, I was taught in school and as an undergraduate. However, I was not so arrogant that I didn't see the standards as documents to help me be a better teacher. As I progressed through my years in the classroom and later at the state education agency, I began to see the real importance of these documents and the effect they had. In this section, I will discuss several challenges that drive the need for these documents and why they are so important to the goals of scientific literacy and citizenship. As I see it, there are several challenges connected to the need for standards, but at the end of the day, they all come back to equity and giving every child an opportunity to be successful in life. The recommendations I will focus on separately are (1) the need for every student to be able to think critically and scientifically in a highly competitive and global society, and (2) the need to guarantee every student has the same opportunities to succeed in life.

THINKING CRITICALLY AND SCIENTIFICALLY IN A COMPETITIVE AND GLOBAL SOCIETY

A major issue facing science education is that we need to provide all students with the ability to think critically about everything in their lives, from daily phenomena to scientific issues that affect them. Failing to equip all students with scientific skills and content will continue the current trend of deficits in producing a STEM-prepared workforce. We have known about the achievement gap in science for a long time, and we have not addressed the challenge; the gap has continued to expand. Beginning with *A Nation at Risk* (NCEE 1983) and continuing in other publications such as *Rising Above the Gathering Storm* (NAS 2007), there is accumulating evidence that students need more science and math. It took the launch of Sputnik to make the United States focus on the need to generate more scientists and engineers, but it is conceivable we face a greater crisis today. For some of our

younger readers, it may be important to point out why Sputnik was such a big deal. Let me give a brief perspective. For the first time, the United States saw itself as behind another country. However, it wasn't just that we were behind, but that we were behind an adversary that we saw as a threat to not only our standing in the world but also our own national security. I think it is fair to say that the United States had become complacent and felt the education system was working, even though the enactment of the Elementary and Secondary Education Act (ESEA) of 1965 demonstrated an understanding that the system was not an equitable one. With the launch of Sputnik, science and mathematics education was seen as being in a crisis. It was time for us to act as a country to educate our students in science and mathematics to ensure our way of life. Sputnik was a wake-up call of national proportions. With all of that said, I would say we are in a greater state of emergency now, or at least we should be. In a 2014 survey conducted by Change the Equation and Business Roundtable, 126 major companies responded in resounding fashion regarding our current challenges in science and, more broadly, STEM. The survey (Business Roundtable 2014) found that 98% of chief executive officers identified the lack of STEM skills as a major problem for their companies. Furthermore, they said that approximately 60% of their job openings require basic STEM literacy and 42% require advanced STEM knowledge and skills. Particularly important about this survey is the finding that nearly 66% of the job openings requiring STEM skills were in manufacturing and the service industries. Additionally, the survey found that 38% of the companies reported that at least half of their entry-level applicants lack basic STEM literacy and 28% of respondents said that at least half of their new entry-level hires lack basic STEM literacy. This survey is indicative of the need for a science education that addresses content as well as the ability to think critically and scientifically. Science is about both producing more scientists and engineers and the need for workers who have basic skills associated with STEM disciplines. The standards are not about holding any students back; rather, they are intended to provide all students with the ability to become fully functioning contributors to society. Standards are important for guiding teachers, districts, and states toward the content and skills needed to be successful in the 21st century. It is also important to realize that this movement is about each child. The work in the Sputnik era did a lot of great things for our country, but we did not see a drastic increase in our students of color or students from underprivileged socioeconomic backgrounds going into STEM careers, nor was much done to ensure access and opportunity for those students, which resulted in a huge achievement gap both in achievement scores and availability of quality STEM programs.

GIVING EVERY STUDENT AN OPPORTUNITY TO SUCCEED

What are standards really about? Is there a particular part of standards that drew me into wanting to focus my career on standards? I had no plans to leave the classroom. I loved it. I loved, and still love, everything to do with being a classroom teacher. At the core, while I loved science, it was the belief I could make a student's life better that drew me to a broader career in education. As my perspective began to grow beyond the classroom, I began to see how my expectations did not match those of oth-

ers. At that point, the district I taught in still had levels for our high school classes. I began to notice that I had different expectations in my classroom depending on the "level" I was teaching. I thought I was making decisions for the benefit of my students, but I came to realize that my students would do what I expected them to do. I also noticed that some things were more important than others, and, oddly enough, they were concepts from the *NSES* and *Benchmarks*—the contemporary standards.

When I moved to the Georgia Department of Education (GaDOE) as the science supervisor, the inequity of our science expectations became more real than ever before. When I came to GaDOE, only 68% of our students were passing the Georgia High School Graduation Test (GHSGT) the first time. This was a requirement for graduation, so 32% of our students were not eligible for graduation before they had even completed their third year of high school. As bad as that sounds, it is not the full picture of how inequitable science instruction was across our state. Only 36% of African American students were passing the first test, and we had a lower rate of passage from our English language learners. You may ask, "How do standards fit into this?" As I did my own research into why this was happening, I found students who did not do well on the exam also had not been exposed to a rigorous set of science coursework. It was my first experience with the phenomena that I have come to call the less is none phenomena, which means that somehow we have convinced ourselves in the education community that students who do not show success in science and mathematics early and often will somehow get better at it by doing it less. So, students who do not perform well in science and math take fewer courses in the subjects, they have lower expectations placed on them, and they continue to perform at lower levels than those students who have higher expectations placed on them. The only thing these students get more of is remediation. So, we give students who have come to dislike the subject more "drill and kill" to prepare for an exam, and lower expectations really equal little to no actual improvement in performance. I am not saying that all students should expect to become scientists. I am saying that students who are not given the opportunity to perform at high levels lack opportunities for success in the workforce after they graduate—if they graduate! This is, from my perspective, a grand challenge.

The STEM pipeline does not have a leak; it's a gusher. A lot of attention is given to the lack of preparedness that causes the losses inside the pipeline. The issue for science education standards is the entry into the pipeline. I would submit that the bigger challenge is not the "leak," but that there are not enough students going into the pipeline to begin with. We live in a society where understanding science and STEM is a necessity for success. The Change the Equation study found that STEM skills are a ticket into the 21st-century job market. Over the next five years, employers expect to replace nearly one million employees because those individuals lack basic STEM skills; stated another way, employers will need a million employees with basic STEM skills and an additional 600,000 with advanced skills (Business Roundtable 2014). Who will fill these positions? The era of "soft skills" is over and has been replaced with STEM skills. Low-wage and low-skill jobs are on the way out. The days of simply standing on an assembly line and being able to support a family are in short supply. Today's jobs are requiring problem-solving skills with very specific skill sets. Welding, robotics, health care, and manufacturing are more and more reliant on understanding technology and the science and math principles behind this technology. Employees need to understand the properties of materials, how to code to program robots in manufacturing, and solution chemistry in health care. While this is certainly a competitiveness issue for our country, as I have discussed in the previous section, it is an issue that educators must embrace as a means to support our students in their lives as workers and citizens. Yes, as educators, we affect the lives of our students while they are in our classes, and that cannot be underestimated or devalued. However, we are also preparing them for a world we have yet to experience. That world is not for the few or the privileged—it is for all. Education is the great equalizer, but only if every student is given the same chance. ***Standards are not about teaching every student the same way; they are about making sure that all students receive the same opportunities.***

A FEW FINAL THOUGHTS ON MY RECOMMENDATIONS

I realize the two points I focused on in this discussion may be what people would expect me to say. I would like to reiterate the issue of equity. From the beginning of the development of *A Framework for K–12 Science Education*, the issue of scientific literacy was always viewed through the lens of equity. It is important in the 21st century for all students to be prepared for the life they choose. In the development of the *NGSS*, it was not an accident that our focus was *all standards, all students*.

The standards have the ability to influence the equilibrium that exists in contemporary education. As with all systems at equilibrium, education will only change with a stress. The *NGSS* were meant to be that stress. The *NGSS* should affect the other components of the education system, such as education policy, instruction, and assessment. If this reform is meant to come to fruition, all of these areas must be viewed through the lens of the *NGSS*. All the major components of the education system will need to change to ensure that these standards are for all students. Only if this is done will all students have the same opportunities for success beyond school.

CONCLUSION

This chapter reviewed several recommendations for contemporary science education. They are challenges we identified based on our professional experiences. You, no doubt, have your own list of recommendations. Those recommendations are the opportunities for your leadership.

REFERENCES

Abd-El-Khalick, F. 2014. The evolving landscape related to assessment of nature of science. In *Handbook of research on science education, volume II*, ed. N. G. Lederman and S. K. Abell, 621–650. New York: Routledge.

Alberts, B. 2012. Failure of skin-deep learning. *Science* 338: 1263.

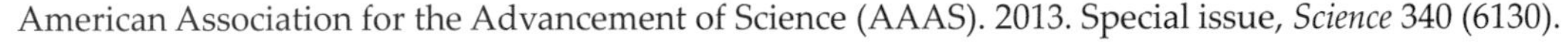

American Association for the Advancement of Science (AAAS). 2013. Special issue, *Science* 340 (6130).

Boyer, E. L. 1990. Civic education for social responsibility. *Education Leadership* 48 (3): 4–7.

Business Roundtable. 2014. *Change the equation*. Washington, DC: Business Roundtable.

Butts, F. R. 1980. *The revival of civic learning: A rationale for citizenship education in American schools.* Bloomington, IN: Phi Delta Kappa Educational Foundation.

Butts, F. R. *Education Week.*1991. Citizenship as a National Educational Goal. December 11.

Carr, P. 2016. National Assessment of Educational Progress (NAEP) 2015 Science Results. Washington, DC: National Center for Education Statistics.

City, E., R. Elmore, S. Fiarman, and L. Teite. 2009. *Instructional rounds in education: A network approach to improving teaching and learning.* Cambridge, MA: Harvard Education Press.

Driver, R., J. Leach, R. Millar, and P. Scott. 1996. *Young people's images of science.* Philadelphia, PA: Open University Press.

Elmore, R. F. 1996. Getting to scale with good educational practice. *Harvard Educational Review* 66 (1): 1–26.

Hall, G., and S. Hord. 2014. *Implementing change: Patterns, principles, and pot holes*. 4th ed. New York: Pearson

Hinton, M. 2016. Science gains seen at 4th, 8th grades. *Education Week* 36 (12): 6.

Hitt, D. H., and P. D. Tacher. 2016. Systemic review of key leader practices found to influence student achievement: A unified framework. *Review of Educational Research* 86 (2): 531–569.

Lederman, N., and J. Lederman. 2014. Research on teaching and learning of nature of science. In *Handbook of research on science education. Vol. II*, ed. N. G. Lederman and S. K. Abell, 600–620. New York: Routledge.

National Academy of Sciences (NAS). 2007. *Rising above the gathering storm: Energizing and employing America for a brighter economic future.* Washington, DC: National Academies Press.

National Commission on Excellence in Education (NCEE). 1983. *A nation at risk: The imperative for educational reform.* Washington, DC: U.S. Deptartment of Education.

National Research Council (NRC). 2012. *A framework for K–12 science education: Practices, crosscutting concepts, and core ideas.* Washington, DC: National Academies Press.

NGSS Lead States. 2013. *Next Generation Science Standards: For states, by states.* Washington, DC: National Academies Press. *www.nextgenscience.org/next-generation-science-standards*

Omenn, G. 2006. Grand challenges and great opportunities in science, technology, and public policy. *Science* 314 (5806): 1696–1704.

Rutherford, F. J., and A. Ahlgren. 1989. *Science for all Americans.* New York: Oxford University Press.

Tyack, D., and L. Cuban. 1995. *Tinkering toward utopia: A century of public school reform.* Cambridge, MA: Harvard University Press.

CHAPTER 24

THE SEMINAR'S CONCLUSION AND YOUR BEGINNING

The science education community needs your leadership. Science educators must address the gaps in knowledge and abilities across race and gender and provide learning experiences that will prepare students for college and careers. The K–12 system of science education has the responsibility to make its contribution to America's future.

We need distributed leadership to achieve the lofty goals we just described. Up until this point in the seminar, we have had the responsibility of presenting the challenges we face and providing you with perspectives that will help you as a leader. Now, as we are at the end of the seminar, we must say that it is your turn to assume leadership.

In this chapter, we conclude with our personal perspectives.

CHAPTER 24

A PERSONAL PERSPECTIVE

NATIONAL STANDARDS AND THE CHALLENGES OF LEADERSHIP

Stephen Pruitt

Challenges? You mean there are challenges with implementing standards, and even standards developed by literally thousands of people across the country, including some of the best minds in science *and* education? Surely this cannot be. Of course, I am being facetious. We have addressed several issues throughout the book, and you have heard from the man who has worked on these issues since the 1990s, Rodger Bybee. I will return to this point because the fact that Rodger has been in this field and has remained committed and passionate about science education for so long should give us hope.

We have two more large challenges ahead that we have not discussed specifically. We have hit on both, but I think they are important to address specifically. First, we must communicate to all involved with education an understanding of the intent and depth of the standards from the perspective of teachers, administrators, and the community. Second, we must manage the change to allow acceptance of the idea that the standards—taught properly—will make a difference for students.

COMMUNICATION

Communication is perhaps the biggest and toughest challenge of them all. Why? I think simply because educators are not trained in communicating and do not always think of it first. I do not mean to sound critical; it is a simple fact that we are not schooled in the fine art of communications in a public sense. We take our methods courses and perhaps a speech class or two, but the actual science of communicating big ideas and concepts to the public is not in our course of study. We leave this task to communication experts. This is a mistake because these experts do not know the hot buttons or sometimes even the audience they are trying to reach. At the same time, leaving all communications up to educators is a mistake too. There needs to be a convergence. During the development of the *Next Generation Science Standards* (*NGSS*), communication was a key factor in everything we did. I was very lucky to have colleagues such as Sandy Boyd and Chad Colby at Achieve to help us stay focused on our communications. We tried to be strategic with every decision we made, every announcement, and really every move.

So, why is communication so critical, and why do we need a plan developed by program staff and communications staff? In short, program staff can get so into the content that they sometimes cannot

see why everyone does not see the content as they do. They need an outside voice to help develop and hone the message to be succinct and clear. One thing that is absolutely critical to understand: One must do all one can to control one's own message. As I like to say, in the absence of information, people make stuff up. Some of my greatest frustration during the development process that carried over into adoption was the idea that if we did not have a problem at that moment, we should not start a controversy, but should just wait until there was an issue to communicate. That idea is simply a recipe for disaster, as it allows others to formulate a message and you lose the opportunity to inform and educate.

One of the first and most needed communications is to make sure people know what standards even are. We have all seen the Facebook posts on what people call standards when in reality they are curricula (and often not aligned curricula, I might add). As we have discussed in prior chapters, standards are devoid of instructional direction. They embody the content students should be able to apply, but many people both in and out of education do not understand the difference between standards and curricula. As part of education, standards become part of the political discussion. It will be critical to have information out so that it can be reviewed and used as an evidence base. Finally, communication about standards must be written in a way that speaks to the community. The communication about the standards should include some of the actual standards language so that people can see it, but the "wraparound" material must be such that non-educators can see why the standards are so important and what the goals are. As much as I hesitate to say it, discussing three-dimensional learning with a non-educator is not helpful. Frankly, in the early days, I am not sure it was meaningful to educators. However, what does seem to speak to people are explanations that standards prepare students to understand the world by focusing on the phenomena and concepts they experience every day; allow students to experience, explain with evidence, and critically think about phenomena around them; or give students an opportunity to do things with the science they learn that can explain phenomena in their lives and prepare them to be more successful in their endeavors after high school. Also, here is a little advice as you consider communications: No matter how much you feel an obligation to educate people on the controversial issues associated with science education, I would advise you to never lead with these controversies. Doing so could very well distract from your bigger message of the good that standards will do for students. These controversies can be worked in, and should be, but do not lead with them.

Now, some of you reading this may be thinking, "Oh, he is speaking to state education agencies, districts, and school personnel." Actually, I am talking to teachers just as much. I went into this earlier, but it bears repeating: Educators must be active participants in the communication around standards and education as a whole. When a chief state school officer, which I was, or others who work at those levels advocate for things such as standards, it appears to be a protection of turf or an indication that this is the "stuff we are supposed to do." Education leaders have to make the specific attempt to expand their worldviews and be leaders in the community and the policy world as well. It is a decision, not an accidental occurrence. It is not always comfortable, and being an education

leader is hard. But, as we have discussed, we have the right to stand before anyone because we stand with children, we are teachers, and people should accept our expertise. Remember, #NeverJust.

MANAGING THE CHANGE

We covered this before, but I want to take a slightly different approach. We have discussed the importance of educators, both teachers and administrators, understanding the intent of standards. The community and general populace need to understand as well. Again, it goes back to deliberate communications, but there needs to be a focus. Because of how most of us learned science, there is an expectation of what content should be in the standards. Even teachers get caught in the trap of looking for topics. The *NGSS* were based on the underlying concepts of science. For instance, people look for bonding and assume there is no chemistry because they do not see the words *ionic* and *covalent*. The intent and depth come in with understanding that bonding is dictated by distance between atoms and nuclei, energy, and fields. But if we are simply looking for the rudimentary, we will not find it. The biggest opponents to this way of thinking believe that memorizing terms or being able to simply state laws somehow leads to a deeper application. Well, research displayed in *A Framework for K–12 Science Education* pretty clearly shows that is not the case. But still, teaching this new way is tough, and understandably so. Not only are teachers uncomfortable with change, but parents and policy makers are as well. As a result, the standards can come under greater fire. Standards are controversial enough, and lacking an understanding of the change is a major contributing factor.

If I may, I will share a short anecdote that illustrates the point. In 2006, Pluto was changed from a full-fledged planet to a dwarf planet. I was at the Georgia Department of Education as the Science Program Manager. It was such an exciting time. We were in the midst of implementing the first new set of science standards in more than a decade, and this was a great example of why we moved to conceptual standards. Rather than have students memorize the planets as had been done in the past (this is also true in the *NGSS*), students will learn about the solar system, Earth's place in the system, why Earth supports life, and the roles that gravity and force play in the system. Students will still discuss planets and all of the other heavenly bodies in the solar system, but there would be no requirement to memorize the order, especially since the order technically changes with a couple of the planets. What a great time to be a teacher! It was a time to discuss the nature of science and the dynamics of how science changes based on acceptance of new evidence. Around this time, I received a request from a reporter to discuss this topic. I spent more than half an hour discussing this great event with him. In fact, I have always joked that I dispersed incredible quotes about science education that day that I was sure would be an inspiration to all. The reporter, however, was aghast that I was not sending out notification to the state teachers that Pluto would no longer be on the state test. I told him it already wasn't on there. We went back and forth, and he simply could not get past the idea that kids were not memorizing the planets anymore. He simply could not get past this because, as he said, "That is how I learned it, and it must be valuable." I said, "What is important here is not that kids can name the planets in order, but they understand their interaction and how science itself works." Well,

after all this time and some great quotes on my part, the reporter asked, as we were about to hang up, "Will students still make models of the solar system for science fairs?" I quipped, "Probably, but now they only need eight balls instead of nine." That was the only quote that got into the article. The point is I really did not do a good job helping him understand the science; he had no context. The real magic to three-dimensional learning is that it provides that context, but we must make it a priority to help others see the intent of the standards. Having said all of that, this is perhaps the toughest part of change. Acceptance is fragile and change is unsettling. It will take the leadership of our incredible educators to make this change.

A NEW HOPE BASED ON YOUR LEADERSHIP

I cannot end with just the challenges. I need to end with what I believe is key to overcoming these and any other challenges to a quality education, science and otherwise.

A large part of this book revolves around the idea of leadership. As educators, we must step up to the plate to really move our country forward in science education. I would add a special message to those of you reading this book who are in preservice positions or early in your teaching career. Make the decision to approach education using a leadership and scientific approach. That is to say, decide from the beginning you will be a leader in your department, your building, your district, and your state. Through that role, you will become a leader in the country. You will not be popular at times, but remember that you joined this profession to help kids and not to be popular among adults. With regard to the scientific approach, require evidence of what is going on in education. Do not take people's word for something. I know that sounds awful, but it is necessary to do your own research before making a judgment. I have been asked many times how I went from a high school chemistry teacher to leading the biggest science education initiative in the country for almost 20 years to a chief state school officer. My answer is simple: I do my own research and remember why I do the job.

Earlier, I alluded to my co-author and his role. Rodger has seen so much change in science education. In fact, I would say he has had a hand in most of it. He is far too humble to say this, but I will. Rodger is inspirational. Why? Because he never lost hope for better science education for our nation's students. Trust me when I say that it would be easy to become cynical and lose hope that real change can come. The truth is that there have been times in my career when I have felt my own hope fade and allowed the thought of "Can we really do this?" enter my mind. But then I work with incredible teachers like the ones I get to work with in Kentucky or with Rodger and other titans of science education, and I am reminded that we *can* do this. The fact that these titans of science education—such as Rodger, Joe Krajcik, Jim Pellegrino, Brian Reiser, Helen Quinn, Cary Sneider, Ramon Lopez, Melanie Cooper, Peter McLaren, Brett Moulding, and more—have not given up but continue to show leadership in their own communities, states, and the whole country inspires me to believe we can do the right thing for our students. It will take time, and it won't be perfect, but our students will be better off. I would guess if you ask these folks, they would tell you there were things that didn't quite go as planned or maybe didn't get as far as they had hoped, but they made a difference.

Right now, many of you reading the book may be thinking that these folks are different; they aspired to do what they do or they were given a special opportunity. These folks were once the people I read about and admired from afar. Now, I consider them friends. I did not aspire to be where I am, nor was anything given to me. What we have in common is a passion to make sure our students are scientifically literate and ready for a world that changes by the second. If you don't believe me, ask them. You will see all of us made a decision to embrace the hope of a scientifically literate world and education system. My students (those I taught previously and those in school now) inspire me to do the job of improving education. My students in Kentucky and around the country need education leaders to stay committed and passionate about improving education. The work will never end. The titans of science education inspire me to know that I need to stay strong, and my hope would be that every science educator embraces that hope, picks up their own torch, and helps us overcome whatever challenges the principle that each child deserves a quality education.

A PERSONAL PERSPECTIVE

HAVE THE COURAGE TO BE A LEADER

Rodger Bybee

In this seminar, Stephen and I have presented many ideas and recommendations for individuals in all aspects of science education. Some challenges can be described as grand; others as important but not grand; and most as small but significant. How one perceives challenges depends, of course, on the personal context. The challenges of classroom teachers of science are not the challenges of developing new state standards for science. I present this example to underscore the point that there are individuals in many and varied components of the science education community, and they all have challenges that must be addressed. Confronting the need for change within your personal contexts is the first connection to leadership.

You may also note that providing leadership in the different contexts of science education requires varied knowledge and abilities. We have done our best to provide you with historical and contemporary foundations that will help you understand and address the challenges you will face.

In this personal perspective, my final statement in the seminar, I discuss something that is essential. You have choices as you confront the opportunities to lead. The realization that you have choices also means there may be questions, doubts, concerns, anxiety, and maybe even fear about making the choice to do something new and different. Here is my point: It takes courage to overcome the concerns, make the choice, and provide leadership.

Courage, as generally understood, is the quality of mind that enables one to face challenges with confidence, resolution, and perseverance. The courage I am describing is neither physical, such as that required of individuals in military combat, nor that discussed by philosophers such as Kierkegaard or Camus; rather, it is the courage to investigate new forms of teaching practices, design curriculum programs, and develop policies, depending on your educational context.

Being a science educator does require some leadership and, subsequently, courage. Today's education system is experiencing radical changes with new state standards, new federal legislation (such as the Every Student Suceeds Act [ESSA], and new priorities with the Trump administration. Science education does require courageous people who understand the historical and contemporary foundations, appreciate the need for reform, and can provide leadership.

You likely will not have to demonstrate the courage of Abraham Lincoln, Martin Luther King, Jr., or war veterans. The courage we talk about in this seminar involves trying standards-based approaches in your classroom, arguing for a contemporary curriculum program in your school, presenting initiatives for professional development to administrators and school boards, joining committees to develop new assessments, and contributing to science education professional organizations.

I cannot end this essay without mentioning the courage it will take to defend the integrity of science against those who question or deny biological evolution or climate change.

In our careers, Stephen and I have demonstrated the courage to be leaders. Now, in closing, we ask you to provide leadership in the ways and means appropriate to your situation.

I wish you the courage to be a leader.

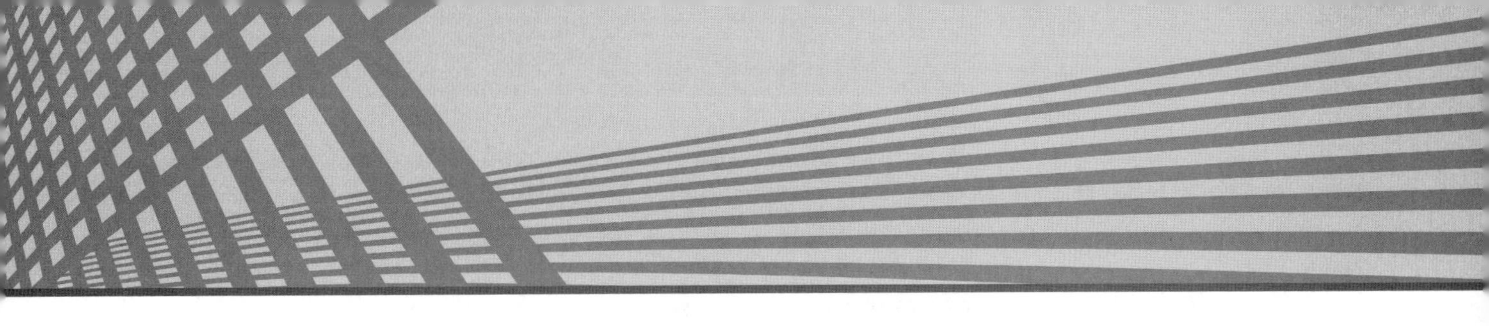

APPENDIX A

HISTORICAL PERSPECTIVES, 1635–1965

Table A.1: Early Education (1635–1751)

GOALS	Religious indoctrination; educational goals linked with Calvinist theocracy
SCHOOLS AND CURRICULUM	Latin Grammar School (Boston Latin School, 1635) private, college-preparatory, no articulation with common school curriculum, little or no scientific orientation, limited chiefly to study of Latin and Greek and the scriptures. Religious explanations for natural phenomena stressed.
TEACHING PRACTICES	Catechistic; indoctrination of youth in faith, formal discipline stressed.
INSTRUCTIONAL MATERIALS	No formal texts, writing of classical authors where available, use of scriptures.
STUDENT POPULATION	Male, usually restricted to well-to-do, able students, only for a very small percentage of population.
TEACHERS' PREPARATION	Usually members of clergy, no training in scientific and practical fields.

Note: All tables in the appendix are adapted from Lacey, A. 1966. *Guide to science teaching in secondary schools.* Belmont, CA: Wadsworth.

Table A.2: Emergence of Science Education (1751–1821)

GOALS	Goals still largely religious but increased emphasis on practical and utilitarian outcomes, preparation for a variety of needed skills of a scientific nature (e.g., industrial, commercial, navigational, agricultural, and surveying).
SCHOOLS AND CURRICULUM	Because of its practical emphasis, the Academy (Philadelphia, 1751) replaced Latin Grammar School, little orientation with the common school. Academy marked beginning of organized instruction in science, but no standard curriculum. Courses included: natural philosophy (i.e., physics), chemistry, astronomy, zoology, botany, geology, physical geography, navigation, agriculture, and surveying.
TEACHING PRACTICES	Lecture, memorization and recitation stressed less emphasis on religious understanding of natural phenomena, beginnings of object lessons.
INSTRUCTIONAL MATERIALS	Some specimens and printed materials used, materials provided by individual instructors, no formal texts.
STUDENT POPULATION	Boys, girls, coed, more heterogeneous but still exclusive.
TEACHERS' PREPARATION	Usually no formal preparation in science, no minimum standards, both lay and clergy taught, increasing influence for lay teachers.

Table A.3: Science Education Is Established (1821–1870)

GOALS	Utilitarian, disciplinary, religious, objectives of high school science teaching were to instruct in practical arts and citizenship.
SCHOOLS AND CURRICULUM	Origin of high school in Boston (1821), later called Boston English High School, practically oriented with little or no articulation with elementary school, beginning of graded classes in secondary schools, elements of arts and sciences, beginning studies in natural philosophy (i.e., physics), and addition of practical mathematics, navigation, surveying, and astronomical calculations. By 1852, fourth year included courses in astronomy, geology, mechanics, engineering, and higher mathematics; after 1860, some high schools offered botany, meteorology, mineralogy, physiology, and physiography. Biological sciences became listed as biology. Botany usually offered as premedical course. Most schools offered one course in Earth sciences, usually geology, geography, or meteorology.
TEACHING PRACTICES	Largely descriptive, lectures and discussions, if textbooks available, method was largely assigned reading and recitation. By 1850's textbook method dominant, in early high school, laboratory method almost unknown, with some use of object lessons and specimens to verify texts; experiments, if used, were crude and inaccurate. By 1865 provision made for laboratory instruction in many high schools. Mental discipline and automatic transfer were theories of learning.
INSTRUCTIONAL MATERIALS	High School texts were largely abridgments of college textbooks. Secondary science textbooks first appeared in the 1830s and 1840s (e.g., Comstock's *Outlines of Geology*, 1839; *Elements of Chemistry*, 1844; and *Gray's Chemistry: Continuing the Principles of the Science, Both Experimental and Theoretical.)*
STUDENT POPULATION	Small percentage of population attended secondary school. In early part of era, pupils largely male and from aristocratic class, in latter part schools were more coeducational.
TEACHERS' PREPARATION	In early part of era teachers generally poorly prepared, usually no formal preparation in science, many trained for clergy. Beginning of normal schools improved situation after 1830s. Later college graduates and those with skill and experience in science field began to teach, in general, qualifications uneven, no minimum standards for teaching.

Table A.4: College and Committee Policies Influence Science Education (1870–1910)

GOALS	Doctrine of formal disciplines influenced this period, religious emphasis declined, informational and utilitarian aims maintained prominence. The goals of science teaching were to develop faculties of reasoning, observation, and concentration in accordance with faculty psychology.
POLICIES	National policies (Committee of Ten, 1893; Committee of Fifteen on Elementary Education, 1893; Committee on College Entrance Requirements, 1895) emphasized need for beginning science earlier than high school. Science as requirement for college admission spurred standardization of curriculum.
SCHOOLS AND CURRICULUM	First public high schools developed, Kalamazoo Decision (1872) set precedent for public, tax-supported secondary education, articulation between elementary and secondary schools increased. Science curricular reform due largely to changes in science instruction in colleges and universities. In 1872, Harvard science faculties became a major influence in shaping high school science curricula by establishing physics (and later chemistry) as college entrance requirements. Secondary science courses subsequently became abbreviated college courses. Other significant curricular developments: physics considered most important science, chemistry next; biological science offerings increased. Biology courses were actually separate, one-year, short courses in botany, zoology, and human physiology with little integration. Earth sciences consisted mainly of geology, which declined in importance after 1900.
TEACHING PRACTICES	Because of faculty psychology, methodology emphasized teaching of science as formal disciplines. Individual differences in abilities, backgrounds, and interests were generally ignored. Laboratory instruction became popular, established first in chemistry, later in physics and biological sciences. Laboratories were narrow, formal, and specialized.
INSTRUCTIONAL MATERIALS	Greatly increased use of specimens and laboratory equipment. Because of increased activity in college science textbook writing, high school texts, (i.e., abridgments of college texts) multiplied. Latter part of period was pre-eminently era of laboratory manuals, almost 40 appearing between 1890 and 1900 in biology alone.

Table A.4. *(continued)*

STUDENT POPULATION	Largely coed, increasingly democratic and less class-determined after 1872. A small percentage of population attended high school. The influence of college-dominated secondary school science diminished popularizing science. After 1890 emphasis was on techniques of specialized laboratory procedures and initiatives requiring certain number of exercises as prerequisite for college admission. Enrollments in physics and chemistry dropped; in 1890 22% of students were enrolled in physics, in 1910, 15%. There was a similar decline in chemistry.
TEACHERS' PREPARATION	Expansion of secondary education was marked by growing demand for college and university-trained teachers, including those in science fields; thus, an increase in normal schools, development of teachers colleges, introduction of teacher education into liberal arts colleges and universities, and development of in-science educational programs for teachers occurred. National polices were significant factors in accelerating development of advanced training for science teachers. Supply of trained teachers remained inadequate.

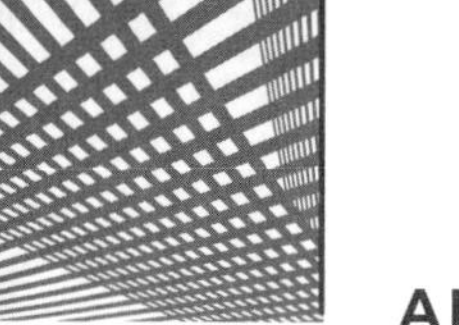

Table A.5: School Reorganization and Development of General Science and the Junior High School (1910–1930)

GOALS	Goals not consistent from school system to school system; some beginnings of life-centered objectives due to growth of new junior high school and development of integrated courses in biology.
POLICIES	In 1920 NEA Commission on the Reorganization of Secondary Education published a report on reorganization of science. Committee urged that sciences be organized and taught in manner that would contribute to Seven Cardinal Principles: (1) health, (2) command of fundamental process, (3) worthy home membership, (4) vocation, (5) citizenship, (6) worthy use of leisure, and (7) ethical character. Committee stated objectives of each of the sciences for the principles. General science emphasized vocation, health, environment, and citizenship; biology the environment; and chemistry and physics preparation for college.
SCHOOLS AND CURRICULUM	Establishment of junior high schools between 1910 and 1928. General science offered only at grade 9, but later at grade 7. Biology became relatively constant at 10th grade; physics and chemistry alternated between grades 11 and 12. With growth of general science, corresponding growth in students studying biology. Significant development of unified concept of sciences, and decline in specialized sciences.
TEACHING PRACTICES	Abandonment of faculty psychology and influence of psychology of learning as propounded by Edward L. Thorndike, due to junior high school movement, tendency toward more correlation and functional content in biology. Few changes in methods of teaching chemistry and physics.
INSTRUCTIONAL MATERIALS	From 1910-1920 textbooks were compendia of various sciences, with materials loosely tied together. Some experimental work for demonstration and laboratory, but written largely from orientation of authors with little influence of psychology of learning. After 1920, some orientation of general science and biology to environment of pupils; however, studies reveal texts usually emphasized factual materials and devoted little attention to principles and generalizations of science. Materials for chemistry and physics were still very much controlled by college science departments.

Table A.5. *(continued)*

STUDENT POPULATION	General Science—507,000 in 9,845 schools Biology—393,391 in 7,686 schools Chemistry—204,000 in 7,346 schools Student population continued to grow in general science and biology, and a drop in percentage enrollment in chemistry from 12.6% in 1922 to 11.1% in 1927, and in physics from 14.6% in 1910 to 6.9% in 1928.
TEACHERS' PREPARATION	Significant increase in number of secondary teachers enrolled in education departments of colleges and universities in preparation for science teaching. However, no significant movement toward training of teachers for junior high general science; many teachers of chemistry and physics found to be poorly prepared.

Table A.6: Progressive Policies and Science Education (1930–1945)

GOALS AND POLICIES	Many elaborate lists of goals were prevalent. Seven Cardinal Principles continued to influence science education. Other policy statements included: National Society for the Study of Education, 31st Yearbook, *A Program for Science Teaching*, 1932. Central objective was "building of learning exercises and experiences around broad principles which are fundamental to the understanding of Nature." Commission on Secondary School Curriculum of Progressive Education Association, *Science in General Education*, 1938. Science should contribute to (a) personal living, (b) immediate personal-social relationships, (c) social-civic relationships, (d) economic relationships, and (e) reflective thinking. Although policy statements were published, studies showed implementation was inconsistent and rare.
SCHOOLS AND CURRICULUM	A leveling off in growth of junior high schools and general science extended to 7th grade. As more state departments of education published courses of study, advanced general science courses were proposed and established for grades 11 and 12 to take place of chemistry and physics, which had declined rapidly. Biology grew as constant basic science in high school grades. During World War II utilitarian science courses (e.g., electricity, aviation, and photography) were added to the high school curriculum.
TEACHING PRACTICES	Methods of teaching general science showed influence of contemporary psychology in selection of subjects relating to pupils' activities, interests, and environments. Few attempts were made to correlate teaching materials between elementary and high schools. Methods of teaching biology were largely taxonomic and descriptive; in chemistry and physics, no significant changes in methods. Most significant new development was the limited implementation of project method of teaching.
INSTRUCTIONAL MATERIALS	Increasing emphasis on materials with civic or social orientation. Most state departments of education published new courses of study and a significant number of new general science and biology textbooks were written. Materials for chemistry and physics remained relatively unchanged.
STUDENT POPULATION	Period marked by fact that between 1930 and 1945 the high school became common school. By 1945 approximately 75 per cent of high school age students were attending school.
TEACHERS' PREPARATION	Teachers in secondary schools were poorly prepared; also, many adequately prepared teachers assigned to diverse tasks that reduced their effectiveness. During World War II, many science teachers were in the armed services or working in other fields.

Table A.7: Post-World War II: New Policies for Preparation of Science Teachers (1945–1955)

GOALS	Goals for science teaching in democratic society widely accepted in American education. Most generally accepted goals placed emphasis on functional aspects of science: facts, concepts, and principles that are functional, skills used in problem solving, attitudes, appreciations, and interest having implications for functional learning.
POLICIES	Meaningful policy statements appeared in significant publications: *Education for American Youth*, from Education Policies Commission, 1944; *General Education in a Free Society* from Harvard Committee on General Education, 1945; *Science Education in American Schools* from National Society for the Study of Education, 1947; *Education for All American Youth-A Further Look* from Educational Policies Commission of National Education Association and American Association of School Administrators, 1951; *Science in Secondary Schools Today,* from National Association of Secondary School Principals, 1953.
SCHOOLS AND CURRICULUM	Science curricula lacked articulation. General science was firmly established in junior high school, with advanced general science offered in 11th and 12th grades. Biology was typically offered in grade 10; chemistry and physics typically in the 11th and 12th grades. Particular developments during latter stage included: more courses to meet needs of more pupils, particularly those gifted in science, increasing number of courses in Earth science, more attempts of correlating science with other curricular areas, greater attention given to elementary school science, and increased activities, such as science fairs.
TEACHING PRACTICES	Teaching science emphasized presentations of content by teachers and textbooks. Instruction included memorization of facts, concepts, and applications of science to daily living. The functional application of science was an important strategy.
INSTRUCTIONAL MATERIALS	Textbook publishers began to compete seriously for authors, and many improved texts completed for all science areas, and laboratory manuals began to lose some of "cookbook" style. National Science Teachers Association completed comprehensive study on science facilities. Suppliers of scientific materials and equipment improved products for effective utilization by science teachers. Increased funds from governmental agencies and private foundations for improving science teaching facilities.

Table A.7. *(continued)*

STUDENT POPULATION	Science became an imperative for all high school pupils; number of pupils attending school approached 80 percent, with approximately 30 percent going on to college. All pupils took two years of general science, approximately 75 percent studied biology, about 33 percent of 11th graders studied chemistry, and approximately 15 percent of 12th graders studied physics.
TEACHERS' PREPARATION	This era was characterized by a "deplorable" state of science teaching and committees proposing the improvement of science teachers' preparation. A study of significance was a report of American Association for the Advancement of Science Cooperative Committee on Science Teaching, published in *School Science and Mathematics*, February 1946, entitled "The Preparation of High School Science and Mathematics Teachers." The report identified three interrelated problems. • Science (and mathematics) teachers are not properly trained for their beginning teaching assignments. • Not enough able men and women are attracted to science teaching because salaries are low and the job lacks dignity. • The high school science curriculum needs reorganization, and the equipment for science teaching needs modernizing. Committees made the following recommendations: • Establish a policy of certification in closely related subjects within broad area of sciences and mathematics. • Devote approximately one-half of prospective teachers' four-year college programs to courses in sciences (and mathematics). • Grant certificates to teach general science in 7th, 8th, and 9th grades on the basis of broad preparation including courses in all subjects concerned with general science. • College and certification authorities should work toward five-year program for preparation of high school science teachers. • Curriculum improvement in small high schools go hand in hand with improvement in teacher preparation. In 1953, the National Science Foundation (NSF) began a program for improvement of "Education in the Sciences" by financing of two summer institutes for secondary school science teachers. In 1954, the National Science Teachers Association established a national program for rewarding meritorious services of science teachers in a program called "Science Teacher Achievement Rewards" (STAR).

Table A.8: A Revolution in School Science Programs (1955–1965)

GOALS	After 1955 there was a re-evaluation of goals in the rapidly changing space, atomic, and nuclear age. Although functional goals continued, there was a distinct shift to intellectual disciplines. Increasing emphasis was placed on processes of science and how scientists work toward answers to questions about natural phenomena. A primary goal was to help pupils understand and develop deeper appreciations of structure and strategies of science, and to instill abilities to learn new knowledge rather than memorizing isolated facts or principles of science.

SCHOOLS AND CURRICULUM	Era characterized by revolutionary curricular changes in science, the most in American education. Factors contributing to revolution:
	• Funds made available for curricular studies by National Science Foundation (NSF) and private foundations.
	• Involvement of professional research scientists and faculties of college and university science departments, in conjunction with high school science teachers and university science educators.
	Important developments included:
	• Establishment of K–12 science curriculum with increased articulation from kindergarten to university levels.
	• Curriculum revisions in
	  • Physics (Physical Science Study Committee), and Project Physics
	  • Biology (The Biological Science Curriculum Study)
	  • Chemistry (The Chemical Bond Approach) and The Chemical Education Materials Committee (CHEM Study)
	  • Earth Science (The Earth Science Curriculum Project)
	• Junior High School Science Project (The Princeton Junior High School Science Project)
	• Elementary School Science (Elementary School Science Project), Science Curriculum Improvement Study (SCIS), Science A Process Approach(S-APA), and Elementary Science Study (ESS)
	• Shifting of scientific concepts previously reserved for college study to secondary and elementary levels.
	• Articulation of elementary and high school science.
	• Establishment of advanced placement science courses in secondary schools.

Table A.8. *(continued)*

TEACHING PRACTICES	The following innovations implemented in varying degrees: • Team teaching • Media • Programmed instruction • Science laboratories • Process approaches in teaching (science as inquiry, discovery, and investigation)
INSTRUCTIONAL MATERIALS	Radical changes of facilities available for teaching. Significant changes included: ratification of National Defense Education Act (NDEA) by Congress in 1958 which provided funds to be distributed to states for improvement of science equipment, increased involvement of National Science Foundation in financing purchases of equipment and facilities, and increase in funds from private foundations and business for purchases of scientific equipment and supplies.
STUDENT POPULATION	Increase in secondary school enrollments across the country; increase in enrollment in 1960s caused by increased birth rates following World War II. Significant trends included: • Increase of pupils enrolling chemistry and physics courses • Pupils studying biology at 9th grade level • Pupils taking advanced science courses
TEACHERS' PREPARATION	Supply of qualified teachers short of numbers needed. Following trends were significant: • Increase from two National Science Foundation Summer Institutes for Secondary Science and Mathematics in 1953 to 412 institutes with about 21,000 teachers supported by NSF in 1963 and 1964. • National Science Foundation established other programs for science teacher improvement, including academic-year institutes, summer fellowship programs, research participation, and in-service institutes. • In 1961, National Association of State Directors of Teacher Education and Certification, in conjunction with committee of American Association for the Advancement of Science, published "Guidelines for the Preparation of Teachers of Science and Mathematics." • Colleges and universities tailor specific programs for secondary school science teachers. • College and university science and education departments offer special training for teachers of new curriculum designs established by PSSC, BSCS, CHEM, CBA, and IPS.

INDEX

Page numbers printed in **boldface** type indicate tables or figures.

INDEX

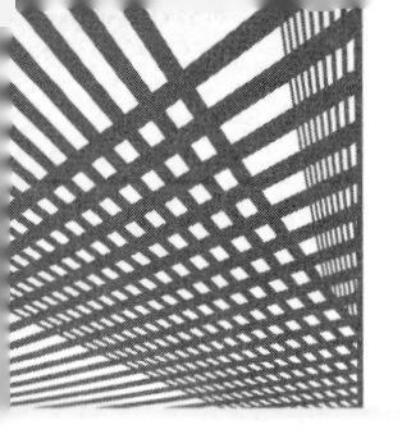

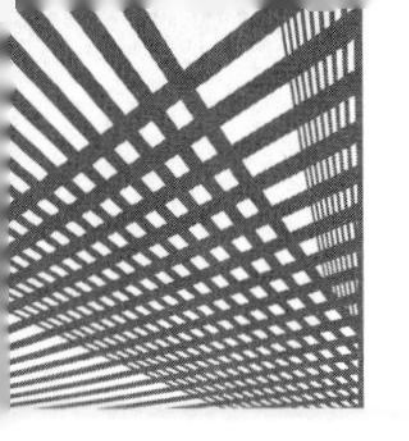

INDEX